INSECTS AND OTHER TERRESTRIAL ARTHROPODS: BIOLOGY, CHEMISTRY AND BEHAVIOR

MOTHS: TYPES, ECOLOGICAL SIGNIFICANCE AND CONTROL METHODS

INSECTS AND OTHER TERRESTRIAL ARTHROPODS: BIOLOGY, CHEMISTRY AND BEHAVIOR

Additional books in this series can be found on Nova's website under the Series tab.

Additional E-books in this series can be found on Nova's website under the E-book tab.

INSECTS AND OTHER TERRESTRIAL ARTHROPODS:
BIOLOGY, CHEMISTRY AND BEHAVIOR

MOTHS: TYPES, ECOLOGICAL SIGNIFICANCE AND CONTROL METHODS

LUIS CAUTERRUCCIO
EDITOR

Nova Science Publishers, Inc.
New York

Library of Congress Cataloging-in-Publication Data

Moths : types, ecological significance, and control methods / editor, Luis Cauterruccio.
p. cm.
Includes bibliographical references and index.
ISBN 978-1-61470-626-7 (hardcover : alk. paper) 1. Moths--Classification. 2. Moths--Morphology. 3. Moths--Ecology. 4. Moths--Control. I. Cauterruccio, Luis.
QL542.M68 2011
595.78--dc23

2011024422

Published by Nova Science Publishers, Inc. † New York

CONTENTS

PREFACE

This book presents topical research from across the globe in the study of the types, ecological significance and control methods of moths. Topics discussed include the moth Steniscadia poliophaea(noctuidae) as as potent enemy of young mahogany trees in Amazonian forests; pest management of the microlepidoptera in fruit production and its economic significance; moth sex-pheromone production; the moth reproductive physiology and natural enemy pressure; sublethal effects of pesticides on exposed moths and their unexposed progen and the genetics of interactions among moths, their host plants and enemies in Crimean oak forests.

Chapter 1 - Most insects are specialists of their host plant. The existence of *Lepidoptera* species is dependent on the site that the female choosestolay hereggs since the hatching larvae are less mobile. Among the different plant stimuli that act at this behavioral step, chemical stimuli are dominant. The most frequently studied stimuli are volatile and non- volatile secondary metabolites that act at a distance or in contact with the plant. Very few studies have been carried on plant primary metabolites at the leaf surface. They should provide information about the physiology of the plant and its nutritional value. These ubiquitous components were previously thought to be of no interest for insect-host plant specific species when selecting a negg-laying site. Soluble carbohydrates at the leaf surface are photosynthates and their quantities vary according to the photosynthesis rate throughout the day. They come from the apoplast, pass through the cuticle and partially re-penetrate into the plant. Their presence at the leaf surface follows a dynamic equilibrium between the inside and outside of the plant, which coincides with carbon as similation and metabolite translocation from the plant. A selective permeability of the cutic letocarbohydrates, passing throughs tomata whose numbers vary according to the plant species and their plant-specific distribution on the leaf surface provide the semolecules with specific, partly genetically-based characteristics. For the model chosen, *Malus domestica*Borkh, and its worldwidepest,*Cydia pomonella* L.(*Lepidoptera, Tortricidae*), many studies have shown the influence of volatilecomponents,trichomesandwaxcomponentsonhostselection.Solublecarbohydrates andsu gar alcohol sontheleafsurfaceshould, however,provideinformationboth about plantphysiology and host plant specificity.Ablendofsixmetabolites-sucrose, D-glucose,D-fructose,sorbitol, quebrachitol,myo-inositol-replicatingthosethatexistontheleaves of the hostandnon-hostplants studied wastested on *C. pomonella*egg-layingbehaviorandlinkedtoplant status. Accept ance and egg-laying stimulation were associated with the composition of the blend. The sorbitol specific to woody rosaceousspecies and D-fructose within the blend has a considerable influence on female behavior. The variation of *C. Pomonella* responses in relation to acceptance oregg- layingstimulation according to ratio changes in a single chemical or chemicalgroup withinthe blendrevealsa flexibility of its behavior.Phytosanitary productsused for protection of apple or chards modify the leaf surface metabolite blend compositionand,subsequently,*C. pomonella*egg-laying. Among the compounds included intheircommercialformulation,thesugarswere studied. Aquantityof10ppmand0.1ppmof sucroseor D-fructose, respectively, applied to apple trees ino rchardsinduces modifications ofthe sugar alcoholswithintheblend, providing gustatorycuesthatreduce*C. pomonella*egg-layinganddamages.The responses to D-fructoseofbothinsects andplantsmay indicate itsimportancein the signaling path ways of plantresistanceto Lepidoptera. Theimplicationof this knowledgewithinanewappleorchardprotectionstrategyagainst*C. pomonella*ispresented.

Chapter 2 - The super-family Noctuoidae is the most species-rich of Lepidoptera, and many appear to be specialized herbivores. Yet little is known about their abundance and ecological significance in diverse forests of the tropics. In this chapter we briefly review these two aspects in the context of diversity maintenance (Janzen-Connell hypothesis), and present the case of the South American moth *Steniscadia poliophaea*. This species feeds only on expanding leaf and stem tissues of seedlings and saplings of the prized timber tree, big-leaf mahogany (*Swietenia macrophylla*). We synthesize published research, observational reports, and anecdotal evidence about *S. poliophaea*'s life history, ecology, and impact on host mahogany populations across southern Brazilian Amazonia. This moth plays an important role in suppressing the early recruitment and growth, and hence potential local dominance, of the fast-growing *S. macrophylla*. We doubt this moth plays a contributing role in structuring local adult densities of *S. macrophylla* in Central America and Mexico where it has not been reported to occur. We compare the ecological significance of *S. poliophaea* to the better known shoot-boring moth, *Hypsipyla grandella* (Pyralidae) that is a major pest in mahogany plantations throughout the Neotropics. Finally, we consider implications of these findings for host-competition and control in the recovery and sustainable management of threatened *S. macrophylla* populations in logged and unlogged South American forests. Moth herbivores in general, and the Noctuoidae in particular, warrant further investigation as potential drivers of Janzen-Connell effects on trees in species-rich tropical forests.

Chapter 3 - One of the leading concerns of pest control in modern fruit production, and for both fruit quality assurance and environmental preservation, has been how conventional control methods affects biodiversity and how they can be altered to mitigate pesticide side effects in all aspects. This chapter discusses the significance of economically important microlepidoptera-moth species in fruit production and is mostly focused on their Integrated Pest Management (IPM). Microlepidoptera is a cluster of moth families commonly known as the 'smaller moths'. Since the group is characterized by polyphyletic diversity this is not, from a taxonomical standpoint, a restrict definition albeit commonly used to group small moth species which in most cases display similar life cycles and habits that are not found in larger Lepidoptera (i.e. butterflies). An overview of the current status of representative codling moths, tortrix, Gelechiidae and leaf-roller moths including: *Cydia pomonella*, *Grapholitha molesta*, *Anarsia lineatella* and *Adoxophyes orana* are presented. The detailed habits and bionomics are documented from prior studies and compared to older and latest references. The work proceeds by the description of numerous control methods and tactics that are currently used in IPM and as part of the wider framework of Integrated Fruit Production (IFP). The development of forecasting models based on degree-days, as well as the development of Economic Injury levels and Thresholds as decision tools to determine the optimal treatment time for biorational insecticides and insect growth regulators is presented. Efforts are also made to discuss and weight constrains of the 'Economic Injury Level concept' to be applicable on a realistic basis in fruit orchards. The major properties of biorational chemical compounds and biological control agents (i.e. bacteria and parasitic nematodes) and possible side effects on beneficial species are short reviewed. Novel control methods such as matting disruption, the attract and kill and push and pull strategies are briefly outlined with the view to be developed and incorporated in future IPM programs on a regular basis to control fruit moths. Finally, actual facts and challenges such as pesticides resistance and restrictions due to the implementation of the latest European Union council directives for pesticides are also discussed.

Chapter 4 - Pheromones are chemicals emitted to send messages to individuals of the same species. Much of the research on chemical communication systems in insects has focused on moths in the order Lepidoptera, which is the second largest insect order with well over a hundred thousand described species. Most of the hundreds of species studied have been found to use a long-distance chemical communication system for attracting mates and the most widely explored are the sex-pheromones of female moths. The discovery of pheromones and their binding proteins have impacted Lepidopteran biology, and neural encoding, processing and integration of olfactory signals from mates are areas in which Lepidoptera continue to serve as important models. Moreover, the exploitation and the use of molecular techniques in the post genomics era have led to many advances in several aspects of moth pheromone research *viz.* the elucidation of biosynthetic pathways; the identification of key enzymes therein; the regulatory physiology of pheromone biosynthesis; the role of G-protein coupled receptors in the initiation and inhibition of these pathways and the role that pheromones play in the speciation process. Reproductive behavior in moths relies on the synchronization of various environmental and physiological events that influence the timing of sexual activities between the males and females. Receptivity in most female moths is broadcasted by the release of a unique blend of fatty acid-derived volatile sex-pheromones when they extrude their pheromone glands thereby assuming typical calling behavior. This behavior occurs only at specific times of the photoperiod, typically during the night (scotophase) and only then is sex-pheromone biosynthesis initiated. The insect's neuroendocrine system is a major regulator of many physiological functions including mating-behavior. Environmental and internal signals such as age, photoperiod, temperature, mating history and host plant volatiles signal the neuroendocrine system to induce downstream events affecting sex-pheromone production. The release of a pheromone-biosynthesis activating neuropeptide (PBAN) into the hemolymph up-regulates the biosynthesis of fatty acid derived compounds. Once the female sex-pheromones are emitted, the males perceive and orient towards the source of the volatile. On reaching the females, males of several species display their hair-pencil complexes. The hair-pencil complexes also contain pheromonal compounds that are related structurally to the female sex-pheromones. These compounds play an essential role for successful mating. Mating affects subsequent female reproductive behavior and the production and release of sex-pheromone is suppressed. This phenomenon has been attributed to the transfer of seminal peptides to the female during copulation. In this review I will provide background on the biosynthesis of sex-pheromones and will focus on elements of regulation with particular focus on the characterization and mode of action of the Pheromone-Biosynthesis Activating Neuropeptide (PBAN) and its receptor. I will delve on inhibitory processes such as the outcome of female receptivity after mating and the action of seminal peptides such as Sex Peptide on its receptor. With our increasing understanding of the regulatory physiology of reproductive behavior we encounter several avenues that could be utilized for the disruption of female receptivity and prevention of subsequent successful mating events, avenues that I will explore as possibilities in the future development of new mating control or disruption strategies for field application against moth pests.

Chapter 5 - Life history theory attempts to define the rules controlling female reproductive effort, the trade-off between fecundity and egg size and the associated trade-off between egg size and offspring performance. In the tortricid moth *Lobesia botrana* Den. and Schiff., egg size is a highly labile trait depending on several proximate – environmental – factors, which correlates positively with offspring performance in adverse habitats but not in favorable ones. Host plant and its environment could then modulate the trade-off between egg size and larval performance and thereby between egg size and female fitness. The ultimate – adaptive – effect of host plant on egg size and related reproductive variables was investigated by comparing thirteen field populations of *L. botrana* derived from a cultivated (*Vitis vinifera* L.) and a wild (*Daphne gnidium* L.) host plant. We selected these hosts because of their ecological, historical and economic connotations. Daphne is considered the ancestral host plant of the moth and hence has been colonized since ancient times, while vine has been recently colonized on a historical scale, in spite of moth is currently the major vine pest worldwide. Results showed that larger females produced larger eggs, were more fecund, lived longer and had a greater reproductive effort irrespective of host plant. Heritability estimates supported the occurrence of heritable variation for egg size and fecundity, but not for longevity. Unlike with female fecundity and longevity, selective pressures imposed by host plant affected significantly the size of eggs, daphne females producing smaller eggs than vine females. Host-mediated egg size differences were consistent despite the huge variation among populations within host plants, which was interpreted because of local adaptation processes under a weak gene flow. Results overall do not support however the expected trade off between egg size and number since fecundity did not differ significantly between host plants. Host-mediated selective forces driving egg size of *L. botrana* and their related reproductive and ecological trade-offs are finally discussed from an evolutionary perspective regarding host plant quality, insect-plant relationships, moth reproductive physiology and natural enemy pressure.

Chapter 6 - Pesticides affect not only target species but also non-target organisms that live in exposed areas. Particular species, populations and individuals vary in terms of their resistance; some may survive pesticide exposure while others are more vulnerable. Moth larvae in particular are important pests, but due to various factors affecting exposure, some that are exposed may reach the mature (imago) stage and produce offspring. Most research on pesticides has focused on their lethal effects, especially ovicidal and larvicidal activity on developmental stages. Data concerning sublethal effects are limited, due to the huge number and vast range of such effects and the fact that until fairly recently the complexity of these systems was not realized. These effects can now be classified in ascending categories, starting from basic biological levels (genetic, biochemical), through more complex phenomena (cellular, histological, and physiological), and on to the levels of single organism and whole population (malformations, behavioral anomalies, reproductive alterations, reduced resistance to endogenous and exogenous stress, etc.). It is now becoming clear that toxic effects (developmental alterations, decreased survival of larvae or pupae) may not be limited to the exposed generation but may reach into subsequent generations as well, and in their natural environments may subject the animals to intense selective pressure and risk of extinction, leading in turn to serious disturbances within foodwebs. The authors review current knowledge concerning the abovementioned sublethal effects of pesticides on moths. Among these, reproductive malfunctions are the most frequently reported. Usually, survival, fecundity (number of eggs), fertility (hatching) and integrity of egg structure are reduced in a dose-dependent manner, although there are cases in which the insecticides do not obviously affect fertility. For example, the chitin synthesis inhibitors, novaluronon and diflubenzuron, and a nicotinoid insecticide, imidacloprid, do not affect the fecundity of *Cydia pomonella, Platynota idaeusalis* and *Tryporyza incertulas*, respectively. According to the literature, frequently observed decreased fertility may be caused by various factors, including suppression of sperm and/or egg production, impairment of those germ cells that are produced, sublethal genetic abnormalities within these cells, or lethality in the embryonic stage. In the insecticide-exposed populations, subsequent generations may lay small eggs, and larval and pupal development may be affected too. Newly hatched larvae of moths may show morphologic defects and feeding disabilities, while adult moths show unusual choices for egg laying sites, compared to control (unexposed) moths or moths exposed to other insecticides. Such aberrant effects are especially important in the case of alkaloids and other natural components showing biological activity against moths. Some research groups report a particularly interesting phenomenon, reduction in egg numbers laid by untreated females mated to treated males or vice versa. This can be an early stage of so-called "assortative mating," as in cases in which conspecific insects do not mate freely if males and females are raised on different diets. In this review we discuss, in light of their known chemical and biological activities, sublethal effects of pesticides in different insects, and we speculate on the possible ecological significance of these effects.

Chapter 7 - Any organism interacts with an environment in accordance to its genetic constitution, raises unique requirements to the environment and in its own manner interacts with representatives of other species. Thus, intra-population genetic variation should strongly influence integrative ecological pattern of the population in the community or ecosystem. This thesis is a basic principle of our investigation on ecological ties among moths and other species, with which they interact in forest ecosystems. Oak leaf roller and, gypsy moths strongly contribute to a pathological situation in the Crimean forests. Application of allozymes, random and specific DNA markers has shown that the consequences of interaction among the moths, their host plants, parasitoids and micro-pathogens strongly depend on their genetics, raising ties among gene pools of their populations. Evolution of the ties during long co-existence of the moths with host plants and enemies forms natural mechanisms of their regulation. These mechanisms have a great perspective in respect to their application in agricultural environments, changing aim of pest control from pest elimination to control of its evolution towards decrease in its damaging capacity. Long co-existence of moths and pathogenic viruses forms feedbacks between their genomes. Nuclear polyhedrosis viruses, for instance, can regulate apoptosis processes in host insect cells to increase the efficiency of the host resources exploitation. Application of relatively short, artificially created DNA fragments of antiapoptosis genes of nuclear polyhedrosis virus on its host insect, gypsy moth, increases in its death rate. These fragments may serve as DNA insecticides. We have found that external application of solution with two short single strand iap-3 gene fragments of

Lymantria dispar multiple nuclear polyhedrosis virus causes reliably high mortality of gypsy moth caterpillars. DNA insecticides may become alternative means of control methods for phyllophagous insects, including the oak leaf roller, and gypsy moth.

Chapter 8 - North-eastern part of India is considered to be a hot spot of bio-diversity. As far as the lepidopteran species are concerned, the region sustains a large variety of butterfly and moth. Among the moths inhabiting the area, a large number of silk moth species including *Antheraea perny, Samia Cynthia ricini, Antheraea roylei, Antheraea pruthi, Antheraea assamensis* etc. are prominent. Besides these, many species of Atlas moths, which produce 'Fagura' silk, have been recorded from North-east India. The diversity of moth species in the region is due to its unique topography, climatic conditions, vegetations etc.

Chapter 9 - Traditional control methods for lepidopteran pests in grapes and hops often involve foliar treatments with broad-spectrum organophosphates or pyrethroids. However, as many of these high-risk chemicals face increased regulation, growers have shifted toward using softer insecticide programs including reduced risk insecticides and biopesticides, which are typically more effective against early instar larvae than on later instar larvae. Attract-and-kill technology was explored as a reduced-risk alternative control strategy against lepidopteran pests of grapes and hops. A feeding attractant (1:1 acetic acid:3-methyl-1-butanol (AAMB)) derived from fermented molasses attracts a large number of noctuid species, including some that are pests of grapes and hops. Attract-and-kill experiments using bait stations with AAMB and a pyrethroid insecticide mixed with PTFE grease were conducted in Washington State vineyards during the summers of 2003 to 2005. Bait stations were hung in wine and juice grape vineyards in replicated 2-ha plots at three treatment levels, 0, 25, and 125/ha. Moths were monitored with feeding attractant, pheromone (spotted cutworm only), and light traps throughout the season. Post-deployment, climbing cutworm numbers in feeding attractant traps in bait station-treated plots were significantly reduced in wine grapes in 2003 and in juice grapes in 2005, while numbers of cutworms usually increased in pheromone and light traps. In 2009 and 2010, targeting the hop looper (*Hypena humuli* (Harris), bait stations with feeding attractant and pyrethroid insecticide were deployed in three Washington hop yards in replicated 2-ha plots at three treatment levels, 0, 25, and 150/ha. Moth populations were monitored with AAMB-baited traps throughout the season. Larvae on hop plants were sampled biweekly during the season. In two of the hop yards during July 2010, trap captures of hop looper moths were significantly reduced in bait station-treated plots compared with control plots in July, approximately one month after deployment of bait stations. For hop looper and other lepidopteran larvae, numbers collected were variable, and no effects due to bait stations were detected. Bioassays involving field-collected noctuid moths exposed to fresh and field-aged insecticide-treated bait stations showed that field residues of bifenthrin, fenpropathrin, and β-cyfluthrin maintained toxicity against noctuids as far out as 9 weeks post-deployment. Based on these data, the use of AAMB-based bait stations has potential as a reduced-risk method to control pest cutworms in grapes and hop looper in hops.

In: Moths: Types, Ecological Significance and Control... ISBN: 978-1-61470-626-7
Editor: Luis Cauterruccio, pp. 1-38

Chapter 1

SUGARS ON LEAF SURFACES USED AS SIGNALS BY THE INSECT AND THE PLANT: IMPLICATIONS IN ORCHARD PROTECTION AGAINST *CYDIA POMONELLA* L. (*LEPIDOPTERA, TORTRICIDAE*)

***S. Derridj* [1], *N. Lombarkia* [1], *J. P. Garrec* [2], *H. Galy* [3] *and E. Ferré*[3]**

[1] INRA, UMR 1272: Insect Physiology, 78000 Versailles, France.
[2] INRA, UMR 1137: Ecologie et Ecophysiologie Forestières, 54000 Nancy, France.
[3] ANADIAG SA 16 rue Ampère, 67500 Haguenau, France.

ABSTRACT

Most insects are specialists of their host plant. The existence of *Lepidoptera* species is dependent on the site that the female chooses to lay her eggs since the hatching larvae are less mobile. Among the different plant stimuli that act at this behavioral step, chemical stimuli are dominant. The most frequently studied stimuli are volatile and non-volatile secondary metabolites that act at a distance or in contact with the plant.

Very few studies have been carried on plant primary metabolites at the leaf surface. They should provide information about the physiology of the plant and its nutritional value. These ubiquitous components were previously thought to be of no interest for insect-host plant specific species when selecting an egg-laying site.

Soluble carbohydrates at the leaf surface are photosynthates and their quantities vary according to the photosynthesis rate throughout the day. They come from the apoplast, pass through the cuticle and partially re-penetrate into the plant. Their presence at the leaf surface follows a dynamic equilibrium between the inside and outside of the plant, which coincides with carbon assimilation and metabolite translocation from the plant. A selective permeability of the cuticle to carbohydrates, passing through stomata whose numbers vary according to the plant species and their plant-specific distribution on the leaf surface provide these molecules with specific, partly genetically-based characteristics.

For the model chosen, *Malus domestica* Borkh, and its worldwide pest, *Cydia pomonella* L. (*Lepidoptera, Tortricidae*), many studies have shown the influence of volatile components, trichomes and wax components on host selection. Soluble carbohydrates and sugar alcohols on the leaf surface should, however, provide information both about plant physiology and host plant specificity. A blend of six metabolites - sucrose, D-glucose, D-fructose, sorbitol, quebrachitol, myo-inositol - replicating those that exist on the leaves of the host and non-host plants studied was tested on *C. pomonella* egg-laying behavior and linked to plant status. Acceptance and egg-laying stimulation were associated with the composition of the blend. The sorbitol specific to woody rosaceous species and D-fructose within the blend has a considerable influence on female behavior. The variation of *C. pomonella* responses in relation to acceptance or egg-laying stimulation according to ratio changes in a single chemical or chemical group within the blend reveals a flexibility of its behavior. Phytosanitary products used for protection of apple orchards modify the leaf surface metabolite blend composition and, subsequently, *C. pomonella* egg-laying. Among the compounds included in their commercial formulation, the sugars were studied. A quantity of 10 ppm and 0.1 ppm of sucrose or D-fructose, respectively, applied to apple trees in orchards induces modifications of the sugar alcohols within the blend, providing gustatory cues that reduce *C. pomonella* egg-laying and damages. The responses to D-fructose of both insects and plants may indicate its importance in the signaling pathways of plant resistance to Lepidoptera. The implication of this knowledge within a new apple orchard protection strategy against *C. pomonella* is presented.

Keywords: Host selection; acceptance; egg-laying; gustatory cues; leaf surface; carbohydrates; sugar alcohols; plant resistance.

INTRODUCTION

The process of insect-host plant selection from a distance to contact with its surface includes a number of sequential behavioral steps, which are primarily responses to chemical cues. Plant chemical constituents comprise an inexhaustible diversity and vary with distance from the plant, within plant sites and over time. Most herbivore insects are plant species-specific. A plant species should be "non-host resistant" for a majority of insects, and plant selection should be more the result of avoidance than of attraction. The volatile compounds resulting from secondary plant metabolism are the focus of the large majority of host plant selection studies reviewed by Bernays and Chapman (1994) and Schoonhoven et al. (2008).

In general, the range of plants accepted by females to lay their eggs is smaller than the diet breadth of the larvae. It is likely that there is a selective and ecological advantage for plants to limit their infestation by an early signaling on females when they deposit their eggs outside the plant. To explain host plant selection for egg-laying and feeding, the generalized sequences described devote relatively little attention to gustatory cues that are used on the plant surface. Finch and Collier (2000) showed that the continuum between host selection and host acceptance for laying eggs is not the rule among specialist pest insects of cruciferous plants. They observed "inappropriate and appropriate landings" when insects were presented with *Brassica oleracea* intercropped with *Trifolium subterranean.* Only appropriate landings stimulated egg-laying. Based on observations of *Delia* fly egg-laying, these authors proposed a new hypothesis to explain host plant selection. Gravid females follow behavior patterns that include landing, running on cabbage leaves and stem surfaces, followed by probing the soil where they lay their eggs. Host plant selection occurs in three successive phases governed by volatile cues, visual cues (color) and non-volatile cues. It is now accepted that host and site selection for egg-laying relies mainly on vision and associative learning involving optical and contact-chemosensory cues that promote time and energy optimization of host selection behavior (Schoonhoven et al. 2005). Insects are adapted to plant surface exploration before egg-laying or feeding by antennal contact, palpation with maxillary palps and styloconic sensilla, walking, standing, drumming, nibbling, and ovipositor scanning, all of which are associated with gustatory sensillae (Southwood 1986; Städler 1986, Chapman and Bernays 1989). The few demonstrations of the influence on insect egg-laying behavior of gustatory non-volatile cues after landing are partly linked to difficulties encountered in this type of study, including: (i) identification of the compounds with which the insect is actually in contact; (ii) appropriate tests to demonstrate the effects of molecules on the behavior; (iii) verification of their detection at the level of peripheral sensory systems and their nervous integration; and (iv) association of chemical stimuli of host selection to the physiological and nutritional needs and detoxification abilities of the insect species.

Leaf Surface Epicuticular Waxes and Insect Plant Recognition

The plant surface is covered by the cuticle, which is a continuous extracellular membrane (Riederer 1991). It has many functions due to the fact that it is the interface for the biotic interactions that it controls: plant transpiration, loss and uptake solutes, exchange of gases and vapors, transport of lipidic substances, water and particle repellence, attenuation of photosynthetically-active and UV radiation, mechanical containment and plant development (Riederer and Müller 2006).

To explain insect/plant relationships with the plant surface, studies have primarily focused on the structural and chemical composition of the upper cuticular layer formed by the epicuticular waxes whose structure (Holloway 1982; Jeffree 1986, 1996, 2006; Barthlott 1990, 1998; Koch et al. 2009) and chemical composition (Baker 1982) are highly variable with plant species (Jeffree 1986), abiotic environmental conditions (Schütt and Schuck 1973; Faini et al. 1999; Giese 1975; Shepherd et al. 1995) and biotic factors (Jetter and Shäffer 2001; Jetter 2006). This lipidic wax layer is an amorphous film that emerges from crystal structures consisting exclusively of aliphatic compounds. A review of different methods for extracting epicuticular wax components shows that changes in extraction methods using organic solvents had consequences on the compounds obtained (Riederer and Schneider 1989). Timing and extraction methods cause difficulties in interpreting relationships. A cryo-adhesive method (Jetter et al. 2000; Ensikat et al. 2000) using gum arabic made it possible: (i) to isolate epicuticular waxes without any intra-cuticular wax contamination; and (ii) to reproduce wax crystals (physical information) in association with their chemical composition.

Different extraction methods and the components extracted and investigated on insects were reviewed by Eingenbrode and Espelie (1995), and then by Müller in 2006. Results on the effects of wax components on insects are mostly limited to studies of correlations between the fractions extracted by various solvents and behavioral insect responses, compared to the solvents that are used to extract them. The chemical components of wax have therefore been shown to play a role in host plant recognition and resistance (Woodhead and Chapman 1986; Eigenbrode et al. 1991; Espelie et al. 1991; Bodnaryk 1992: Adati and Matsuda 1993; Yang et al. 1993; Juniper 1995; Brooks et al. 1996; Espelie 1996; Powell et al. 1999; Spencer et al.

1999; Brennan et al. 2001; Jones et al. 2002; Rapley et al. 2004; Steinbauer et al. 2004; Maher et al. 2006; Steinbauer et al. 2009).

An additional difficulty to perform a demonstrative activity of epicuticular components is the study of the peripheral electrophysiological response from insect gustatory receptors stimulated by non-water soluble lipid components. An improvement in the methods used should make it possible to test these chemicals in the near future (Lacaille et al. 2007; Hiroi 2008).

Leaf Surface Secondary Chemicals as Insect Stimuli outside Epicuticular Wax Components

On the plant surface and outside the compounds of the epicuticular waxes, insects may be in contact with substances derived from secondary metabolism that come from the internal tissues and cross through the cuticle. They can be extracted from leaf epicuticular waxes or collected above them. It is postulated that water and solutes can flow through two parallel and independent diffusion paths across the plant cuticle: (i) lipophilic paths in the polymer matrix (Wattendorf and Holloway 1984; Jeffree 1996; Schlegel et al. 2005); and (ii) a reticulum of polysaccharide microfibrils branching out and stretching through the cuticular membrane and reaching the outer surface (hydrophilic pores) (Schönherr and Schreiber 2004; Schreiber 2006). The predominance of one path and variations among plant species is still a debate. The cuticle covering the guard cells has higher water permeability than the cuticle of the epidermis (Maier-Maercker 1983, Schlegel et al. 2006). Stomata also contribute to the explanation of cuticle permeability (Eichert and Burkhardt 2001; Schlegel and Schönherr 2002, Schlegel et al. 2005; Eichert and Goldbach 2008; Eichert et al. 2008; Mac Gregor et al. 2008). Their densities may vary according to leaf side and plant species (maize: 52-68 mm^{-2}; tomato: 12 -130 mm^{-2}; potato: 50-160 mm^{-2}).

The affinity of the solvent used to extract molecules, as in the case of wax, may change the range of chemicals collected. The molecules collected could then be artifacts as a result of the extraction method. Spraying an acidic solution (0.50 mM per liter of sulfuric acid) makes it possible to collect picomoles of pyrrolizidine alkaloids from *Senecio jacobaea* per cm^2 of leaf surface. Their chemical spectrum is larger than when using only water (Vrieling and Derridj 2003).

Some secondary plant metabolites were thereby described as strong leaf surface stimuli. This is the case of alkaloids, as well as phytosterols (Holloway 1971), large amounts of which are also found in the plasma membrane (Borner et al. 2005), and of free fatty acids with C16 and C18, which are very abundant in the tissues. Glucosinolates were described as egg-laying and larval stimulants and probably play a role in the host acceptance of *Lepidoptera* such as *Pieris brassicea* (Van Loon et al. 1992), *Plutella xylostella* (Spencer et al. 1999), *Hellula undalis* (Mewis et al. 2002), and *Diptera* such as *Delia radicum* and *D. floralis* (Roessingh et al. 1992). The "Cabbage Factor Identification" was also extracted from epicuticular waxes and stimulates *D. radicum* egg-laying (Hopkins et al. 1997; Roessingh et al. 1997). As revealed by extraction using gum arabic (Müller and Riederer 2005; Reifenrath et al. 2005), glucosinolates are, however, probably not present in epicuticular waxes, and Städler and Reifenrath (2009) explain their activity on insects through penetration of their sensillae into the wax layers or through stomata.

Progress Supporting the Study of Primary Metabolites on Leaf Surfaces as Plant Cues for Insects

This bibliographic review describes the obstacles encountered when studying secondary metabolites from the plant surface as cues for insect host selection. New knowledge of cuticle permeability and better separation techniques of its compounds should help to more effectively demonstrate their effects on insects. It was previously revealed (Tukey 1970) that sugars, sugar alcohols, pectic substances, amino acids, phenolic substances, gibberelins and vitamins were found in the throughfall of plants and were leached from their surfaces. The occurrence of water-soluble metabolites in the phylloplane has been reported for many plant crop species to elucidate their role in its colonization by epiphytic microorganisms (Morris and Rouse 1985). Primary metabolites found in all green plants were considered as nutrients, with no impact on plant evolution or, consequently, on insect specialist host selection. Nevertheless, in 1959, Thorsteinson included simple carbohydrates (sugars) that occur in free form in plants, in the class of token stimuli (now referred to as allelochemicals) and feeding stimulants as secondary substances.

It is now accepted that substances from both primary and secondary metabolism within a plant interact as stimuli in host plant recognition and feeding-site selection (Kennedy 1953-1958). Sugars interact as stimuli with several groups of substances. When added to sinigrin, sucrose and not fructose elicits a feeding response in *Leptinotarsa* (Thorsteinson unpublished).

Peripheral gustatory neuron cells may be specialized to detect primary and/or secondary metabolites (Albert 1980; Panzuto and Albert 1997; Roessingh et al. 1999; Städler 2002; Schoonhoven and Van Loon 2002; Green et al. 2003). Carbohydrates and amino acids are detected by separate cells in *Lepidoptera.* Three to five neuron cells located at the base of the gustatory sensillae allow the detection of water, carbohydrates and salts. Little is still known about how the nervous system integrates gustatory cues at the level of the thoracic and sub-esophageal ganglia (Rogers and Newland 2003). This is another reason why taste has been little investigated in comparison to olfaction for which the processing of peripheral inputs within the central nervous system (CNS) has been widely studied.

Means Used to Reveal Cue Activity of Leaf Surface Primary Metabolites in Insect-Host Plant Selection

To show that primary metabolites at the leaf surface can act as plant cues for egg-laying females, their relationships with this behavior must be demonstrated. If they are capable of playing a role in host plant selection as potential host, resistant or non-host, they must be at least partly genetically-based. To test these hypotheses, the study focused on: (i) sugars at the leaf surface. Besides their general interest in relation to plants and insects, they have many practical and technical advantages compared to the other molecules studied so far. They can be collected using water alone, their synthetic products can be easily acquired for behavioral studies, and their electrophysiological activity can be easily recorded; (ii) cuticle permeability to soluble carbohydrates and sugar alcohols; (iii) the specialist codling moth *Cydia pomonella* L. (*Lepidoptera:Tortricidae*) and its host, the apple tree *Malus domestica* Borkh (Rosaceae), as a plant/insect model. The non-nectarivore females do not feed during their egg-laying period that lasts a few days, there is therefore neither the influence of female feeding on their fertility nor on hemolymph composition, which could influence gustatory perception and egg-laying behavior (Bernays and Simpson 1982). Hatching larvae have a low ability to move from the egg-laying site to their feeding site. Therefore, when females select plants and organs for laying their eggs, the chemical cues need to be adapted to the future of their progeny. The codling moth *C. pomonella* L. is an apple pest worldwide and is found in various climatic zones. It prefers apple and pear over walnut, quince, crab apple and hawthorn (Balachowsky 1966). Females lay eggs separately, one-by-one, and not in groups, on foliage close to the fruit so that the neonate larvae can locate and penetrate into the fruit for feeding until their last larval stage (Geier 1963). In temperate climates, females generally lay their eggs early in the season on the upper side of the corymb leaves (leaves that surround inflorescences and fruits) or of bourse shoot leaves. The latter is a vegetative shoot that contributes to the growth of fruits through an additional supply of photosynthates (Abbot 1960). As the season progresses, the proportion of eggs laid on the fruits gradually increases and the leaf side chosen may change depending on the *M. domestica* cultivar (Audemard 1976; Blomefield 1997).

Carbohydrates on Leaf Surface

Physical Evidence

The visualization of soluble carbohydrates at the leaf surface is possible using a method that makes water-soluble hexoses, i.e., D-glucose, D-fructose, xylose and mannose, visible *in situ* at the leaf surface. The major difficulties to overcome are to avoid inducing leaching by water or disrupting the cuticular permeability by an organic solvent and to characterize very small quantities of substances (ng per cm²). The method consists in deep-freezing leaf samples (with an area of 4 cm² each) in nitrogen paste (-196°C) immediately after they have been excised, and then slowly removing water over 24 hours by freeze-drying at -40°C to avoid molecular movement. Leaves are then dipped into a methanol solution saturated with barium at -10°C for one hour, leading to the formation of non-soluble sugar-barium complexes. Barium is then substituted for silver in an alcoholic solution of silver nitrate for better visualization of silver-monosaccharide forms by a scanning electron microscope (SEM Philips EM 525) filled with a microprobe to enable microanalysis by X-ray spectroscopy (OXFORD microanalysis unit) (Fiala et al. 1993). Hexose distribution was compared for leaves removed two hours before sunset (insect egg-laying period) from *Zea mays* L. and *Cichorium endiva var. latifolia* grown in greenhouses and from *Prunus laurocerasus L.* grown outdoors. Precipitates of monosaccharide-silver complexes were observed in different densities and distributions. The leaf surface of *C. endiva* is covered by monosaccharides and shows the most extensive allocation without any concentrations above the epidermal cell wall junction. On maize leaf surfaces, hexoses are scattered in aggregates located primarily at the bottom of the anticlinal epidermal cell walls, around the stomata and over secondary veins where different trichomes are found. Within silver-monosaccharide granulates the presence of different precipitates of inorganic ions K^+ and Ca^{2+} can be also detected. Craters of 0.2-0.4

micrometers in diameter can also be observed. Precipitates are very scattered on *P. laurocerasus* and areas where none are present can be observed. The granulates are smaller than those observed on maize and no concentrations above the epidermal cell wall junction could be found (Fiala et al. 1993; Derridj 1996). Schlegel and Schönher (2002) observed sugar leaking around the stomata, on anticlinal epidermal cell walls and at the base of trichomes for several plant species. Hexose distributions on leaf surfaces may differ among plant species and induce physical variations in their contact with the insect when it lands and walks around on the surface.

Collection Method and Chemical Analyses

The water jet allows good access within epicuticular wax crystalline structures. To avoid inducing leaching, the duration and spraying pressure have to be defined according to plant species or organs. In general, only a few seconds (less than one minute) are necessary. In the case of several plant species, spraying the leaves with ultra-pure water for 20 s at a pressure of 17 L/min makes it possible to collect quantities of ng of soluble carbohydrates, sugar alcohols, free amino acids and organic acids per cm^2 of leaf area (Fiala et al. 1990, 1993; Derridj et al. 1996 a and b; Derridj 1996; Soldaat et al. 1996). Leaves were positioned at a slope of 60° and the foliar surfaces were sprayed from a distance of 20 cm with approximately 10 mL of ultra-pure water per 100 cm^2 of leaf area. Fruit surfaces were sprayed with 10 mL of ultra-pure water per 300 cm^2. Collection of internal leaf and fruit fluid is avoided by sealing the wounded part of the plant in liquid paraffin (44-46°C). Nitrogen gas is used as a carrier gas to spray water since it has neutral activity on metabolite release and no reaction with them. Washings are filtered (0.25-µm Millipore filter) immediately after collection to remove impurities and epiphytic microorganisms that naturally live on leaf surfaces (Andrew and Harris 2000; Andrews et al. 2002). The quantities of metabolites collected by this method were thus greater and more reproducible than when leaves were simply dipped in water, and provided estimations of relative quantities encountered by the insect after landing (Fiala et al. 1990). The methods used for chemical analyses of leaf surface metabolites require a low detection threshold (picomoles and nanomoles per sample including several leaves). Chemical analyses of soluble carbohydrates were carried out on derivatives (sylilated products) by gas chromatography coupled with a flame ionization detector (FID), Delsi Nermag DN 200 apparatus, making it possible to define relative quantities rather than absolute values (Lombarkia and Derridj 2002). Technical progress today in high performance liquid chromatography (HPLC) leads to good results without derivation of molecules, as well as more accurate quantities. This was not the priority, however, since quantitative approximations already existed due to the distribution of carbohydrates on the leaf surface, to the collection method and to the amount actually perceived by the insect upon contact with the leaf surface. The quantities at the maize leaf surface are very low and variable, about a few mg m^{-2} (ng cm^{-2}), ranging from 10^{-6} to 10^{-5} mol m^{-2}, whereas free amino acids were detected at 10^{-6} mol m^{-2} (Derridj et al. 1989).

Photosynthetic Origin

The biochemical composition of the leaf surface can be of exogenous origin (dust, pollen, productions of epiphytic microflora and insects) or of endogenous origin (from the plant). To verify the photosynthetic origin of carbohydrates at the maize leaf surface, experiments were carried out on maize grown in a phytotronic chamber when photosynthesis reached a high constant level within the whole plant at the late tassel stage. Filtration of the air, the use of a UV ray germicide tube and controlled relative humidity (< 95%) inside the phytotronic chamber limited the presence of exogenous substances and epiphytic microorganisms on the leaf surfaces. The median part of the eighth maize leaf was pulsed for 30 min with pure CO_2 containing 90% $^{13}CO_2$. As early as the end of the pulse, the proportions of ^{13}C-labeled soluble carbohydrates vs. non-labeled carbohydrates increased at the leaf surface up to a maximum at between 3 to 6.5 hours after the pulse, followed by a decrease during the dark period. No visible variations of proportions of ^{13}C-labeled carbohydrates could be registered simultaneously, either in the apoplast (free space) or in the leaf tissues. Photosynthesis led to the enrichment of labeled carbohydrates at the leaf surface probably originating from the apoplast. The decrease could be explained by continuous diffusion of unlabeled sugars, as well as the re-entry of labeled sugars into the plant (Derridj et al. 1996b).

Eighty ng cm^{-2}(4.4 µmole of labeled molecules) the quantity chosen to detect radioactivity when only 1% of D-glucose and D-fructose labeled with ^{14}C could penetrate, were deposited in 10 droplets of 0.22 µl per leaf (12th position) two hours before dark on the upper leaf side of 35-40 leaf endive stage under controlled conditions. Penetration into the plant was observed at a maximum rate before the droplets dried (between 9 to 15 min), and 18.8 ± 4.8 % of D-Glucose and 14.5±4.7% of D-fructose penetrated over 24 hours after a slow gradual increase. Re-penetrations of sugars vary depending on the carbohydrate molecule, plant species, growth stage and environmental conditions.

Quantities of D-fructose, D-glucose and sucrose collected from the leaf surface of *C. endivia* var. *latifolia* or *Z. mays* on plants cultivated in the greenhouse showed a progressive rise from 9 a.m. until 1 p.m. (solar time) with a maximum situated between 11 a.m. and 4 p.m., depending on the carbohydrate and plant species, followed by a progressive fall until reaching the low values observed 24 hours earlier. Amounts of soluble carbohydrates change throughout the day, coinciding with carbon assimilation and metabolite translocation. There are quantitative differences between the three carbohydrates, sucrose, D-fructose and D-glucose, which are permanently maintained throughout the day. There is a period during the day when the greatest difference is observed, between 1 p.m. and 4 p.m. i.e. on maize ear leaf at the early tassel stage and young lettuce leaf, the proportion of each carbohydrate was respectively 45, 37 and 18% on maize, compared to 14, 64 and 21% on *C. endiva*.

The small population of epiphytic organisms (10^3 bacteria cm^{-2}) cannot explain the dynamics of leaf surface sugars observed here. However, under natural conditions, plant species-specific microflora in the phylloplane could contribute to the dynamic variations in relation to abiotic factors and, particularly, to high air residual humidity.

Selective Cuticular Permeability

The permeability of isolated astomatous cuticles from *Prunus laurocerasus L.* to soluble carbohydrates was studied with respect to their quantities generally found in the apoplast. The permeability to sucrose, D-glucose and D-fructose is selective. They passed through the cuticle at decreasing rates expressed by the cuticle permeance: 21.7, 4.13 and 2.34 nms^{-1}, respectively (Stammitti et al. 1995). Permeance to fructose and sucrose was similar (n=6 to 8, P=0.790). Using the same method, permeance to sorbitol of isolated astomatous cuticles of *Malus* sp., *P. laurocerasus L.*, *Ilex aquifolium* L. and *Pyrus communis* L. showed that the cuticle of *Malus* sp. is the most permeable to sorbitol, with 20.5 nms^{-1} of permeance vs. the other ones, which had similar values of 2.41, 2.55 and 2.01, respectively. The permeance of isolated cuticles to D-Glucose ranged from 0.1 nm s^{-1} for *Euonymus japonicus* leaf (Kannan 1969), to 200 nm s^{-1} for *Malus sylvestris* fruit (Goodman and Addy 1963). The same permeance to D-glucose and sucrose was found for *E. japonicus* leaf and *Lycopersicum esculentum* fruit: 0.3 nm s^{-1} (Kannan 1969). Neither the water solubility nor the molecular mass of these molecules could explain these selective passages through the cuticles. The impact of the quality of the *L. laurocerasus L.* cuticular membrane was investigated by modifying its content by chloroform rinsing. The cuticle selectivity to D-glucose and D-fructose was maintained (Stammitti et al. 1995). It is hypothesized that the selective permeability to carbohydrates is primarily due to hydrophilic pores of the cuticle inside the cutin (not destroyed by chloroform) and is mainly responsible for carbohydrate ratios at the plant surface. Abiotic factors that play a physical role in the internal lipidic structure of the cuticle probably have a greater influence on the quantities of each molecule than on their ratios at the plant surface. An increase in humidity from 2 to 100% increases the flow of water by a factor of 2 to 3 (Schönherr and Merida 1981; Schreiber et al. 2001).

Two interesting selective properties of cuticular permeability can be observed: between molecules for a cuticle from a plant species and between plant species for one molecule. This provides primary metabolites with specific traits at the leaf surface. There is no direct relationship between the composition of the leaf surface and that of the apoplast because of the selective molecule passing through the cuticle. Sucrose dominates in the apoplast but crosses the cuticle more slowly than hexoses. It thus appears that the composition of the surface blend is closer to that of the cell composition, which contains nearly 50% of hexoses (unpublished).

Variations of carbohydrate composition at the plant surface were observed according to plant and leaf ages, sides and leaf zones, organs, plant species, cultivars and time of the year. Interestingly, at the leaf surface of maize at the vegetative growth stage, the composition of sugars varied with leaf positions and leaf zones, whereas the composition of tissues was similar. The female therefore probably obtains more information on the leaf functions and sugar gradient from the basal area to the apical ones on the basis of sugar ratios and leaf surface quantities than from contact with leaf tissues (Fiala et al. 1990; Derridj et al. 1999). To link insect behavior to metabolite cues present at the leaf surface, plant sites and time of insect activity are parameters that therefore must be considered. For example, to explain crepuscular *Lepidoptera* behavior, it is necessary to collect leaf surface metabolites when sunset begins.

Leaf Surface Sugars and *C. Pomonella* Behavioral Responses

It has already been shown that sugars from plant surfaces are involved in adult host selection mechanisms for egg-laying (Fiala et al. 1993; Derridj et al. 1996a; Roessingh et al. 2000; Städler 2002; Maher et al. 2006). Study methods nevertheless need to be more precise and extensive.

A Sugar Blend Activity

Protocol used: To demonstrate the activity of leaf surface sugars on *C. pomonella*, the same protocol for behavioral assays was followed throughout all the experiments described. Acceptance and egg-laying stimulation responses of gravid females were examined under no-choice conditions. Females used in the experiment had been laying eggs on neutral substrate (without any substances) for two days before the test. Each isolated gravid female (without a male) was then confined to a small cylindrical cage of 11 cm in diameter and height, lined at the top, bottom and along the walls with dried nylon cloth impregnated with whole corymb leaf surface water washings of *M. domestica* cultivars or a synthetic blend of six sugar components found in the washings. At least three replicates of ten single females were monitored at three different days. The impregnated nylon cloths were given to the females one hour before the onset of darkness. Egg-laying response was observed after 63 min (60 min of light and 3 min of darkness) of contact with the substrate. For the control (nylon cloth impregnated with ultra-pure water), 50% of the females laid their eggs after that time. A distinction was made between "egg-laying acceptance", which is the number of females that laid at least one egg on the artificial substrate at that time, and "egg-laying stimulation", which is the number of eggs deposited per female that laid at least one egg (Lombarkia and Derridj 2002). After 20 min of darkness, all females laid their eggs, not providing any information about the acceptance step. On trees, egg-laying occurs within minutes after landing (Wildbolz 1958) and depends on the landing organ: 299.57 ± 89.44 s on fruits and 67.4 ± 21.95 s on leaves (Lombarkia and Derridj 2002).

C. pomonella behavioral responses to sugars: There is an impressive collection of inorganic and organic substances in the leaf surface washes, some of which are not stable. It was postulated that sugars and especially sugar alcohols specific to ligneous rosaceous plants and generally well detected by moths should be important for plant acceptance of *M. domestica*. The major function of sugar alcohols is to store energy. Other possible functions include osmoregulation, protection of plants from desiccation and frost damage. D-sorbitol and quebrachitol are the most specific constituents. The first one is the most abundant and allows a classification between families. Like sucrose, it is translocated in the phloem from source to sink. *Myo*-inositol is one of the stereoisomers of inositol. It is present in all plants and is an element found at the crossroads of plant metabolism. It is found in a free state in walnut leaves (Miller 1973) a secondary *C. pomonella* host.

The blend of three soluble carbohydrates and three sugar alcohols - D-glucose, D-fructose, sucrose and D-sorbitol, quebrachitol and *myo*-inositol – was considered. This one has effects on female acceptance and egg-laying stimulation (Lombarkia and Derridj 2002,

2008). A good correlation between the blend of these sugars and insect behavior was observed. This surface component blends (referred to as "blend") were maintained throughout the experiments. The same blend components of the apple fruit (A) and of the corymb leaf (L) from the Golden delicious cultivar differed in sugar alcohol quantities. Acceptance and egg-laying stimulation were reduced when removing one of the two chemical groups from blend A. This was only observed on blend L after removing the soluble carbohydrate group. This implies that acceptance and egg–laying are not only influenced by soluble carbohydrates that are similar in both blends A and L, but also by ratios between the two groups (Lombarkia and Derridj 2002).

The study of the influence of each component from the blend was then pursued by removing metabolites one-by-one from the A and L blends (Lombarkia and Derridj 2002). Sorbitol and, to a lesser degree, the other sugar alcohols are necessary to maintain blend acceptance. D-fructose, sorbitol and, to a lesser degree, *myo*-inositol stimulate the egg-laying properties of the blend. Sorbitol therefore plays a role at both behavioral steps. The removal of D-glucose and quebrachitol had no effect on either acceptance or on egg-laying stimulation (Lombarkia and Derridj 2002).

Within the sugar blend, chemical groups and four single components influence C. pomonella acceptance and egg-laying.

C. Pomonella Host Selection and Leaf Surface Sugar Blend

Landings on Host and Non-Host *Malus* Species

Malus floribunda Baugène *(M.f.*B) is often grown in French apple orchards to cross-pollinate *M. domestica.* It is a *C. pomonella* non-host plant that is not subject to damage in orchards. Landing and egg-laying of gravid females (15 to 20 two-day-old females) were observed on two-year-old specimens of these plants. The plants were grown in containers until the young fruit stage and given to females under no-choice controlled conditions. Observation time was fixed at one hour of darkness from 5 to 6 p.m., the period during which more than 33% of females lay eggs on the tree. On both *Malus. sp*., females landed in the same proportions and generally on the upper side of the corymb leaves (44%) and fruits (25%). After landing, 60% of the females generally laid eggs on the bourse shoot leaves of *M. domestica* vs. 0% on *M.f.*B. Four main sequential behaviors of females were observed: exploring the site by walking, stopping, ovipositor scanning while walking or stopping. Ovipositor scanning was the most discriminating behavior between host and non-host rosaceous plants. More scanning and more locomotion took place on host vs. non-host plants (speed: 0.55 ± 0.08 vs. 0.32 ± 0.03 cm/sec) (Figs. 1A and B). Host and non-host plant status could not be distinguished at the landing step; ovipositor scanning of the landing site surfaces was associated with host plant acceptance and egg-laying stimulation.

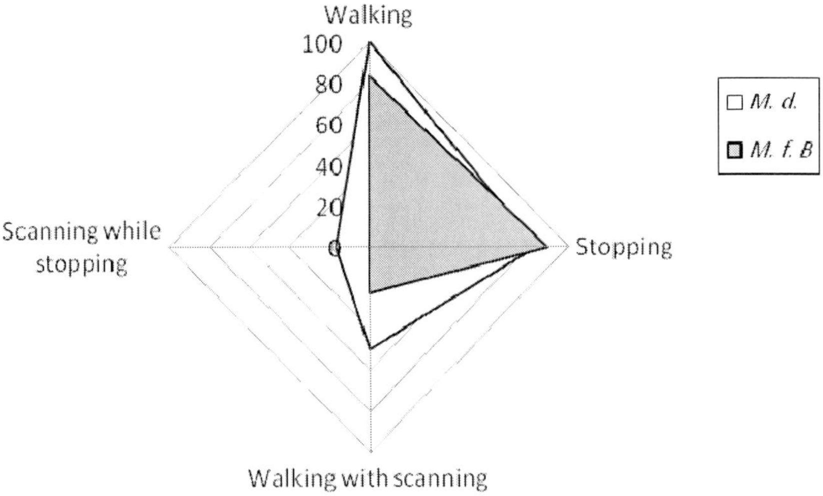

a

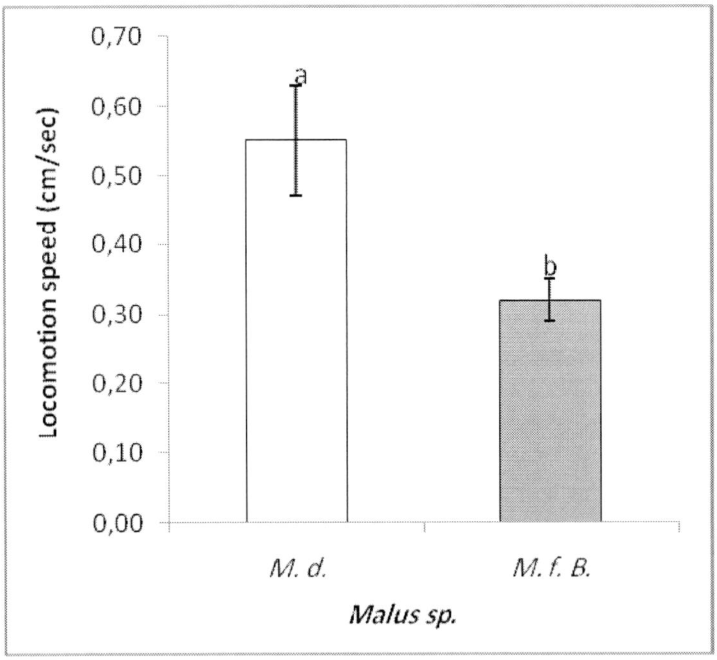

b

Figure 1. (A and B) Egg-laying behavior sequences and speed locomotion of *C. pomonella* females after landing on rosaceous tree host *M. domestica (Reine des Reinettes* cultivar*)* and non-host *M. floribunda* Baugène (*M.f.*B).

Constitutive Variations of the Leaf Surface Sugar Blend and *C. Pomonella* Hosts

C. pomonella females land on both host and non-host rosaceous plants (see above). A blend of soluble carbohydrates and sugar alcohols influences host acceptance and egg-laying on *M. domestica*, its main host. Could this blend also be related to non-host or resistance within rosaceous and hosts in other plant families?

Non-host rosaceous plants: Chemical analyses on the upper corymb leaf sides of both host and non-host *Malus sp.* reveal considerable differences in relation to data gathered from *M. domestica*, compared to the *M.f.*B blend (Derridj et al. 1999). Opposite ratios were observed between the two chemical groups: 61.68 ± 11.88 vs. 29.88 ± 2.70 of soluble carbohydrates, and 38.32 ± 11.88 vs. 70.12 ± 2.70 of sugar alcohols on *M. domestica* vs. *M.f.*B, respectively. Quantities of metabolites were much higher for *M.f.*B (ratio: 8) (Fig. 1C). *C. pomonella* acceptance and egg-laying stimulation were drastically reduced for the *M.f.*B blend vs. the *M. domestica* blends (Fig. 1D). These results suggest that the non-acceptance for egg-laying on *M.f.*B could be related to opposite ratios between the two chemical groups within the blend and/or very large quantities of sugar alcohols. This confirmed observation of trees under controlled conditions (see above). Lack of damage generally observed on *M.f.*B in orchards could be explained by its resistance to egg-laying.
Females land on *M.f.*B leaves whose surface blend composition stimulates neither ovipositor scanning nor egg-laying.

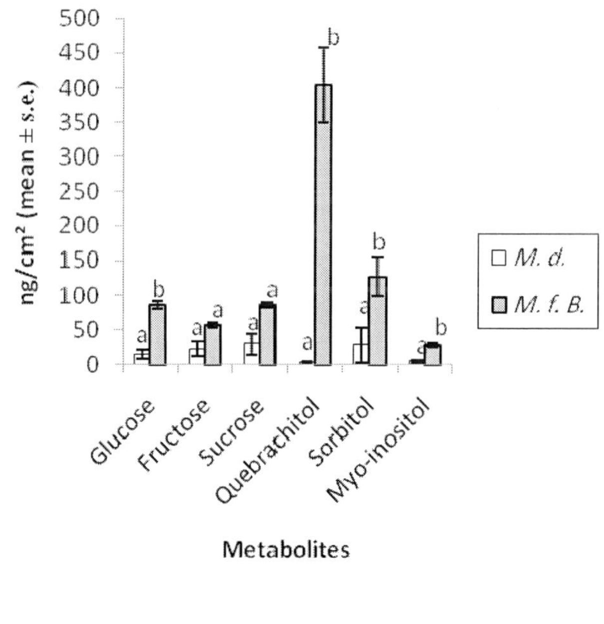

Figure 1. (C) The bourse shoot upper leaf surface blend compositions of host *M. domestica* (*M.d.*) and non-host *M. floribunda* Baugène (*M.f.*B) rosaceous plants and *C. pomonella* acceptance and egg-laying.

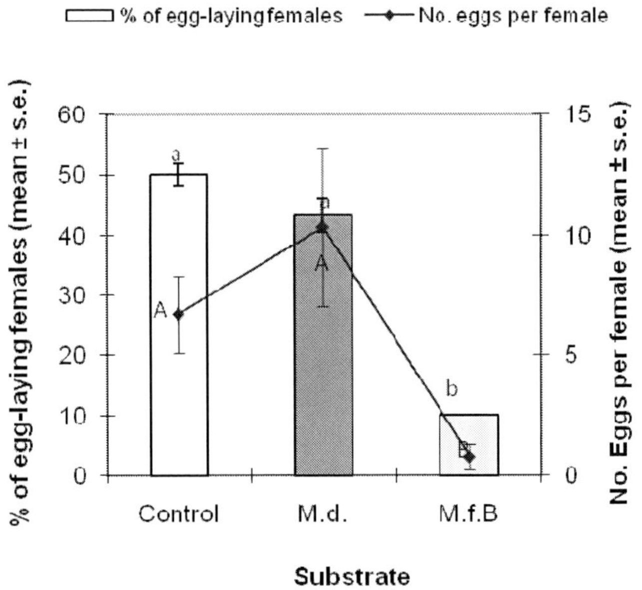

Figure 1. (D) *C. pomonella* acceptance and egg-laying on artificial substrate impregnated with the metabolite blend that reproduces the bourse shoot upper leaf surface composition of *M. domestica* (*M.d.*) and *M. floribunda* Baugène (*M.f.*B) and water control.

Constitutive resistance of M. domestica: Goonewardene and Howard (1989) reported that the apple cultivar E31-10 (= X65-11) is resistant to *C. pomonella* larval damage in the orchard, as well as in the laboratory. Fewer eggs were laid in orchards on X65-11 vs. P5R50A4, a more susceptible cultivar grown in its vicinity (INRA, Gautheron, France). Behavioral assays with blends on nylon cloth showed that the resistance of X65-11 could be associated to avoidance of substrate and egg-laying deterrence due to its blend surface composition. It was less of an egg-laying stimulant than the water leaf surface washings but acted in the same way as observed in orchards, e.g., the resistant cultivar X65-11 blend vs. P5R50A4 blend reduced egg-laying at the same ratio as it was observed in orchard (Lombarkia and Derridj 2008). The ratio between metabolites within the blend (Fig. 2) was greater for the antixenosis properties than for their quantities, which were quite low, e.g., the egg-laying reduction was maintained when the X65-11 blend metabolite quantities were multiplied by 100, 1,000 and 10,000. One small change was observed in acceptance behavior when the blend concentration was multiplied by 10,000, which resulted in a behavioral response equivalent to the one observed for the ultra-pure water control (Lombarkia and Derridj 2008). Leaf surface ratios of the blend could be linked to apple tree cultivar resistance.

The walnut tree Juglans regia (Juglandaceae) as host: The primary original hosts of *C. pomonella* were wild apple trees of the *M. silvestris* group. The human-mediated dispersal of apple tree cultivation explains a relatively recent insect distribution from Eurasia to the entire world. The walnut tree *Juglans regia* L. (*Juglandaceae*) was often cultivated in the same geographic areas as apple trees/*M. domestica,* and walnut became a secondary host for *C.*

pomonella. Codling moth populations worldwide show a lack of genetic variation and have not yet developed premating isolation (Taret et al. 2010). The emission profile of 90 volatile compounds emitted by walnut trees was qualitatively and quantitatively different from that of *M. domestica*. Many of the major and minor compounds, however, are shared by both plant species (Casado et al. 2008). They can vary throughout the season and during the day. A preference for *J. regia* vs. *M. domestica* or vice versa at a given moment could be the result of such variations in the emission of a blend of ubiquitous volatile components (Bruce et al. 2005). In France, between the months of July and September, females from the second flight lay eggs on walnut leaves and mainly on nuts. The surfaces of walnut leaves near the nut, as well as of nuts cultivated in the arboretum of Roquencourt (France), were analyzed in August. Surprisingly, sugar alcohols that are characteristic of woody rosaceous plants, such as sorbitol and quebrachitol, were present in large quantities on leaves and nuts (Fig. 2). The proportions of the sugar alcohol group within the surface blend of the upper leaf sides near the nut and the nut surfaces were quite close to those generally found on the upper side of corymb leaf surfaces of *M. domestica.* Analyses revealed that ratios within metabolite blends were different from those found on *M. domestica.* As for volatiles, it is possible that the blend composition is quite similar for each *M. domestica* and *J. regia* plant, but at different periods of the season, e.g., one plant species could be less accepted than the other at a given moment, and vice versa, which could explain host shifts throughout the season between both plants.

Use of Knowledge in Orchard Protection

Phytochemical products used for orchard protection against bacteria or *C. pomonella* larvae show side effects on the leaf surface metabolite blend composition and on *C. pomonella* egg-laying.

Side Effects of Acibenzolar-S-Methyl (ASM) on the Blend and on *C. Pomonella*

The development of acibenzolar-S-methyl (ASM), an analog of salicylic acid, as an inducer of resistance to the bacteria *Erwinia amylovora* on *M. domestica* revealed induction of defense mechanisms in leaves (Brisset et al. 2000). It modifies plant metabolism and metabolite translocations. Consequently, it should also modify metabolite composition of the plant surface. Observations on *C. pomonella* egg-laying behavior were made 14 days after ASM treatment on two-year-old trees of Golden Delicious apples grown in containers. Females coming from the cage walls to the trees landed primarily on the upper side of the bourse shoot leaves. As a result of the ASM treatment, landings were reduced by 50% compared to non-treated trees. No changes were observed in landings within the trees from site to site, with a maximum observed on bourse shoot leaves and fruit. Without any modifications in their distribution within the trees, the number of eggs per tree was dramatically reduced by 60%. They were mainly found on the upper leaf side of the bourse shoot where 5.75 ± 2.2 eggs per tree and site were recorded, compared to 21.0 ± 4.8 on control trees. No variation was observed in relation to female acceptance (Derridj and Borges

2006). The main chemical modifications due to the treatment were on the upper leaf side of the bourse shoot leaves on which smaller quantities of glucose, sucrose, mannitol, *myo*-inositol and of some free amino acids such as isoleucine, phenylalanine, aspartic acid, glutamine and arginine were observed. The ASM treatment induced a high variability of sucrose and sorbitol quantities within the blend and their ratios were negatively linked.

In addition to blend changes induced by the ASM treatments (Fig. 2), other unknown cues acting at a distance were probably also modified. ASM induced reduction in host attraction and then after landing reduced the egg-laying stimulation. The non-host plants or those resistant (constitutive or induced) to egg-laying have a blend composition in which the fructose or sucrose are found in amounts smaller or equal to those of sorbitol. The opposite is observed otherwise (Fig. 2).

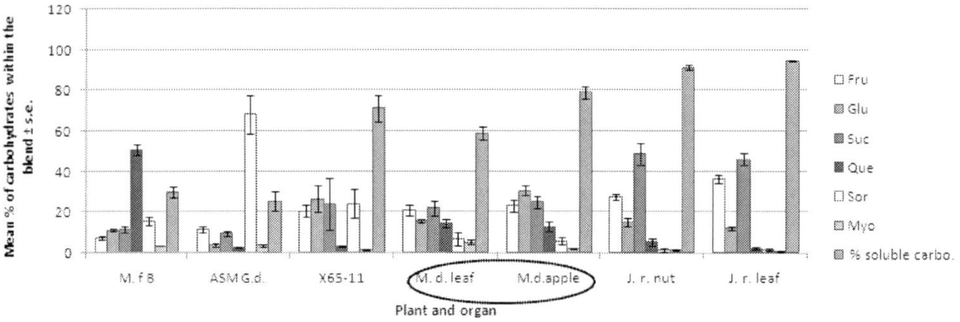

Figure 2. Leaf and fruit surface metabolite blend compositions. The quantities of D-fructose, D-glucose, sucrose, quebrachitol, sorbitol and *myo*-inositol are expressed in % of the blend. Chemical analyses were carried out on the same gas chromatograph by the method described above. Washings were collected on both the leaf side and fruit of the same shoot preferred by *C. pomonella* to lay its eggs, in the afternoon (4 to 5 p.m. solar time), during July and at the beginning of August. The following are presented: the non-host upper leaf sides of *M floribunda* Baugène (*M.f.*B) (n = four samples of six leaves); Golden Delicious (G.d.) upper leaf sides modified by ASM (ASM G.d.) (n = three samples of four leaves); a mean of four commercial cultivars cultivated in an organic orchard near Versailles of the main host *M. domestica (M.d.)* on corymb or bourse shoot leaves with two apples (n for leaves = 11 samples of six leaves, n for apples= four samples of two apples); and the secondary host, *J. regia,* leaves surrounding two nuts (n = four samples of six leaves), nuts, n = four samples of two nuts.

Unexpected Effects of a Formulated Granulovirus on *C. Pomonella* Egg-Laying via the Leaf Surface Blend

Applications of granulovirus against the codling moth, *C. pomonella*, which target neonate larvae before or during initial entry into the fruit, now offer selective control. In an experimental North Italian orchard, this product was used on two *M. domestica* cultivars. An egg-laying preference was observed on Golden Delicious vs. Red Chief, either non-treated or treated with Madex®, a commercial granulovirus product. Unexpectedly, the Madex® treatment dramatically reduced the number of eggs by 52 to 55% vs. non-treated tree from both cultivars. Chemical analyses of the metabolite blend on leaf and fruit surfaces showed modifications of sugar alcohol components on both cultivars when treated with Madex® (Lombarkia et al. 2011). Mimicking the blends on nylon cloth, those from leaf surfaces treated with Madex® reduced acceptance and egg-laying stimulation vs. blends from non-treated leaf surfaces for both cultivars. Observations from the orchard could be reproduced in laboratory assays with artificial surface blends. The granulovirus itself had no direct effect on acceptance and egg-laying stimulationThe reduction of eggs on 'Red Chief' could be primarily explained by a drastic effect on egg-laying stimulation, whereas the reduction on 'Golden Delicious' was partly linked to a lower acceptance. The effects of reduced egg-laying caused by applications of Madex® were associated to biochemical changes in surface blends, depending on the cultivar.

Role of Sugars from the Granulovirus Formulation and Their Effects on *C. Pomonella* Damage Reduction

These leaf surface blend modifications are most probably the consequences of the formulation composition that contains products used to stimulate larval feeding such as sugars (under a secret patent) or adjuvants to protect the granulovirus from UV rays and/or wetting agents for better adhesion to leaf surfaces. When sprayed on the leaves, sugar may directly act on the insect and/or via the plant by penetrating into it and participating in complex sugar signaling pathways, just like endogenous sugars (Gibson 2000; Rolland et al. 2006; Ramon et al. 2008) and, in this way, modify the chemical composition of the leaf surface. This last hypothesis was tested by removing "sugars" from the formulation and comparing the effects of the subtraction in the apple orchard to those of the formulated product and to the activity of sucrose or D-fructose water solutions at 100 ppm on *C. pomonella* damage. Sugar quantities were chosen on the basis of those that would only play a role as a plant signal and not improve larval ingestion. The study was conducted in a commercial orchard managed according to organic farming requirements and located in the region of Montauban (southern France). The orchard was composed of two rows of the apple cultivar Reinette du Canada, planted in 1991. Each elementary plot consisted of five trees. This trial was conducted on the second generation of *C. pomonella*. Damage from the first generation was eliminated by a preliminary assessment conducted before the trial began. Four treatment modes were compared: (i) the commercial Carpovirusine 2000®; (ii) a sugar-free formulation of Carpovirusine 2000®; (iii) D-fructose and sucrose solutions at 100 ppm; and (iv) D-fructose and sucrose at 100 ppm added to the sugar-free Carpovirusine 2000® formulation. Method N°18 of the French Biological Trials Commission (Audemard 1987) was applied in a randomized block design including four replicates. Each block (groups of experimental units) was as homogeneous as possible in equal numbers of treatment modes. Within each block, the terms were distributed randomly and independently of their distribution in other blocks. The number of trees was calculated to be able to observe a minimum of 250 fruits in each elementary plot at harvest. Applications were conducted with a mechanical jet sprayer using a spray volume of 1000 L/ha to ensure effective wetting of the vegetation according to good agricultural practices. The quantities were 0.1 L/100 L for the Carpovirusine 2000®, 10 g/100 L for sucrose and 10 g/100 L for D-fructose. The carbohydrates were applied four times, 9-20 days apart, in the morning. Assessments on infested fruits were performed on a regular basis on the fallen fruits per plot (fruit abscission is often a consequence of *C. pomonella* infestation). At harvest, all fruits were assessed in the central part of each plot. The variable, '% of infested fruit at harvest', is based on the ratio of the total number of infested fruits (fallen and attached) and the total number of assessed fruits (fallen and attached) per plot. Level of damage assessed in the untreated control treatment was presented in its absolute terms '% of infested fruits at harvest' while the results of the treated objects were expressed relatively in % of efficacy using the Abbott efficacy transformation: ABBOTT efficacy = $100 \times \frac{T0 - Tt}{T0}$, T0 = % of infestation in the Untreated control plot , Tt = % of infestation in the treated plot. An analysis of variance followed by a Student-Newman-Keuls test (α=0.05) were used to compare the level of damages recorded in each treatment (Fig. 3). This experiment made it possible to draw the following conclusions: (1) foliar applications of single exogenous carbohydrates could reduce *C. pomonella* damage; (2) fructose was more efficient than sucrose at the doses tested; and (3) the removal of "sugars" from the granulovirus formulation decreases the effectiveness of the commercial product.

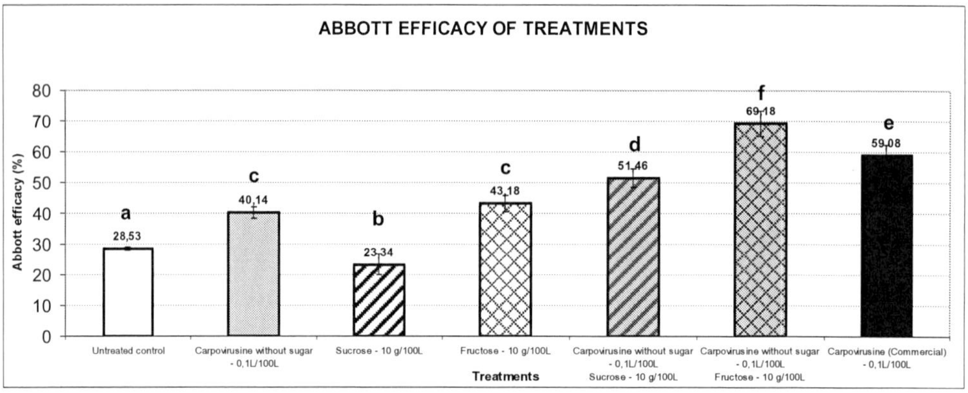

Figure 3. Abbott efficacy of carbohydrate applications on *C. pomonella* apple tree damage associated with the formulation of the granulovirus product and sprayed alone (results of the Untreated control are presented in its absolute terms: '% of infested fruits at harvest').

A NEW CONCEPT FOR ORCHARD PROTECTION AGAINST *C. POMONELLA*

Induced Modifications of the Leaf Surface Metabolome by Foliar Application of Single Sugars

These experiments show that foliar applications of sugars should induce reduction of *C. pomonella* damage on apple trees. In order to explain *C. pomonella* damage reduction, the hypothesis that carbohydrate foliar application had an effect on insect behavior due to modifications in the leaf surface metabolite blend was maintained. The metabolome of the leaf surface was analyzed to broaden the spectrum of metabolites. In the orchard of the ENSP, Potager du Roi (Versailles, France), solutions of sucrose at 10 ppm and of D-fructose at 0.1 ppm were applied to two separate rows of Golden Delicious trees. On one row, there were four alternating plots of four trees, either treated or not treated with the sugar solutions (600 L/ha). The morning treatments were carried out every 20 days throughout the season from

April until the end of August. Two trees in the middle plots were sampled in July for leaf surface chemical analysis. Leaf surface washing samples were collected from 4 to 5 p.m. solar time. Three basal leaves on the bourse shoots were collected at the top half of the tree. Their underside (preferred by insects in this trial) was washed. Three samples were collected on each tree, consisting of two associated shoots, one oriented towards the east and the other towards the west. The metabolome analyses (water soluble metabolites) were carried out after a derivation (sylilation). The method used was based on that of Fiehn (2006). The identification of compounds was carried out with GC-TOF gas chromatography using AMDIS software and quantification with Waters QuanLynx software. The Student t test was used to compare one carbohydrate tree treatment vs. non-treated trees (n=9). A total of 119 water-soluble metabolites were detected with certainty (deconvolution (AMDIS)) on the bourse shoot leaf surfaces. Among the 24 components modified by the 10-ppm sucrose treatment, ten of them were also modified by the 0.1-ppm fructose treatment. Quebrachitol, *scyllo*-inositol, *myo*-inositol, alanine and erythronic acid quantities increased, whereas those of GABA *(*Gamma Amino Butyric Acid*)*, ethanolamine and two undetermined compound decreased (Table 1A).

Table 1.A. Apple leaf surface components in the metabolome (water-soluble metabolites) whose quantities are significantly modified by foliar applications in the orchard of solutions of sucrose (10 ppm=10 μg ml^{-1}) and D-fructose (0.1 ppm=0.1 μg ml^{-1}) every 20 days (five times before analysis) on Golden delicious in July. In italic letters: components common to both treatments (Analyses carried at INRA, Plant Nitrogen Nutrition, 78000 Versailles, Fr)

Sucrose-treated (10 μg/ml		Fructose-treated (0.1 μg/ ml)	
Name	t-test	Name	t-test
quebrachitol 5tms 1859.3/318	0.0009	*1380.5/184*	0.0001
1380.5/184	0.0010	*quebrachitol 5tms 1859.3/318*	0.0002
scyllo-inositol 1949.3/318	0.0013	*GABA 1527.7/304*	0.0005
glyceric acid 1335.5/292	0.0014	*ethanolamine 1269.8/174*	0.0011
Sucrose-treated (10 μg/ml		Fructose-treated (0.1 μg/ ml)	
GABA 1527.7/304	0.0017	*scyllo-inositol 1949.3/318*	0.0012
2-ketogluconate 1779.8/292	0.0017	*1145.1/235*	0.0027
1145.1/235	0.0025	*myo-inositol 2086.7/318*	0.0075
myo-inositol 2086.7/318	0.0047	*alanine 1109.1/116*	0.0095
succinic acid 1320.9/247	0.0047	1530.2/314	0.0155
ethanolamine 1269.8/174	0.0054	*beta-alanine 1430.7/248*	0.0200
alanine 1109.1/116	0.0097	*erythronic acid 1538.6/292*	0.0260
decanoic acid 1459.2/229	0.0108	Diethylene glycol ? 1245.8/117	0.0363
glycerol 1278.1/205	0.0154	1452.9/350	0.0429
raffinose 3355.5/361	0.0164	1247.7/127	0.0464
1433.5/233	0.0188		
arabinose meox 1662.9/307	0.0229		
unk 1918.7/375	0.0248		
nonanoic acid 1368.3/215	0.0260		
beta-alanine 1430.7/248	0.0270		
beta-lactate ? 1148.8/219	0.0287		
erythronic acid 1538.6/292	0.0323		
C19 1.006 lysine 174	0.0335		
oxalic acid 1138.3 /147	0.0353		
betahydroxybutyrate [938] 1162	0.0508		

There was no visible effect on the soluble carbohydrates: sucrose, D-fructose and D-glucose. Modifications observed in these analyses partly concerned the blend and, in particular, the quantity of several sugar alcohols which have in a principal component analyses a high discriminant power, including the sorbitol of which the quantities rather variable show a tendency to be reduced.

Consequences of Induced Blend Leaf Surface Metabolite Changes on *C. Pomonella* Behavior

Females: Behavioral assays on nylon cloths as described above were carried out with these modifications linked to the amounts generally reported for Golden Delicious (Table 1B).

The combined modifications of the blend by sucrose or D-fructose reduced *C. pomonella* acceptance and egg-laying (Fig. 4).

The foliar application of carbohydrates disrupts the information that the female gathers on the plant surface to accept the site and lay eggs.

Table 1.B. Quantities (ng per cm²) and ratios of the blend metabolites (%$_1$ between the six metabolites, %$_2$ within three soluble carbohydrates and three sugar alcohols), found on the upper side of the bourse shoot leaf surface of Golden Delicious (Natural Blend=NB), after sucrose (10 ppm) or fructose (0.1 ppm) solutions inducer treatments AITB (After Inducer Treatment Blend)

	Soluble carbohydrates									Sugar alcohols								
	Glucose			Fructose			Sucrose			Sorbitol			Quebrachitol			*Myo*-inositol		
	ng/cm²	%$_1$	%$_2$	ng/cm²	%$_1$	%$_2$	ng/cm²	%$_1$	%$_2$	ng/cm²	%$_1$	%$_2$	ng/cm²	%$_1$	%$_2$	ng/cm²	%$_1$	%$_2$
NB (x 10^{-2})	67	27	38	72	29	41	38	15	21	15	6	22	32	13	46	22	9	32
AITB (x 10^{-2})	67	23	38	72	25	41	38	13	21	7.5	3	6	64	22	55	44	15	38

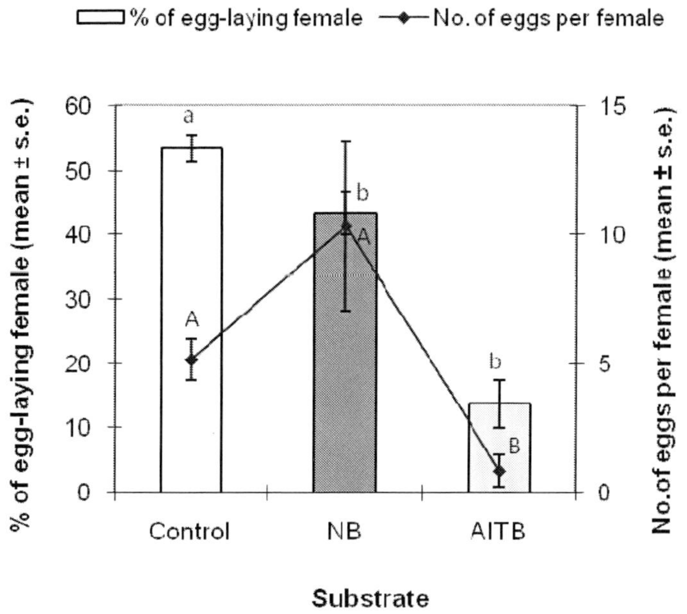

Figure 4. *C. pomonella* acceptance (percentage of egg-laying females) and egg-laying stimulation (number of eggs per female) (n = 4 x 10) on nylon cloths impregnated with: ultra-pure water (control), the man-made blend of six metabolites at similar concentrations that occur on the upper side of bourse shoot leaves of Golden Delicious (NB) and AITB. Different lower case letters indicate significantly different percentages of egg-laying females ($P < 0.05$; χ^2 test), and different upper case letters indicate significantly different numbers of eggs ($P < 0.05$; Mann-Whitney test).

Neonate larvae: The resistance effect on *C. pomonella* female egg-laying is predominant in the reduction of larval damage, but this does not exclude effects on the neonate larvae before penetrating into the fruit.

The influence of the individual metabolites from the blend was studied on nylon cloths. For each sugar alcohol molecule from the blend, 21 two-hour-old single neonate larvae were tested under red light, at 25°C ± 2 and 70% ± 3 RH. Larvae were deposited in the middle of a 10-cm diameter Petri dish, with a nylon cloth at the bottom impregnated with a single metabolite found on leaf surfaces in several concentrations. Observations of several behaviorals were recorded until the larvae reached the edge of the Petri dish and after 10 min when they did not. No significant impact of single blend components and concentrations were observed except qQucbrachitol and *myo*-inositol, which arrested neonate larvae at the highest concentrations (Derridj et al. 1999) (Fig. 5).

Foliar application of carbohydrates in apple orchard protection against the codling moth could be a new element in the urgent search for alternatives to ongoing withdrawal of pesticides or at least to their reduction.

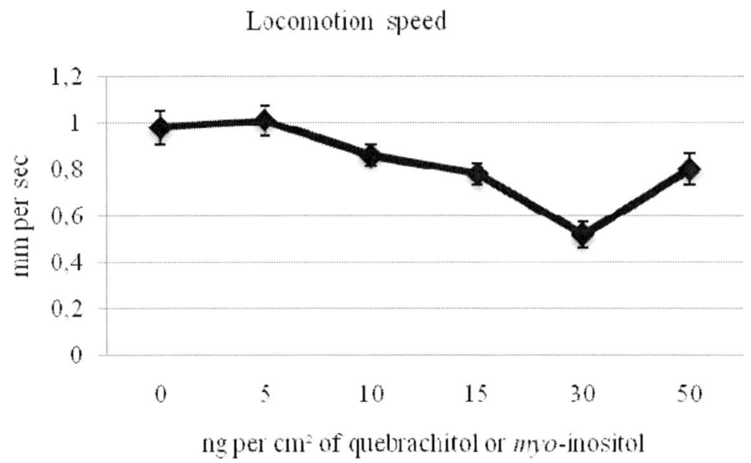

Figure 5. Sugar alcohol concentrations on nylon cloths and *C. pomonella* neonate larval speed locomotion.

Method and Efficiency of Sugar Application for Protecting Apple Orchards from *C. Pomonella*

D-fructose and sucrose were sprayed on *M. domestica* at 10 and 100 ppm in commercial orchards over three years and in three European countries (France, Italy and Greece). Plots were arranged in a randomized Fisher block design. The first application of sugar (1000 to 1500 L/ha) took place 20 days before the maximum egg-laying period of the second generation and was renewed four to five times within a 14-day interval. Sugars and insecticide spray solutions were applied between 4:30 a.m. and 7:45 a.m. Assessments on infested fruits were performed at each application date on the fallen fruits. At harvest, 300 to 500 fruits were assessed in the central part of each plot. The variable '% of infested fruit at harvest' is based on the ratio of the total number of infested fruits (fallen and attached) and the total number of infested fruits (fallen and attached) per plot. The percentages of infested fruits were separately analysed in each trial by a Student-Newman-Keuls means comparison test performed at a 5% significance level. The ANOVA assumption of homogeneity of variance was checked using a Bartlett's chi-square test. In all trials, the Abbott efficacy at harvest of each sugar foliar spray, were 37.2±16.8 % and 40.6±16.0% for D- fructose and sucrose, respectively. No significant differences were observed between the two carbohydrate solutions or between the concentrations applied (Student-Newman-Keuls test; $P < 0.05$). When D-fructose and sucrose were compared to the recommended dose of an organophosphate insecticide, they revealed comparable efficacy (Table 3). The first two digits of the trial number indicate the year; letters indicate the country: FR: France; ITA: Italy; GR: Greece. Results are presented in % of infested fruit at harvest (bold numbers) and in % of Abbott efficacy vs. untreated control damage level (numbers between brackets). IMIDAN 50 WP (Phosmet 50% w/w, WP) at 100 g/100 L (registration dose rate) and at 150 g/100 L: an organophosphate chemical insecticide.

Table 3. Effects of aqueous solutions of sugars and insecticides on *Cydia pomonella* damage in commercial apple orchards

Year and country of the trial	06/FR	08/GR	08/ITA	07/FR (1)	07/FR (2)	07/ITA	07/GR
Apple cultivar	Granny-Smith	Golden Delicious	Sansa	Golden Delicious	Golden Delicious	Golden Delicious	Mondial Gala
Replicate numbers of three trees	4	4	4	4	4	4	4
Sugar and quantity added to insecticides	-	-	-	Sucrose 100 ppm	Sucrose 100 ppm	Fructose 10 ppm	Fructose 10 ppm
Untreated control	**22.26** a	**24.93** a	**44.35** a	**40.67** a	**1.02** a	**1.75** a	**37.58** a
Fructose 10 ppm		**16.61** a (33.25)	**17.58** b (60.04)			**1.28** a (36)	**30.32** b (19.52)
Sucrose 10 ppm		**17.11** a (31.34)	**16.27** b (63.33)	**28.06** b (30.66)	**0.65** b (38.09)		
Sucrose 100 ppm		**19.63** a (21.23)		**25.11** bc (37.98)	**0.33** c (61.9)		
Sucrose 1000 ppm	**16.19** b (24.14)						
IMIDAN 50 WP 100g/100L				**23.66** bc (41.55)	**0.16** c (80.95)	**0.63** a (54.63)	**20.17** cde (46.53)
IMIDAN 50 WP 150g/100L						**0.52** a (65.29)	**14.55** e (60.81)
IMIDAN 50 WP 100g/100L+ Sugar				**15.75** d (61.34)	**0.08** c (92.86)	**0.73** a (45.13)	**18.46** de (50.66)

Foliar application of low quantities of carbohydrates is a promising method based on inducing resistance in apple tree against its major insect pest. The resistance is an antixenosis against females and probably also against neonate larvae. This type of resistance is rarely subject to resistance selection by plant breeders partly due to difficulties involved in estimating the number of C. *pomonella* eggs on trees.

CONCLUSIONS

To understand host plant selection for egg-laying by *C. pomonella*, it is necessary to follow the behavioral steps from plant attraction at a distance to plant and site acceptance after landing. A missing link should provide a partial explanation. The *C. pomonella* host plants may be attractive to insects due to volatile compounds such as (E, E)-α-farnesene, (E)-β-farnesene or a pear ester, (E, Z)-2,4-decadienoate, that affect male and female behavior (Wearing and Hutchins 1973; Hern and Dorn 1999; Yan et al. 1999; Light et al. 2001; Reed and Landolt 2002; Coracini et al. 2004). Witzgall et al. (2005) found a shared ubiquitous volatile blend released from apple, pear and walnut that affects egg-laying. Its high variability suggested a considerable plasticity in the female response to host plant odors. Trichomes were shown to be involved in the initial stages of *C. pomonella* host acceptance on the leaf surface. Their density on each side of the leaf is negatively correlated with the distribution of *C. pomonella* eggs (Blomefield et al. 1997). As a result of their physical nature, epicuticular waxes also influence egg-laying (Elmer 1980).

One chemical group has been ignored when looking for cues at the leaf surface. This group consists of primary metabolites and, in particular, soluble carbohydrates and sugar alcohols. They should, however, be of great interest for the following reasons: (i) given that they can cross through the plant cuticle, they should be excellent indicators of the physiological and metabolic state of the plant and organ; (ii) insect evolution has maintained their perception by gustatory sensillae present on many parts of their bodies outside of the mouth parts; (iii) they provide information on the nutritional value of the plant to females that do not feed on them; (iv) due to their selective passage through the cuticle, they also provide indications as to the plant species and organ specificity; and (v) non-volatile secondary metabolites appear to have a greater importance in the epidermis and mesophilic parts than on the plant surface. The fact that they are water-soluble molecules makes it possible to collect them more easily without any destruction of waxes and with few artifacts and to measure their perception and integration into the insect's nervous system.

They are freely phloem-mobile inside the plant and have a physiological role in plant metabolism and carbon transport. Outside the plant, on its surface, sugar quantities are the result of a continuous balance between the output from the apoplast and the partial subsequent penetration of the molecules into the plant. They are present on the leaf surfaces of plant species already studied (if not all plants), e.g., *Allium porrum*, *Eucalyptus sp.*, *Glycine max, Helianthus annuus*, *Malus sp., Medicago sativa, Picea glauca, Prunus laurocerasus*, *Senecio sp., Spinacia oleracea*, *Triticum* spp., *Vitis sp., Zea mays*, etc. They are of photosynthetic origin and have a photosynthetic rate, are transported throughout the plant, and accumulate in organs or parts of them, and are characteristic of physiological and nutritional plant states, leaf age and side.

The selective permeability of the leaf cuticle to the different sugars, the number of stomata by which the water-soluble carbohydrates pass through the cuticle, and leaf surface distributions of hexoses are partly genetically based. This provides these ubiquitous metabolites whose quantities vary with stable ratio characteristics when the insect is selecting its host. The metabolite ratios may vary with the metabolite blend considered (sugars and sugar alcohols in the case presented here) and thus become signals that enable *C. pomonella* to accept the appropriate plant species on which to lay its eggs. A group of six free amino acids among 20 others found on leaf surfaces discriminate four *Senecio* species, including the oligophagous Lepidoptera *Tyria jacobeae,* to deposit their eggs on (Soldaat et al. 1996). Through plant species specificity of the permeability properties of the cuticle, carbohydrate ratios can therefore reach a referent stability that the insect can keep as an image in centralized (still unknown) nervous centers for plant recognition (Popp et al. 2005).

C. pomonella behavioral responses to the sugar blends are quite precise, showing flexibility due to the influence of cues ranging from subtle ones consisting of a single component to that of one chemical group consisting of three components of soluble carbohydrates or sugar alcohols. Chemical subtractions from the blend, e.g., D-glucose or quebrachitol, could have almost no effect, whereas that of D-fructose and sorbitol has very high ones. The extent of consequences on acceptance and/or egg-laying and the intensity of female disturbance could vary depending on the chemicals concerned. Variations of the site or plant acceptance, e.g., leaf side, organ, cultivar, plant species, are probably more distinctly related to variations in the ratios among the blend components than to their quantities. The ratios between fructose or sucrose and sorbitol within the blend are closely related to host plant selection for egg-laying. These effects can be compared to those observed on the ratio-specific odor recognition of pheromones and plant volatiles (Bruce et al. 2005). Although relationships between the blend surface composition of the plant and insect acceptance and egg-laying are well related, it is likely that other chemical stimuli occur at the plant surface and add to the complexity of signals that influence behavior. In contrast, the mannitol and malic acid plant species-specific compounds of *Malus sp* are little used by females for recognition purposes (personal data).

Finding similar blend ratios between the two chemical groups on *M. domestica* and *J. regia* confirms the hypothesis of the blend activity on the recognition of a plant as host by *C. pomonella*. This is another example among rosaceous species and *Lepidoptera* where knowledge of the leaf surface metabolite composition and, particularly, that of sugar alcohols may help to explain host shift (Hora and Roessingh 1999; Roessingh et al. 1999).

The relationships consistently found between the metabolite blend tested without any volatiles in *C. pomonella* egg-laying behavioral assays and observations in orchards probably reveal either a narrow link between volatile blends and gustatory metabolite blends or the predominance of gustatory cues over volatile cues. On all host or non-host plants reported here, landing sites are similar. They include the corymb and the bourse shoot leaves during the season and then, gradually, the fruit itself. It is assumed that the choice is based at a distance on volatile compounds that are related to organ functions. The contact with the plant is then crucial because it makes it possible to differentiate the host plant from the non-host plants on which the females will lay their eggs. This supports the hypothesis of Finch (2000) that focuses on plant landing and on non-volatile compounds to explain insect egg-laying. The hypothesis that insects accept the plant site and lay eggs only on the basis of secondary metabolite cues seems also difficult to sustain, knowing that so many cues may be perceived after landing. Trichomes, waxes, volatile and non-volatile compounds, secondary and primary metabolites may all influence plant and site acceptance for egg-laying sequentially and/or simultaneously. More studies are required to differentiate between or to link plant cues, both at the same time or sequentially, to understand the flexibility of insect host selection.

Insect responses to the blend permit an adaptation to the physiological changes of the plant. Females follow the sugar translocations. They show a remarkable adaptation by responding to this sugar blend that permits egg-laying in the most well-adapted location throughout fruit growth for the future of their offspring (Whetter and Taper 1963; Chong and Taper 1971; Wallart 1980; Vemmos 1995). Sugar alcohol quantities in relation to sucrose in both sink and source tissues are phloem-mobile and shift as the leaf ages and environmental conditions change (Loescher 1987). Sorbitol accounts for 60 to 90% of the carbon exported from the leaf to the fruit (Brown and Hening 1996). Recent progress in research on the metabolism and transport of sorbitol also shows that it can play a role in biotic and abiotic plant responses, e.g., reactions to pathogen infections and tolerance to environmental stress (Kanayama 2009). Responses to the balance between sucrose and sorbitol shown by the ASM plant treatment may also provide clues for insects as to the state of stress of the plant in relation to environmental conditions (Merchant et al. 2006).

The permeability of the cuticle remains determinant in the presence and ratio of a metabolite on the leaf surface. It is suggested that various plant families including monocotyledons and gymnosperms can synthesize sorbitol, but in smaller quantities than in rosaceous species (Kanayama 2009). Thus, the presence of sugar alcohols on walnut surfaces could be due to its freely mobile forms and/or a high permeability of the cuticle as the result of being a deciduous species less well adapted to environment stress (Kirsh et al. 1997).

The adaptation of females when selecting their host plant is also shown by the energy savings due to the speed of gathering and integrating the information received. *C. pomonella* shows a reduced time frame at several behavioral stages: (i) detection of plant metabolite cues from the leaf surface without penetrating into the plant tissues; (ii) sequentiality between behavioral events, e.g., stimulation of ovipositor scanning, followed by egg-laying stimulation; and (iii) responses to photosynthates and their ratios that provide complex information on plant physiology, carbon transport, reaction to stress, nutritive value, plant species specificity, etc., at the same time. "There is clearly a selective advantage for the insect to recognize its goal from the wrapping, the packaging, and on the other hand for the plant to warn or mislead before its integrity is breached" (Southwood 1996).

The plant also has an evolving interest in limiting and preserving its integrity by providing signals at its frontier with the environment. In the case of *C. pomonella* and *Malus sp*., it appears that any change in the leaf surface metabolite blend is connected to the behavior of the pest. The ratio between the two groups of compounds is associated with a rosaceous non-host (*M.f.*B), and the ratios between blend components are related to a cultivar resistance or induced of *M. domestica*. The resistance provided by the leaf surface blend is primarily an antixenosis, i.e., acting on insect behavior. It can limit any wound that induces costly energy defense mechanisms.

Factors external to the plant may affect leaf surface composition and contribute to the emission of cues. Host-specific non-pathogen associated epiphytic microorganisms can induce leaking of metabolites from plants (Knoll and Shreiber 1998) and/or produce them (Georgiou et al. 1992). Their possible contributions could also help to better understand insect host plant range (Wilson and Lindow 1994; Mercier and Lindow 2000).

The application of phytosanitary products may change the composition of the leaf surface blend. Among the compounds of the commercial formulations of many phytosanitary products are those introduced to improve the adhesion to the surface and its penetration through the cuticle. This could possibly lead to a risk of modification of the cuticular permeability and, therefore, of the composition of the blend and its effects on pest behavior.

Other compounds from the formulation such as sugars are introduced into the commercial products as phago-stimulants for more effective ingestion by larvae. What is particularly surprising is that the exogenous foliar application of a single sugar can induce changes in the blend composition, thus inducing plant resistance to the pest. Application of quantities of 10 ppm of sucrose or of 0.1 ppm of D-fructose to leaves in the morning, quantities that can usually be found on apple leaves in the evening, induces changes in the sugar alcohol group several days later. In this way, antixenosis is induced before the larva attacks. This induced resistance differs from the induced defense by elicitors that amplifies defense while wounding. It nevertheless supports the concept that exogenous sugars such as plant molecules outside the intact plant cell and other disintegrated plant cell contents are perceived by plants as signals of 'damaged self recognition' and therefore trigger systemic, hormone-mediated defense responses (Heil 2009). Exogenous carbohydrates can also penetrate into the plant by stomata (Eichert and Goldbach 2008; Mac Gregor et al. 2008), and into the guard cells as well. The majority of guard cells have chloroplasts, which would therefore provide an ideal and convenient location for sensory or regulatory mechanisms (Lawson 2008). Guard cell aploplastic sucrose can also exert an osmotic effect, which can lead to stomatal closure, acting as a possible signal between the mesophyll assimilation rate and transpiration (Kang et al. 2007). It was postulated that sucrose concentrations near the guard cell regulate gene expression, as has been shown in many other tissues (Baiser et al. 2004).

Sucrose and hexoses (of which glucose has been the most studied) play dual functions in gene regulation as exemplified by the up-regulation of growth-related genes and the down-regulation of stress-related genes (Roland et al. 2006; Ramon et al. 2008). Prominent functions of glucose such as cellular signaling in specific regulatory pathways that modulate plant growth and development strangely did not appear as significant plant cues for *C. pomonella* host selection within the surface blend.

In contrast, fructose is a major leaf surface cue for Lepidoptera. *O. nubilalis* and *C. pomonella* detect fructose better than glucose or sucrose. Fructose is a positive signal for them to lay their eggs. Among two maize hybrids grown under the same conditions, variations of D-fructose from tissues in the range of 40 to 25% were registered in the ear leaf tissues, depending on the plant hybrid and growth stage. This trait could be transmitted from inbred lines to hybrids, and low quantities of fructose contributed to resistance to *O. nubilalis* Hbn egg-laying (Derridj et al. 1990) and through foliar applications, it can induce maize resistance to *O. nubilalis* egg-laying (personal data). It also induces apple resistance to *C. pomonella* by modifying the sugar alcohol composition of the blend. Its output and input through the plant cuticle could possibly protect the plant from insects. In the same way, foliar application of a fructose analog, 2, 5-Dihydroxymethyl-3, 4-dihydroxypyr-rolidine (DMDP), induced a root resistance in tomato against nematodes (Birch et al. 1993). Although fructose has remained unexplored in plants and animals, this does not mean that it does not play an interesting role. Fructose signaling appears to interact positively with abscisic acid (ABA) signaling via hormone biosynthesis, whereas it is probably antagonized by ethylene signaling (Young-Hee and Sang-Dong 2011). A transcription factor has recently been identified but the sensor has yet to be found (Wind et al. 2010). The question then arises as to whether or not this is a form of self-protection of the plant due to the cuticular permeability to D-fructose and its involvement in particular signaling pathways within the plant. The dual activity of fructose on insects and plants should be the result of a long co-evolution. It would be interesting to extend the study of fructose signalization pathways and genes in the plant and their evolution within plant species (Büttner 2007) to understand the meaning of fructose detection by insects and its signaling role in host plant selection. It was also recently shown in mammals that dietary fructose was implicated in rat cell signaling perturbation and metabolic syndromes such as insulin resistance, obesity, type 2 diabetes, and high blood pressure. Fructose probably has a signaling function of general interest in the living world that we are just beginning to discover in a number of different kingdoms.

Knowledge of relationships between leaf surface chemical cues and *C. pomonella* host selection should be used to find new solutions in engineering or plant breeding to reduce its impact on crops and particularly on *M. domestica*. It can also help assess the risk of changes in the host when introducing new cultivars or crops. By inducing damage reduction through the application of sub-doses of sugars for both a period of several years and under different climates, it is possible to confirm the robustness of our assumptions on the basis of leaf surface studies. Vigilence is still of utmost importance when inducing apple tree resistance to *C. pomonella* through carbohydrates so as not to expand the host range of the codling moth that could prefer non-treated plants, or to induce a host shift from another insect to apple trees. Moreover, direct or indirect effects on auxiliary and parasitic organisms cannot be completely excluded. One must keep in mind that pesticides may be able to modify plant surface signals. This can be applied to any product that can elicit plant reactions (Boerth et al. 2008) and/or alter cuticular permeability (Baur 1999; Schreiber 2006).

Plant surface chemical signalization could greatly benefit from additional ecological knowledge through studies of cuticular permeability to water-soluble metabolites and relationships with biotic and abiotic factors and epiphytic microorganisms. These results open new possibilities for insect-host plant selection mechanisms and new crop management methods.

ACKNOWLEDGMENTS:

We would like to thank Philippe Couzi for his ongoing technical help at INRA Versailles, as well as F. Combe and C. Iorriatti and their staff at the INRA Gotheron and IASMA Research Stations for providing the possibility of working in their experimental orchards. For their contribution to experiments, we would like to thank A. Borges (ASM), E. Cochet (chemical analyses of commercial apple cultivars and *J. regia*), I. Cavana (neonate larval behavior), F. Moulin (ENSP) for his four years of work in Le Potager du Roi, (Versailles), G. Clément for the metabolome analyses at INRA, PTSCV (Versailles), Y. Lespinasse (INRA Angers), for sharing his knowledge on apple trees, and J. Dasque from CIREA (Bordeaux) who was at the origin of the research on apple tree and *C. pomonella* relationships.

REFERENCES

Abbot, D. L. (1960). The bourse shoot as a factor in the growth of apple fruits. *Annals of applied biology, 48*, 2, 434-438.

Adati, T. and Matsuda, K. (1993). Feeding stimulants for various leaf beetles (*Coleoptera: Chrysomelidae*) in the leaf surface wax of their host plants. *Applied Entomology and Zoology, 28*, 319-324.

Albert, P. J. (1980). Morphology and innervation of mouthpart sensilla in larvae of the spruce budworm, *Choristoneura fumiferana* (Clem.) (Lepidoptera Tortricidae). *Canadian Journal of Zoology, 58*, 842-851.

Andrews, J. H. and Harris, R. F. (2000). The ecology and biogeography of microorganisms of plant surfaces. *Annual Review of Phytopatholy, 38*, 145-180.

Andrews, J. H., Spear, R. N., Nordheim, E. V. (2002). Population biology of *Aureobasidium pullulans* on apple leaf surfaces. *Canadian Journal of Microbiology,48*, 6, 500-513.

Audemard, H. (1976). Etude demoecologique du carpocapse (*Laspeyresia pomonella* L.) en verger de pommiers de la basse vallée du Rhône. Possibilités d'organisation d'une lutte intégrée. Thèse d'Etat. University F. Rabelais. Tours. 365p.

Audemard, H. (1987). Biological and integrated control in apple, pear, and apricot orchards. *Entomophaga, 32*, 59-72.

Baiser, M., Hemmann, G., Holman, R., Corke, F., Card, R., Smith, C., Rook, F., Beva, M.W. (2004). Characterization of mutants in *Arabidopsis* showing increased sugar specific gene expression, growth and development. *Plant Physiology, 134,* 81-91.

Baker, E. A. (1982). *Chemistry and morphology of plant epicuticular waxes*. In D. F. Cutler, K. L. Alvin and C. E. Price (Eds.), *The Plant Cuticle,* (Academic Press, 139-165), London.

Balachowsky, A. S. (1966). Entomologie appliquée à l'agriculture. Tome II Lépidoptères. Masson et Cie., Paris. Vol. 1, 653-743.

Barthlott, W. (1990). *Scanning electron microscopy of the epidermal surface in plants*. In D. Claugher (Ed.), *Scanning Electron Microscopy in Taxonomy and functional Morphology*, (Clarendon Press, Clarendon Press, 69-94). Oxford.

Barthlott, W., Neinhuis, C., Cutler, D., Ditsch, F., Meusel, I., Theisen, I. and Wilhelmi H. (1998). Classification and terminology of plant epicuticular waxes. *Botanical Journal of the Linnean Society, 126*, 237-260.

Baur, P. (1999). Surfactant effects on cuticular penetration of neutral polar compounds: dependence on humidity and temperature. *Journal of Agricultural and Food Chemistry, 47*, 2, 753-761.

Bernays, E. A. and Chapman, R. F. (1994) *Host-Plant Selection Behavior of Phytophagous Insects*, Chapman and Hall, New York.

Bernays, E. A. and Simpson, S. J. (1982). Control of food intake. *Advances in Insect Physiology, 16*, 59-118.

Birch, A.N.E., Robertson, W.M., Geoghegan, I.E., McGavin, W.J., Alphey, T.J.W., Philips, M.S., Fellows, L.E., Watson, A.A., Simmonds, M.S.J., Porter, E.A. (1993a). DMDP- A plant derived sugar analogue with systemic activity against plant parasitic nematodes. *Nematologia, 39*, 521-535.

Blomefield, T. L., Pringle, K. L. and Sadie, A. (1997). Field observation on egg-laying of codling moth, *Cydia pomonella* (Linnaeus) (*Lepidoptera*: *Oleutreutidae*), in an unsprayed apple orchard in South Africa. *African Entomology, 5,* 319-336.

Bodnaryk, R. P. (1992). Leaf epicuticular wax, an antixenotic factor in *Brassicaceae* that affects the rate en pattern of feeding of flea beetle, *Phyllotreta cruciferae* (Goeze). *Canadian Journal of Plant Science, 72*: 1295-1303.

Boerth, D. W., Eder, E., Stanks, J. R., Wanek, P., Wacker, M., Gaulitz, S., Skypeck, D., Pandolfo, D. and Yashin M. (2008). DNA adducts as biomarkers for oxidative and genotoxic stress from pesticides in crop plants. *Journal of Agriculture and Food Chemistry*, *56*, 6751–6760.

Borner, G. H. H., Sherrier, D. J., Weimar, T., Michaelson, L. V., Hawkins, N. D., MacAskill, A., Napier, J. A., Beale, M. H., Lilley, K. S. and Dupree, P. (2005). Analysis of detergent-resistant membranes in *Arabidopsis*. Evidence for plasma membrane lipid rafts. *Plant Physiology, 137*, 104-116.

Brennan, E. B., Weinbaum, S. A., Rosenheim, J. A. and Karban, R. (2001). Heteroblasty in *Eucalyptus globulus* (*Myricales: Myricaceae*) affects egg-laying and settling preferences of *Ctenarytaina eucalypti* and *C. spatulata* (*Homoptera: Psyllidae*). *Environmental Entomology*, *30*, 1144-1149.

Brisset, M. N., Cesbron, S., Thompson, S. V. and Paulin, J. P. (2000). Acibenzolar-S-Methyl induces the accumulation of defense related enzymes in apple and protects from the fire blight. *European Journal of Plant Pathology,* 106, 6, 529-536.

Brooks, J. S., Williams, E. H. and Feeny, P. (1996). Quantification of contact egg-laying stimulants for black swallowtail butterfly, *Papilio polyxenes*, on the leaf surfaces of wild carrot, *Daucus carota. Journal of Chemistry and Ecology*, *22*, 2341-2357.

Brown, P. H. and Hening, H. (1996). Phloem mobility of boron is species dependent: evidence for phloem mobility in sorbitol-rich species. *Annals of Botany, 77,* 5, 497-505.

Bruce, T. J. A., Wadhams, L. J. and Woodcock, C. M. (2005). Insect host location: a volatile situation. *Trends in Plant Science, 10*, 6, 269-274.

Büttner, M. (2007). The monosaccharide transporter (-like) gene family in *Arabidopsis*. *FEBS Letters* 581, 2318–2324.

Casado, D., Gemeno, C., Avilla, J. and Riba, M. (2008). Diurnal variation of walnut tree volatiles and electrophysiological response in *Cydia pomonella* (*Lepidoptera : Tortricidae*). *Pest Management Science, 64,* 736-747.

Chapman, R. F. and Bernays, E. A. (1989). Insect behavior at the leaf surface and learning as aspects of host plant selection. *Experientia*, *45*, 215-222.

Chong, C. and Tager, C. D. (1971). Daily variations of sorbitol and related carbohydrates in *Malus* leaves. *Canadian Journal of Botany, 49*, 173-177.

Coracini, M., Bengtsson, M., Liblikas, I. and Witzgall, P. (2004). Attraction of codling moth males to apple volatiles. *Entomologia Experimentalis et Applicata, 110*, 1-10.

Derridj, S. (1996). *Nutrients on the leaf surface*. In C. E. Morris, P. C. Nicot P.C. and C. Nguyen (Eds.), *The Aerial and plant surface microbiology*, (Plenum Press, 25-41). New York.

Derridj, S., Anglade, P., Fiala, V. and Panouillé, A. (1990). Perspectives d'utilisation des critères biochimiques de la feuille dans la sélection de maïs défavorable à la ponte de pyrale (*Ostrinia nubilalis* Hbn.). International Conference on Pest in Agriculture 2/1990-12.04/ Versailles, FRA, Paris: ANPP (1990), 3, 1209-1216.

Derridj, S. and Borges A. (2006). Apple tree resistance against an insect pest by an elicitor (ASM). Investigations by analyses of the leaf surface metabolites of the tree sites. IOBC workshop on Induced resistance in plants against insects and diseases *IOBC/wprs, 29 (8), 9-13*.

Derridj, S., Boutin, J. P., Fiala, V. and Soldaat, L. L. (1996a). Composition en métabolites primaires de la surface foliaire du poireau: étude comparative, incidence sur la sélection de la plante hôte pour pondre par un insecte. *Acta Botanica Gallica*, *143*, 125-130.

Derridj, S., Cabanat, I., Cochet, E., Couzi, P., Lombarkia, N. and Wu, B.R. (1999). Incidence des métabolites présents à la surface des organes du pommier sur le comportement de *Cydia pomonella* (Lepidoptera Tortricidae) ; A.N.P.P. 5ème conférence internationale sur les ravageurs en agriculture. Montpellier, 7-8-9 décembre, II, p279-286.

Derridj, S., Gregoire, V., Boutin, J. P. and Fiala, V. (1989). Plant growth stages in the interspecific oviposition preference of European corn borer and relation with chemicals present on leaf surface. *Entomologia Experimentalis et Applicata*, *53*, 267-276.

Derridj, S., Wu, B.R., Stammitti, L., Garrec, J. P. and Derrien, A. (1996b). Chemicals on the leaf surface, information about the plant available to insects. *Entomologia Experimentalis et Applicata, 80*, 197-201.

Eichert, T. and Burkhardt, J. (2001). Quantification of stomacal uptake of ionic solutes using a new model system. *Journal of experimental Botany, 52*, 771-781.

Eichert, T. and Goldbach, H. E. (2008). Equivalent pore radi of hydrophilic foliar uptake routes in stomatous and astomatous leaf surfaces – further evidence for a stomacal pathway. *Physiologia Plantarum, 132*, 491-502.

Eichert, T., Kurtz, A., Steiner, U. and Goldbach, H. E. (2008). Size exclusion limits and lateral heterogeneity of the stomacal foliar pathway for aqueous solutes and water – suspended nanoparticles. *Physiologia Plantarum, 134*, 151-160.

Eigenbrode, S. D. and Espelie K. E. (1995). Effects of plant epicuticular lipids on insect herbivores. *Annual Review of Entomology, 40*, 171-194.

Eigenbrode, S. D., Espelie, K. E. and Shelton, A. M. (1991). Behavior of neonate diamondback moth larvae *Plutella xylostella* L. on leaves and on extracted leaf waxes of resistant and susceptible cabbages. *Journal of Chemical Entomology, 17*, 1691-1704.

Elmer, A, Hagley, C, Bronskill, J. F. and Ford, E. J. (1980). Effect of the physical nature of leaf and fruit surfaces on egg-laying by the codling moth, *Cydia pomonella* (*Lepidoptera*: *Tortricidae*). *Canadian Entomologist, 112*, 5, 503-510.

Ensikat, H. J., Neinhuis, C. and Barthlott, W. (2000). Direct access to plant epicuticular wax crystals by a new mechanical isolation method. *International Journal of Plant Sciences, 161*, 143-148.

Espelie, K. E. (1996). *Integrated case study: effects of maize leaf epicuticular lipids on insect pests*. In G. Kerstiens (Ed.), *Plant Cuticles: An Integrated Functional Approach*, (BIOS Scientific Publishers, 223-230), Oxford.

Espelie, K. E., Bernays, E. A. and Brown, J. J. (1991). Plant and insect cuticular lipids serve as behavioral cues for insects. *Archives of Insect Biochemistry and Physiology, 17*, 223-233.

Faini, F., Labbe, C. and Coll, J. (1999). Seasonal changes in chemical composition of epicuticular waxes from the leaves of *Baccharis linearis*. *Biochemistry Systematic and Ecology, 27*, 673-679.

Fiala, V., Boutin, J. P., Barry, P. and Derridj, S. (1993). Les métabolites de surface foliaire (phylloplan) : présence et rôle dans les relations plante-insecte. *Acta Botanica Gallica, 140*, 207-216.

Fiala, V., Glad, C., Martin, M., Jolivet, E. and Derridj, S. (1990). Occurrence of soluble carbohydrates on the phylloplane of maize (*Zea mays* L.): variations in relation to leaf heterogeneity and position on the plant. *New Phytology, 115*, 609-615.

Fiehn, O. (2006). Metabolite profiling in *Arabidopsis*. *Methods of molecular biology, 323*, 439-47.

Finch. S. and Collier, R. H. (2000). Host-plant selection by insects - a theory based on 'appropriate/inappropriate landings' by pest insects of cruciferous plants. *Entomologia Experimentalis et Applicata, 96*, 2, 91-102.

Fraenkel, G. S. (1959). The raison d'être of secondary plant substances. *Science, 129*, 1466-1470.

Geier, P. W. (1963). The life history of codling moth, *Cydia pomonella* L. (*Lepidoptera, Tortricidae*) in the Austrian Capital Territory. *Australian Journal of Zoology*, *11*, 323-367.

Georgiou, G., Lin, S. C. and Sharma, M. M. (1992). Surface-active compounds from microorganisms. *Bio/technology*, *10*, 60-65.

Gibson, S. I. (2000) Plant sugar response pathways. Part of a complex regulatory web. *Plant physiology*, *124*, 1532-1539.

Giese, B. N. (1975) Effects of light and temperature on the composition of epicuticular wax of barley leaves. *Phytochemistry*, *14*, 921-929.

Goodman, R. N. and Addy, S. K. (1963). Penetration of excised apple cuticular membranes by radioactive pesticides and other model compounds. *Journal of Phytopathology*, *46*, 1-10.

Goonewardene, H. F. and Howard, P. H. (1989). E7-47, E7-54, E29-56, and E31-10 Apple Germplasm with Multiple Pest Resistance. *HortScience. 24*, 1, 167-169.

Green, P. W. C., Stevenson, P. C., Simmonds, M. S. J. and Sharma, H. C. (2003). Phenolic compounds on the pod-surface of pigeonpea, *Cajanus cajan*, mediate feeding behavior of *Helicoverpa armigera* larvae. *Journal of Chemistry and Ecology*, *29*, 811-821.

Harborne, J. B. and Baxter, H. (1993). *Phytochemical Dictionary. A Handbook of Bioactive Compounds from Plants*, Taylor and Francis, London.

Heil, M. (2009). Damaged-self recognition in plant herbivore defence. *Trends in Plant Science, 14*, 7, 357-363.

Hern, A. and Dorn, S. (1999). Sexual dimorphism in the olfactory orientation of adult *Cydia pomonella* in response to α-farnesene. *Entomologia Experimentalis et Applicata, 92,* 63-72.

Hiroi, M., Tanimura, T. and Marion-Poll, F. (2008). Hedonic Taste in *Drosophila* Revealed by Olfactory Receptors Expressed in Taste Neurons. *PLoS ONE 3(7): e2610. doi:10.1371/journal.pone.*

Holloway, P. J. (1971). *The chemical and physical characteristic of leaf surfaces.* In T. F. Preece and C. H. Dickinson (Eds.) , *Ecology of Leaf Surface Microorganisms*, (Academic Press, 39-53). London.

Holloway, P. J. (1982). *Structure and histochemistry of plant cuticular membranes: an overview.* In D. F. Cutler, K. L. Alvin and C. E. Price (Eds.), *The Plant Cuticle*, (Academic Press, 1-32), London.

Hopkins, R. J., Birch, A. N. E., Griffiths, D. W., Baur, R., Städler, E. and McKinlay, R. G. (1997). Leaf surface compounds and egg-laying preference of Turnip root fly *Delia floralis*: the role of glucosinolate and nonglucosinolate compounds. *Journal of Chemical Ecology*, *23*, 629-643.

Hora, K. H. and Roessingh, P. (1999). Egg-laying in *Yponomeuta cagnagellus*:the importance of contact cues for host plant acceptance. *Physiological Entomology, 24*, 109-120.

Jeffree, C. E. (1986). *The cuticle, epicuticular waxes and trichomes of plants, with reference to their structure, functions and evolution.* In B. Juniper and R. Southwood (Eds.), *Insects and the Plant Surface*, (E. Arnold, 23-64). London.

Jeffree, C. E. (1996). *Structure and ontogeny of plant cuticles.* In G. Kerstiens (Ed.), *Plant Cuticles: An Integrated Functional Approach*, (BIOS Scientific Publishers, 33-82). Oxford.

Jeffree, C. E. (2006). *The fine structure of the plant cuticle.* In M. Riederer and C. Müllller (Eds.), *Biology of the plant cuticle, Annual Plant Review, 23, 11-126.* Blackwell Publishing, Oxford.

Jetter, R. (2006). *Composition of plant cuticular waxes.* In M. Riederer and C. Müllller (Eds.), *Biology of the plant cuticle*, Annual Plant Review, 23, 145-181. Blackwell Publishing, Oxford.

Jetter, R., Schäffer S. and Riederer, M. (2000). Leaf cuticular waxes are arranged in chemically and mechanically distinct layers: evidence from *Prunus laurocerasus* L.. *Plant Cell and Environnement, 23*, 619-628.

Jetter, R. and Schäffer, S. (2001). Chemical composition of the *Prunus laurocerasus* leaf surface. Dynamic changes of the epicuticular wax film during leaf development. *Plant Physiology, 126*, 1725-1737.

Jones, T. H., Potts, B. M., Vaillancourt, R. E. and Davies, N. W. (2002). Genetic resistance of *Eucalyptus globulus* to autumn gum moth defoliation and the role of cuticular waxes. *Canadian Journal of Forest Research, 32*, 1961-1969.

Juniper, B. E. and Southwood, T. R. E. (1995). *Waxes on plant surfaces and their interactions with insects.* In R. J. Hamilton (Ed.), *Waxes: Chemistry, Molecular Biology and Functions*, (The Oily Press, 157-176). West Ferry, Dundee, Scotland.

Kanayama, Y. (2009). Physiological Roles of Sugar-alcohols in Horticultural Crops. *Journal of the japanese society for horticultural science, 78,* 2, 158-168.

Kang, Y., Outlaw, W.H. Jr, Anderson, P.C., Fiore, G.B (2007). Guard cell apoplastic sucrose concentration-a link between leaf photosynthesis and stomatal aperture size in aploplastic phloem loader *Vicia faba* L.. *Plant, Cell and Environment, 30,* 551-558.

Kannan, S. (1969). Penetration of iron and some organic substances through isolated cuticular membranes. *Plant Physiology, 44*, 517-521.

Kennedy, J. S. (1953). Host plant selection in *Aphididae. Transactions of the 9th International Congress of Entomology (Amsterdam), 2, 110-113.*

Kennedy, J. S. (1958). The experimental analysis of aphid behaviours and its bearing on current theories of instinct. *Proceeding 10th International Congress of Entomology. Montreal 2, 397-404.*

Kirsch, T., Kaffarnik, F., Riederer, M. and Schreiber, M. (1997). Cuticular permeability of the three tree species *Prunus laurocerasus* L., *Ginkgo biloba* L. and *Juglans regia* L.: comparative investigation of the transport properties of intact leaves, isolated cuticles and reconstituted cuticular waxes. *Journal of Experimental Botany, 48,* 310, 1035-1045.

Knoll, D. and Schreiber, L. (1998). Influence of epiphytic micro-organisms on leaf wettability: wetting of the upper leaf surface of *Juglans regia* and of model surfaces in relation to colonization by micro-organisms. *New Phytology, 140*, 271-282.

Koch, K., Bhushan, B. and Barthlott, W. (2009). Multifunctional surface structures of plants: An inspiration for biomimetics. *Progress in Materials Science, 54*, 137–178.

Lacaille, F., Hiroi, M., Twele, R., Inoshita, T., Umemoto, D., Manière, G., Marion-Poll, F., Ozaki, M., Francke, W., Cobb, M., Everaerts, C., Tanimura, T. and Ferveur, J. F.(2007). An Inhibitory Sex Pheromone Tastes Bitter for *Drosophila* Males. *PLoS ONE 2(8): e661. doi:10.1371/journal.pone.*

Lawson, T., Lefebvre, S., Baker, N.R., Morison, J.I.L. and Raines, C. (2008). Reductions in mesophyll and guard cell photosynthesis impact on the control of stomatal responses to light and CO_2. *Journal of Experimental Botany 59,* 3609-3619.

Light, D.M., Knight, A. L., Henrick, C. A., Rajapaska, D., Lingren, B., Dickens, J. C., Reynolds, K. M., Buttery, R. G., Merrill, G., Roitman, J. and Campbell, B. C. (2001). A pear-derived kairomone with pheromonal potency that attracts male and female codling moth, *Cydia pomonella* (L.). *Naturwissenschaften, 88*, 333-338.

Loescher, W. H. (1987). Physiology and metabolism of sugar alcohols in higher plants, *Physiologia Plantarum, 70*, 3, 553–557.

Lombarkia, N. and Derridj, S. (2002). Incidence of apple fruit and leaf surface metabolites on *Cydia pomonella* egg-laying. *Entomologia Experimentalis et Applicata, 104*, 79-87.

Lombarkia, N. and Derridj, S. (2008). Resistance of *Malus domestica* to *Cydia pomonella* egg-laying due to leaf surface metabolites. *Entomologia Experimentalis et Applicata, 128*, 57-65.

Lombarkia, N., Iorriatti, C., Bourguet, E. and Derridj, S. (2011). Unexpected effect of a larvicide, Madex® (granulovirus), on egg-laying of *Cydia pomonella* L. (*Lepidoptera, Tortricidae*) due to changes in chemical signalization on the apple leaf surface. *Entomologia Experimentalis et Applicata*, (submitted).

Loon van, J. A., Blaakmeer, A., Griepink, F. C., van Beek, T. A., Schoonhoven, L. M. and de Groot, A. (1992). Leaf surface compound from *Brassica oleracea* (*Cruciferae*) induces oviposition by *Pieris brassicae* (*Lepidoptera: Pieridae*). *Chemoecology, 3*, 1, 39-44.

Loon Van, J. A. and Schoonhoven, L. M. (1999). Specialist deterrent chemoreceptors enable *Pieris* caterpillars to discriminate between chemically different deterrents. *Entomologia Experimentalis et Applicata, 91*, 1, 29-35.

Maher, N., Thiery, D. and Städler, E. (2006). Egg-laying by *Lobesia botrana* is stimulated by carbohydrates detected by contact chemoreceptors. *Physiological Entomology, 31*, 14-22.

Maier-Maercker, U. (1983). The role of peristomacal transpiration in the mechanism of stomacal movement. *Plant, Cell and environment, 6*, 369-380.

Merchant, A. M., Tausz, S. K., Arndt, S. K. and Adams, M. A. (2006). Cyclitols and carbohydrates in leaves and roots of 13 Eucalyptus species suggest contrasting physiological responses to water deficit. *Plant cell Environment, 29, 2017-2029.*

Mercier, J. and Lindow, S. E. (2000). Role of leaf surface carbohydrates in colonization of plants by bacterial epiphytes. *Applied Environmental Microbiology, 66*, 369-374.

Mewis, I. Z., Ulrich, C. and Schnitzler, W. H. (2002). The role of glucosinolates and their hydrolysis products in egg-laying and host-plant finding by cabbage webworm, *Hellula undalis*. *Entomologia Experimentalis et Applicata, 105*, 129-139.

Miller, L. P. (1973). *Phytochemistry*. The process and products of photosynthsesis. Van Nostrand Rienhold Company, vol. 1, p168-175.

Morris, C. E., and Rouse, D. I. (1985). *Role of nutrients in regulating epiphytic bacterial populations*. In C. E. Windels and S. E. Lindow (Eds.), *Biological control of the phylloplane*, (The American Phytopatologist Society, pp 63-82). Minnesota.

Müller, C. (2006). *Plant-insect interactions on cuticular surfaces*. In M. Riedererand C. Müllller (Eds.), *Biology of the plant cuticle, Annual Plant Review, 23, 398-422*. Blackwell Publishing, Oxford.

Müller, C. and Riederer, M. (2005). Plant surface properties in chemical ecology. *Journal of Chemical Ecology, 31,* 11, 2621-2651.

Panzuto, M. and Albert, P. J. (1997). Different sensitivities of the sugar receptor of the lateral styloconic sensillum in fourth and sixth-instar larvae of the spruce budworm *Choristoneura fumiferana*. *Entomologia Experimentalis et Applicata, 82*, 335-340.

Popp, C., Burghardt, M., Friedmann, A. and Riederer, M. (2005). Characterization of hydrophilic and lipophilic pathways of *Hedera helix* L. cuticular membranes: permeation of water and uncharged organic compounds. *Journal of experimental Botany, 56*,421, 2797-2806.

Powell, G., Maniar, S. P., Pickett, J. A. and Hardie, J. (1999). Aphid responses to non-host epicuticular lipids. *Entomologia Experimentalis et Applicata, 91*, 115-123.

Plourde, D. F., Goonewardene, H. F. and Kwlek, W. F. (1985). Pubescence as a factor in codling moth, egg-laying, and fruit entry in five apple selections. *Hortscience, 20*, 82-84.

Rapley, L. P., Allen, G. R. and Potts, B. M. (2004). Susceptibility of *Eucalyptus globulus* to *Mnesampela privata* defoliation in relation to a specific foliar wax compound. *Chemoecology, 14*, 157-163.

Ramon, M., Rolland, F. and Sheen, J. (2008). Sugar sensing and signaling. The *Arabidopsis* Book, American Society of Plant Biologists. 22p.

Reed, H. C. and Landolt, P. J. (2002). Attraction of mated female codling moths (*Lepidoptera*: *Tortricidae*) to apples and apple odor in a flight tunnel. *Florida Entomologist, 85*, 2, 324-329.

Reifenrath, K., Riederer, M. and Müller, C. (2005). Leaf surface wax layers of *Brassicaceae* lack feeding stimulants for *Phaedon cochleariae. Entomologia Experimentalis et Applicata, 115*, 1, 41-50.

Riederer, M. (1991). The cuticle as the barrier between terrestrial plants and the atmosphere-significance of growth-structure for cuticular permeability. *Naturwissenschaften, 78*, 201-208.

Riederer, M. and Müller, C. (2006). *Biology of the plant cuticle*, Annual Plant Review, 23. Blackwell Publishing, Oxford, pp 438.

Riederer, M. and Schneider, G. (1989). Comparative study of the composition of waxes extracted from isolated leaf cuticles and from whole leaves of *Citrus*: evidence for selective extraction. *Physiologia Plantarum, 77*, 373-384.

Roessingh, P., Hora, K. H., Van Loon, J. J. A. and Menken, S. B. J. (1999). Evolution of gustatory sensitivity in *Yponomeuta* cartepillars: sensivity to the stereo-isomers dulcitol and sorbitol is localised in a single sensory cell. *Journal of Comparative Physiology* A, *184*, 119-126.

Roessingh, P., Hora, K. H., Ying Fung, S., Peltenburg, A. and Menken, S. B. J. (2000). Host acceptance behaviour of the small ermine moth *Yponomeuta cagnagellus*: larvae and adults use different stimuli. Chemoecology, 10, 41-47.

Roessingh, P., Städler, E., Baur, R., Hurter, J. and Ramp, T. (1997). Tarsal chemoreceptors and egg-laying behaviour of the cabbage root fly (*Delia radicum*) sensitive to fractions and new compounds of host-leaf surface extracts. *Physiological Entomology, 22*, 2, 140-148.

Roessingh, P., Städler, E., Fenwick, G. R., Lewis, J. A., Kvist Nielsen, J., Hurter, J. and Ramp, T. (1992). Oviposition and tarsal chemoreceptors of the cabbage root fly are stimulated by glucosinolates and host plant extracts. *Entomologia Experimentalis et Applicata*, 65, 3, 267-282.

Roger, S. and Newland, P. L. (2003). The neurobiology of taste in insects. *Advances in Insect Physiology, 31*, 141-204.

Rolland, F., Baena-Gonzalez, E. and Sheen, J. (2006). Sugar sensing and signaling in plants: Conserved and novel mechanisms. *Annual Review of Plant Biology, 57*, 675–709

Schlegel, T. K. and Schönherr, J. (2002). Selective permeability of cuticules over stomata and trichomes to calcium chloride. *Acta Horticulturae, 549*, 91-96.

Schlegel, T. K., Schönherr, J. and Schreiber, L. (2005). Size selectivity of aqueous pores in stomateous cuticles of *Vicia faba* leaves, *Planta, 221*, 648-665.

Schlegel, T. K., Schönherr, J. and Schreiber, L. (2006). Rates of foliar penetration of chelated fe (III): role of light, stomata, species, and leaf age. *Journal of Agriculture and Food Chemistry, 54*, 6809-6813.

Schönherr, J. and Merida, T. (1981). Water permeability of plant cuticular membranes: the effects of humidity and temperature on the permeability of non–isolated cuticles of onion bulb scales. *Plant Cell and Environment, 4*, 349-354.

Schönherr, J. and Schreiber, L. (2004). Size selectivity of aqueous pores in astomatous cuticular membranes isolated from *Populus canescens* (Aiton) Sm. leaves. *Planta, 219*, 3, 405-411.

Schoonhoven, L. M., Jermy, T. and Van Loon, J. J. A. (1998). *Host–plant selection when to accept a plant; from physiology to evolution,* Chapman and Hall (eds.), (156-193), London, UK

Schoonhoven, L. M. and van Loon, J. J. A. (2002). An inventory of taste in caterpillars: Each species its own key. *Acta Zoologica Academiae Scientiarum Hungaricae, 48*, 215-263, Supplement 1.

Schoonhoven, L. M., van Loon, J. J. A. and Dicke, M. L. M. (2008). *Insect-Plant Biology*. Schoonhoven, L. M., van Loon, J. J. A. and Dicke, M. L. M. (eds), Ed.2, (Oxford University Press Inc., pp 421) New York.

Schreiber, L. (2006). *Characterisation of polar paths of transport in plant culticles.* In M. Riederer and C. Mülller, *Biology of the plant cuticle, Annual Plant Review, 23, 280-289.* Blackwell Publishing, Oxford.

Schreiber, L. (2006). Review of sorption and diffusion of lipophilic molecules in cuticular waxes and the effects of accelerators on solute mobilities. *Journal of Experimental Botany, 57*, 11, 2515-2523.

Schreiber, L., Skrabs, M., Hartmann, K. D., Diamantopoulos, P., Simanova, E. and Santrucek, J. (2001). Effect of humidity on cuticular transpiration of isolated cuticular membranes and leaf disks. *Planta, 214*, 274-282.

Schütt, P. and Schuck, H. J. (1973). Seasonal variation in the composition of cuticular waxes in *Pinus sylvestris*. *Flora, 162*, 206-214.

Shepherd, T., Robertson, G. W., Griffiths, D. W., Birch, A. N. E. and Duncan, G. (1995). Effect of environment on the composition of epicuticular wax from kale and Sweden. *Phytochemistry, 40*, 407-417.

Soldaat, L. L., Boutin, J. P. and Derridj, S. (1996). Species specific composition of amino acids on leaf surface of four *Senecio* species. *Journal of Chemical Ecology, 22*, 1, 1-12.

Southwood, T. R. E. (1986). Plant surfaces and insects. In B. Juniper and T. R. E. Southwood (eds.), *Insects and the Plant Surface* (Edward Arnold, 1-22). London.

Southwood, T. R. E. (1996). Insect-plant relations: overview from the symposium. *Entomologia experimantalis et applicata, 80*, 20-324.

Spencer, J. L., Pillai, S. and Bernays, E. A. (1999). Synergism in the egg-laying behavior of *Plutella xylostella*: sinigrin and wax compounds. *Journal Insect Behavior, 12*, 483-500.

Städler, E. (1986). Egg-laying and feeding stimuli in leaf surface waxes. In B. Juniper B. and R. R. E. Southwood (Eds.), *Insects and the plant surface* (Edward Arnold, 105-121).). London.

Städler, E. (2002). Plant chemical cues important for egg deposition by herbivorous insects. In M. Hilker and T. Meiners (Eds.), *Chemoecology of Insect Eggs and Egg Deposition* (Blackwell Publishing, 171-204). Berlin.

Städler, E. and Reifenrath, K. (2009). Glucosinolates on the leaf surface perceived by insect herbivores: review of ambiguous results and new investigations. *Phytochemistry Revue, 8*, 207-225.

Stammitti, L., Garrec, J. P. and Derridj, S. (1995). Permeability of isolated cuticles of *Prunus laurocerasus* to soluble carbohydrates. *Plant Physiology and Biochemistry, 33*, 319-326.

Steinbauer, M. J., Davies, N. W., Gaertner, C., Derridj, S. (2009). Epicuticular waxes and plant primary metabolites on the surfaces of juvenile *Eucalyptus globulus* and *E. nitens* (Myrtaceae) leaves. *Australian Journal of Botany, 57*, 474-485.

Steinbauer, M. J., Schiestl, F. P. and Davies, N. W. (2004). Monoterpenes and epicuticular waxes help female autumn gum moth differentiate between waxy and glossy *Eucalyptus* and leaves of different ages. *Journal of Chemical Ecology, 30*, 1117-1142.

Taret, G., Sevilla, M., Wornoayporn, V., AIslam, A., Ahmad, S., Caceres, C., Robinson, A. S., Vreysen, M. J. B. (2010). Mating compatibility among populations of codling moth *Cydia pomonella* Linnaeus (*Lepidoptera*: *Tortricidae*) from different geographic origins. *Journal of Applied Entomology, 134*, 3, 207–215.

Thosteinson, A. J. (1959). Host selection in phytophagous insects. *Annual Review of Entomology, 5*, 193-218.

Tukey, H. B. Jr. (1970). The leaching substances from the plants. *Annual Review of Plant Physiology, 21*, 305-324.

Vemmos, S. N. (1995). Carbohydrates changes in flowers, leaves, shoots and spurs of "Cox's orange pippin" apple during flowering and fruit setting periods. *Journal of Horticultural Science, 70*, 889-900.

Vrieling, K. and Derridj, S. (2003). Pyrrolizidine alkaloids in and on the leaf surface of *Senecio jacobaea* L. *Phytochemistry, 64*, 1223-1228.

Wallart, R. M. A. (1980). Distribution of sorbitol in *Rosaceae*. *Phytochemistry, 19*, 2607-2610.

Wattendorff, J. and Holloway, P. J. (1984). Periclinal penetration of potassium permenganante into mature cuticular membranes of *Agave* and *Clivia* leaves: new implications for plant cuticle development. *Planta, 161*, 1-11.

Wearing, C. H. and Hutchins, R. F. N. (1973). α-farnesene, a naturally occurring egg-laying stimulant for the codling moth, *Laspeyresia pomonella*. *Journal of Insect Physiology, 19*, 1251-1256.

Whetter, J. M. and Taper C. D. (1963). Note en seasonal occurrence of sorbitol (D-glucitol) in buds and leaves of *Malus*. *Canadian Journal of Botany, 41*, 175-177.

Wildbolz, T. (1958). Über die Orientierung des Apfelwicklers bei der Eiablage. *Mitteilunger der Schweizerishen Entomologishen Geselshaft, 31*, 25-34.

Wilson, M. and Lindow, S. E. (1994). Coexistence among epiphytic bacterial populations mediated through nutritional resource partitioning. *Applied and Environnmental Microbiology, 60*, 4468–4477.

Wind, J., Smeeken,s S., Hanson, J. (2010). Sucrose: Metabolite and signaling molecule. *Phytochemistry 71,* 1610-1614.

Witzgall, P., Ansebo, L., Yang, Z., Angeli, G., Sauphanor, B. and Bengtsson, M. (2005). Plant volatiles affect egg-laying by codling moths. *Chemoecology, 15*: 77-83.

Woodhead, S.and Chapman, R. F. (1986). Insect behaviour and the chemistry of plant surface waxes.In B. Juniper, and R. Southwood, *Insects and the Plant Surface* (Arnold, 123-135). London.

Yan, F., Bengtsson, M. and Witzgall, P. (1999). Behavioural response of female codling moth, *Cydia pomonella*, to apple volatiles. *Journal of Chemical Ecology, 25*, 6, 1343-1351.

Young-Hee, C., Sang-Dong, Y. (2011). Signaling role of fructose mediated by FINS1/FBP in Arabidopsis thaliana. PLoS Genetics 7, 1, e1001263, DOI: 10.1371/journal.pgen. 1001263.

Reviewed by: Dr. Martin Heil, Position: Researcher 3C, SNI III. CINVESTAV, Unidad Irapuato, *Plant Ecology Laboratory*, mheil@ira.cinvestav.mx

In: Moths: Types, Ecological Significance and Control... ISBN: 978-1-61470-626-7
Editor: Luis Cauterruccio, pp. 39-74

Chapter 2

THE INTRIGUING CASE OF *STENISCADIA POLIOPHAEA* (NOCTUIDAE): POTENT MOTH ENEMY OF YOUNG MAHOGANY TREES IN AMAZONIAN FORESTS

Julian M. Norghauer[1] and James Grogan[2]

[1] Institute of Plant Sciences, University of Bern, Altenbergrain 21
Bern, 3013, Switzerland.

[2] School of Forestry and Environmental Studies,
Yale University, 195 Prospect Street
New Haven, Connecticut, 06511, U.S.A.

ABSTRACT

The super-family Noctuoidae is the most species-rich of Lepidoptera, and many appear to be specialized herbivores. Yet little is known about their abundance and ecological significance in diverse forests of the tropics. In this chapter we briefly review these two aspects in the context of diversity maintenance (Janzen-Connell hypothesis), and present the case of the South American moth *Steniscadia poliophaea*. This species feeds only on expanding leaf and stem tissues of seedlings and saplings of the prized timber tree, big-leaf mahogany (*Swietenia macrophylla*). We synthesize published research, observational reports, and anecdotal evidence about *S. poliophaea*'s life history, ecology, and impact on host mahogany populations across southern Brazilian Amazonia. This moth plays an important role in suppressing the early recruitment and growth, and hence potential local dominance, of the fast-growing *S. macrophylla*. We doubt this moth plays a contributing role in structuring local adult densities of *S. macrophylla* in Central America and Mexico where it has not been reported to occur. We compare the ecological significance of *S. poliophaea* to the better known shoot-boring moth, *Hypsipyla grandella* (Pyralidae) that is a major pest in mahogany plantations throughout the Neotropics. Finally, we consider implications of these findings for host-competition and control in the recovery and sustainable management of threatened *S. macrophylla* populations in logged and unlogged South American forests. Moth herbivores in general, and the Noctuoidae in particular, warrant further investigation as potential drivers of Janzen-Connell effects on trees in species-rich tropical forests.

1. Diversity of Moths and Trees in Tropical Forests: The Herbivory Nexus

Flowering plants and their insect herbivores dominate Earth's organismal diversity on land (Strong et al. 1984, Price 2002). This antagonistic interaction offers insight into the origins of biodiversity and ecosystem function — herbivory is a major channel for energy flow to higher trophic levels — and is a key feature of natural and managed ecosystems (Fraenkel 1959, Howe and Westley 1988, Huntly 1991, Cyr and Pace 1993, Schoonhoven et al. 2005). The breadth and complexity of plant-herbivore interactions reaches its pinnacle in tropical forests (Nair 2007, Speight et al. 2008, Kricher 2010), which harbor > 50% of the world's terrestrial species on 7% of its land surface (Laurance 1999). There is an urgent need to examine both evolutionary and ecological processes that contributed to such high plant and herbivore diversity (Dethier 1954, Ehrlich and Raven 1964, Thompson 1999), and how these interactions help explain present and predict future community composition, maintenance, and dynamics (Strong et al. 1984, Futuyma and Agrawal 2009).

Among the insects, the order Lepidoptera — the moths and butterflies of the world — is extremely diverse (Scoble 1992). While not as species-rich as the Coleoptera, there are ~160 000 described Lepidoptera species, with up to ~500 000 believed to exist. The primary era of speciation by this group tracked that of the angiosperms in the Cretaceous period (Kristensen et al. 2007). In recent years, more than 800 new Lepidoptera species have been described annually, aided in part by the use of increasingly sophisticated molecular tools (e.g., DNA bar-coding, Hebert et al. 2004) and the revision of phylogenies using computer-intensive methods (Zahiri et al. 2010). Among currently described species, the super-family Noctuoidae dominates with at least ~45 000 species, more than twice that of the next largest clade, the Geometroidae (~21 500 species). This super-family Noctuoidae presently comprises six families: Eribidae, Euteliidae, Oenosandridae, Notodontidae, Nolidae and Noctuidae (Zahiri et al. 2010). A large number of Noctuidae herbivore species, many probably locally rare, have yet to be described from tropical zones (Scoble 1992, Price et al. 1995, Nair 2007), especially from forests sustaining high levels of host diversity, with up to several hundred woody plant species per hectare (Richards 1996, Turner 2001, Kricher 2010). For example, moth diversity (excluding Pyralidae) sampled in a Bornean forest over a 12-month period totaled 1053 species, of which 39% were in the Noctuidae family (Barlow and Woiwod 1989). In a temperate deciduous forest, 27% of moth species (141 of 512) also were Noctuidae (Summerville and Crist 2002). The vast majority of Lepidoptera are herbivores during larval stages (Strong et al. 1984, Scoble 1992, Schoonhoven et al. 2005, Speight et al. 2008).

Owing almost entirely to the negative impact of their larvae on host plants' survival, growth and fruit production, a disproportionate number of Lepidoptera species are of great economic importance (Nair 2007). The family Noctuidae alone accounts for 1034 of 5781 Lepidoptera species considered economically important by Zhang (1994) — followed by the Pyralidae (748 spp.), Tortricidae (687) and Geometroidae (351). According to Barbosa (1993), close to half of all Lepidopteran pest species are Noctuidae, Pyralidae and Tortricidae (51% and 46% for tropical and temperate zones, respectively). In timber tree plantations in the tropics, which have expanded rapidly since the end of WWII (Gray 1972), Noctuidae caterpillars can be formidable defoliators of tree seedlings and saplings (e.g., widespread *Hyblaea peura* on *Tectona grandis*, Gray 1972; *Eligma narcissus* on *Ailanthus* spp. in Asia, Nair 2007). In natural forests, where pest outbreaks are thought to be rare, two documented cases were both by Noctuid larvae feeding on newly flushing leaves, *Eulepidotis superior* and *E. phrygionia* each on a single host tree species (*Quararibea asterolepsis*: Bombacaceae and *Peltogyne gracilipes*: Caesalpiniaceae in Panama and Brazil, respectively; Wong et al. 1990, Nascimento and Procter 1994). Another Noctuid, *Antiblemma leucocyma*, heavily damages young leaves of the small tree *Miconia calvecsens* (Melastomataceae) in Brazil. This plant was the only species *A. leucocyma* was found to eat, suggesting a narrow diet breadth (Badenes-Perez and Johnson 2008).

What traits enable a given insect herbivore capable of feeding on the host plant to become a 'pest'? When agricultural crops or trees are planted as monocultures at high density, they are likely to become evermore susceptible to herbivore enemies whose populations grow at faster rates than under natural conditions (Root 1973, Nair 2007, Speight et al. 2008), at least in part due to lowered mortality factors under artificial conditions (e.g., pathogens, animal predators, parasitoids, etc.; Letourneau et al. 2011). This predictable response suggests indirectly that insect herbivores are density-responsive, able to increase numerically in response to increasing host abundance, and thus probably also highly efficient at locating host plants. This ability may be enhanced if herbivores also have a narrow diet breadth (Bernays 2001) and thus we expect many Lepidoptera pests to be specialists (Barbosa 1993). Yet while there is little doubt that insect herbivores can reduce the fitness of individual plants, whether they also regulate plant populations and communities (Louda and Potvin 1995, Carson and Root 2000) under more natural conditions remains unclear (Crawley 1989, Huntly 1991, Maron and Crone 2006).

Herbivory by insects is especially problematic in forests because young and small tree progeny (seedlings and saplings) are relatively abundant, highly vulnerable to mortality, and exposed to enemies for extended periods because of the many years (20–50) typically needed to reach maturity (Harper 1977, Richards 1996, Swaine 1996, Hanley 1998, Turner 2001). This brings us to a classic hypothesis at the herbivory nexus that is potentially very important for promoting tree species diversity and coexistence in the tropics: the Janzen-Connell (J-C) hypothesis, which states that specialized natural enemies help keep host species rare by weakening or killing juveniles where they are concentrated, typically close to the parent plant, thereby freeing up space and resources for other tree species in the community (Figure 1; Janzen 1970, Connell 1971). The result of this overcompensating form of density-dependence is that parent tree will not be replaced in the same place by one of its progeny when it dies. Instead, as a corollary, the J-C hypothesis predicts enhanced host species fitness and recruitment through 'escape in space' via dispersal away from conspecific individuals (Howe and Smallwood 1982). This in turn can be viewed as a qualitative defense against specialist herbivores, though not likely against generalist herbivores, which are probably deterred more by a suite of chemical and physical defenses (Marquis 2005). In this way, the J-C hypothesis can function as a 'stabilizing force' in the community to prevent competitive exclusion, especially by key potentially or already dominant species, and may explain in part why so many rare tree species pesist in these forests (Janzen 1970, Clark and Clark 1984, Carson et al. 2008, Kricher 2010).

Although Janzen-Connell effects (density- and or distant-dependent plant recruitment and mortality) have increasingly been found in tropical forests (Harms et al. 2000, Queenborough et al. 2007, Comita et al. 2010, Metz et al. 2010; reviewed by Carson et al. 2008 and Zimmermann et al. 2008), their biotic mechanisms remain uncertain. This is because, apart from seeds (Wilson and Janzen 1972, reviewed in Turner 2001), the actual impact of insect enemies on seedlings and saplings in terms of growth and mortality has rarely been identified and quantified in the context of natural or semi-natural forest conditions (Leigh 2004, Carson et al. 2008). And while pathogens are increasingly understood to drive J-C effects under more *shaded* forest conditions (Bell and Freckleton 2006, Bagchi et al. 2010, Mangan et al. 2010), insect herbivores remain woefully understudied by comparison (Janzen 1971, Blundell and Peart 1998, Marquis 2005, Massey et al. 2005).

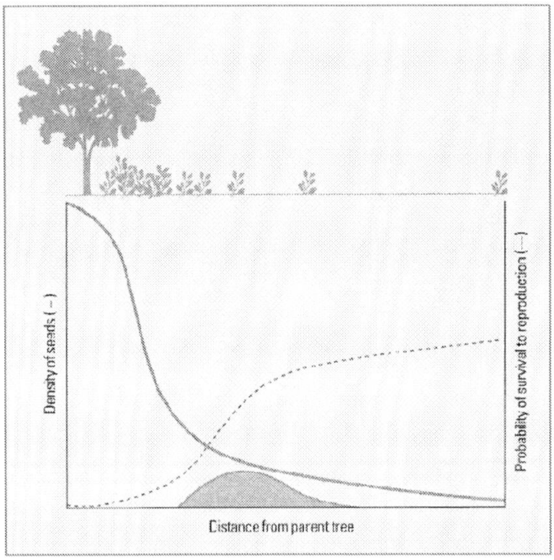

Figure 1. The mechanism underpinning the classic Janzen-Connell hypothesis predicts that the probability of a seed becoming an adult is related to distance from the parent tree. Beneath and near crowns adult recruitment is effectively zero because of overcompensating density-dependent predation and herbivory by biotic enemies upon seeds, seedlings and/or saplings. Although proportionally fewer seeds move farther away from their parent, these locally rare progeny are more likely to escape enemy attacks and avoid being eaten. Also shown is how, for a given species, the interplay between limited seed dispersal and individual susceptibility to host-specific enemy(s) results in a distance from the parent tree where an adult recruitment event is most likely to happen. It is important to bear in mind that in two dimensions, in contrast to seed density (solid line), the relative frequency of seeds usually peaks some intermediate distance from the parent. The mechanism is driven by host-specific enemies, and not intraspecific competition among conspecific seedlings, or asymmetric competition with parents. The graphic is a modified version of the original in Janzen (1970) and reproduced with permission from p. 103 in the book chapter entitled "Plant-insect interactions in terrestrial ecosystems" by SS Strauss and AR Zangerl in *Plant-Animal Interactions: An Evolutionary Approach* (2002), edited by CM Herrera and O Pellmyr and published by Wiley-Blackwell.

Furthemore, insect herbivores generally prefer younger over older plant stages, and vigorous individuals over weakened ones (Feeny 1970, Strong et al. 1984, Price 1991, Hanley 1998). For these reasons, tree seedlings or saplings are expected to be under stronger selective pressure to deploy anti-herbivore defenses to minimize loss of valuable photosynthetic tissues compared to reproductive adults (Hanley 1998, Boege and Marquis 2005). This strategy is thought to be under strong selection in nutrient-poor habitats where it is very difficult to replace plant parts lost to herbivory (Janzen 1974, Coley et al. 1985, Fine et al. 2004). Within plants, expanding young leaves in particular (Harper 1989) — tender and more nutritious than mature leaves (Feeny 1970, Coley 1983, Choong 1996, Coley and Kursar 1996) — are especially vulnerable to leaf-chewing insect herbivores, especially Lepidoptera, many of which are presumably specialized to survive consumption of defensive secondary compounds (Ehrlich and Raven 1964, Coley and Barone 1996, Kursar and Coley 2003, Coley et al. 2006, Kursar et al. 2006, Novotny et al. 2006, Bluthgen and Metzner 2007, Dyer et al. 2007, Richards and Coley 2007, Kursar et al. 2009, Novotny et al. 2010).

2. *STENISCADIA POLIOPHAEA*: A NOCTUIDAE SPECIALIST HERBIVORE

In this broader context of tree species coexistence, one recently studied moth-plant interaction of ecological significance is found in tropical lowland Amazonian forest, between the microlepidopteran *Steniscadia poliophaea* (Noctuidae: Sarrothripinae) and its only known host plant *Swietenia macrophylla* (Meliaceae). The host is better known as big-leaf mahogany, a canopy emergent tree long renowned for its prized timber (Lamb 1966) and currently threatened throughout its natural range (Snook 1996, Kometter et al. 2004, Grogan et al. 2010). Hence our investigation of the moth was admittedly pursued from the viewpoint of its host: we were interested in the patterns of attack and impacts of this moth's larval instars on mahogany populations from both basic ecological and forest management perspectives. What follows is a synthesis of field research pursued to date, mainly in two forests c. 200 km apart in the southeast corner of the state of Pará, Brazil, one unlogged and one logged.

While our scientific understanding of this specific moth-plant interaction is relatively new, *S. poliophaea* was first identified in the early 20th century by Sir George F. Hampson (Hampson 1912, V. Becker *pers. comm.*). Nursery managers in this region became aware of the moth's predations as soon as plantation efforts, requiring large numbers of nursery-grown mahogany seedlings, began in the late 1980s. However, so far as we are aware there has been no further mention of *S. poliophaea* in the scientific literature until recently.

Like all Lepidoptera, the life cycle of *S. poliophaea* has four distinct life stages: (1) egg, (2) larva, (3) pupa, and (4) adult (Scoble 1992). These life stages are intimately linked to mahogany's size, phenology, local abundance and distribution, and capacity for vigorous growth.

S. poliophaea adult females are active at night, laying eggs on young expanding mahogany leaves. These can be either the first new simple leaves of germinating seedlings at ground level (Figure 2.H), or flushing leaves – simple or compound – of established seedlings and saplings shorter than c. 5 m tall (Figure 2.B). Though *S. poliophaea* females may accidentally oviposit on other plant species' leaves, as noted in other Lepidoptera (Thompson and Pellmyr 1991), this herbivore appears to be truly monophagous at the genus (*Swietenia*) level. Feeding trials and *in situ* transferal of early instars to young leaves of neighboring plants in the unlogged forest, called Pinkaití, resulted in no signs of herbivory and the caterpillars' rapid disappearance. In this respect, targeted oviposition by *S. poliophaea* is truly remarkable, insofar as females must search for an ephemeral resource on host plants within a species-rich and structurally complex forest understory. *S. poliophaea* caterpillars do not leave their host mahogany plants, and thus are not 'free-moving' among individual plants as has been documented for some tropical moth species elsewhere (Janzen 2003). The latter are likely more generalist herbivores, capable of feeding on many distant genera within and/or among families. This highlights the well recognized importance of female choice in host plant oviposition for offspring success (Thompson and Pellmyr 1991, Renwick and Chew 1994, Bernays 2001).

Figure 2. Photos showing the life history, feeding behavior, and impact of the small Noctuid moth *Steniscadia poliophaea* on its only known host, the neotropical forest timber tree, big-leaf mahogany (*Swietenia macrophylla*, Meliaceae). Shown in (A) the variation in caterpillar sizes associated with early, mid and late instars; (B) distinctive early feeding patterns on an expanding mahogany leaf; (C, E) late instars feeding from underside of leaves; note the collapsed leaflets from chewed midrib vein and characteristic webbing and frass; (D) an attacked mature leaf that escaped severe damage; (F, K) multiple late instars (~2 cm length) atop defoliated leaf flush of a mahogany sapling; (G) early instars trying to feed on a recently expanded leaf resulting in negligible damage; (H) first leaves of newly germinating seedling; (I) new seedling stem 100% defoliated by a caterpillar – note the new shoot at base; (J) newly established seedling that escaped timely female oviposition, resulting in negligible damage to its leaves (< 5% leaf area eaten); (L) recently expanded leaf flush and older, dark green leaves lower on the stem of a mahogany sapling; (M, N) boat-shaped cocoons of the microlepidopteran moth (~1 cm length); (O) gray-colored adults, ~1 cm length, that emerge after ~8–11 days in the pupae life-stage. Photos A, F, and K were taken at Marajoara and the rest at the Pinkaití forest reserve in the state of Pará, Brazil.

We have only observed *S. poliophaea* eggs, translucent and circular in shape, laid near the petiole on the underside of new leaves and not on their surface. The majority (73%) of microlepidopteran leaf miners in Britain laid eggs in a similar way (Reavey and Gaston 1991). The number of eggs laid on a single leaf is variable, but seems to increase with stem size of the plant and potential surface area available. Since larger leaves can sustain more caterpillars, this suggests that female moths are able to judge the potential size and food resource available to hatching instars, perhaps via a combination of leaf shape and reflectance, and/or chemical and physical sensory information further available upon plant contact (Visser 1986, Renwick and Chew 1994, Schoonhoven et al. 2005). Unfortunately we lack photos of the eggs to complete the life stages illustrated in Figure 2.

The eggs hatch quickly, in 1–3 days. The first pale green-yellow instar is barely visible to the eye (see arrow in Figure 2A). Like most moth larvae, *S. poliophaea* instars eat specific plant parts, feeding almost exclusively on newly flushing leaf tissues (Figure 2A). Tender young mahogany leaves are crucial during the early instar stages because caterpillar jaws are poorly formed. Early instars cannot chew through mature leaf tissues, though we have observed later instars occasionally 'back-feeding' onto mature leaves to complete their life cycle. Thus the timing of oviposition is crucial for larval fitness because as mahogany leaves mature, growing tougher and darker green in color, they become less palatable to *S. poliophaea* larvae.

This restricted chewing ability of *S. poliophaea* is not unlike that reported for caterpillars of the larger Sphingidae moths, whose early instars feed only on new leaves but can consume mature leaves during later, larger instar stages. These species' mandibles are adapted to eating softer 'flimsy' leaves, whereas co-occurring Saturniidae caterpillars are better at chewing through tough mature leaves high in phenolics, especially tannins (Bernays and Janzen 1988). It is notable, and we later argue perhaps no coincidence, that the Costa Rican sphingid moth species are also more specialized in their host diet breadth, being restricted to one or a few closely related species (Janzen 1984 in Bernays and Janzen 1988).

S. poliophaea caterpillars grow rapidly — they must in order to complete larval development before mahogany seedling and sapling leaves mature and become unpalatable. The signs of first and second instar activity are quite distinct: very small holes (< 1–2 mm diameter) appear on expanding leaf blades, accompanied by irregular nibbling along the leaf margins (Figure 2B). As the caterpillars grow larger during mid to late instar stages, and leaves approach full expansion, characteristic frass and webbing appears, and caterpillars often attempt to cut the growing leaf's midrib in one or more places (Figure 2C, E). This crippling of the leaf, which, if successful, causes the leaf to fold over, though not necessarily to fall off because of caterpillar-generated webbing, could serve two functions. First, it may prevent the host plant either from deploying more potent toxins or from transferring signals to hasten leaf maturation against feeding caterpillars. Second, cutting the midrib with subsequent collapse of the leaf onto itself may also provide temporary shelter from potential predators. This behavior has been documented in other Lepidoptera caterpillars in many systems (Dussourd 1993). As in other tree leaf Lepidoptera herbivores (Feeny 1970, Aide and Londono 1989), when *S. poliophaea* oviposition occurs later in the leaf maturation cycle, the small instars are ineffective feeders on nearly mature leaves (Figure 2G), resulting in negligible leaf damage (Figure 2D, J, L).

Unpublished field trials at Pinkaití suggest there are 5–6 instar stages, ending with late instars that are bright yellow with a discernable halo of fine protruding hairs (Figure 2A, F).

By the time *S. poliophaea* caterpillars reach the final instar, they typically have consumed more than 50% and up to 100% of host leaves (Figure 2F, K, I). As mahogany seedlings germinate from seeds in late October and early November, hungry caterpillars may consume all leaf tissue and then move on to seedling stems, stripping them down to the 'bone', leading to certain death even if seedlings have enough reserves to resprout new leaf shoots (Figure 2I). Severe damage occurs especially in the early wet season, and, consistent with the basic premise and predictions of the Janzen-Connell hypothesis, in close proximity to mahogany adult trees or where these are aggregated (Norghauer et al. 2006a,b, 2008a, 2010).

After 10–14 days of intense feeding, final caterpillar instars, c. 1.5–2.0 cm long, spin themselves into boat-shaped cocoons, c. 1 cm long (Figure 2N), either affixed to the leaf undersurface if any leaves remain (Figure 2M) or in the leaf litter near the seedling stem. This pupae stage lasts 9–11 days after which a pale grey moth emerges. We estimate a single generation to require c. 18–22 days in total.

We know little about the population dynamics of this moth herbivore. Sources of mortality at the larval stage likely include viral or fungal pathogen infection of caterpillars, as we have noticed cases of stalled instar development (though not motion paralysis) — not for lack of tender leaf food — and disease symptoms as suggested by discoloration combined with compromised health. We have not yet witnessed parasitism of caterpillars in the field. This third trophic level, in addition to bottom-up forces in the form of leaf toxins sequestered by caterpillars as anti-predator defenses, might play an underappreciated role in the evolution of narrow host-specificity in tropical forest moths (Bernays and Graham 1988, Janzen 1988, Dyer 1995, Dicke 2000). The disappearance of one or more *S. poliophaea* caterpillars from both new mahogany seedlings and large leaf flushes may indicate predation by understory birds or other arthropods (reviewed by Heinrich 1993, Montllor and Bernays 1993), especially ants less deterred by toxic specialists (Dyer et al. 2004); or cannibalism among competing larvae for rapidly diminishing food resources — more common in Lepidoptera than other insect orders (reviewed by Richardson et al. 2010). In addition, herbivore competition for hosts may be limited by territorial behavior via 'host-marking pheromones' that deter oviposition by females arriving later (Damman 1993, Schoonhoven et al. 2005).

Because *S. poliophaea* caterpillars need young mahogany leaves for food, we anticipate that moth populations closely track mahogany phenology. Mahogany seedlings and saplings can flush new leaves 3–4 times during the course of a year in well-lit gaps, whereas in the understory leaf flushing is typically limited to once per year, in the early wet season. The onset of rains marking the transition to the wet season results in a synchronous population-wide leaf flush by host seedlings and saplings. As this first pulse of food becomes available, both in bright canopy gaps and in the shaded forest understory within seed dispersal distance of fruiting mahogany trees, moths demonstrate incredibly effective locational skill, and the caterpillar population growth rate spikes.

S. poliophaea population growth presumably continues with the next leaf flush during November-December, with first generation females searching for mahogany host leaf flushes on a broader spatial scale (Norghauer et al. 2008a), but then begins to decline in the latter half of the wet season (January onward). Leaf flushes increasingly occur out of synchrony as the wet season progresses, possibly because of detrimental effects on host plants caused by prior attacks.

We propose that this forces foraging of second or third generation adults on a wider spatial scale for leaf flushing, possibly by seeking out high-light patches, namely canopy gaps formed by tree or branch-falls, where newly flushing mahogany leaves are more likely to be found (that is, where seedling and sapling growth rates are higher than in the forest understory).

At Pinkaití, for example, mahogany seedlings planted in gaps more than 200 m away from the nearest adult stem — albeit at unnaturally high density (c. 2 m^{-2}) — were eventually located and attacked by *S. poliophaea* in the late wet season (February–April). By the end of the wet season (May), *S. poliophaea* activity is infrequent, presumably curtailed by population reductions driven by bat predation of adults, and/or built-up parasitism or infection loads on caterpillars. With the wet season's last heavy rainfall, a final leaf flush by mahogany can occur, but few of these leaves suffer attacks (Norghauer, unpublished data).

In the dry season (June–August) caterpillars are noticeably absent in the forest. Though population-wide leaf production is diminished by water stress, limited occasionally to only well-rooted seedlings and saplings in gaps, attacks seldom occur. What happens to moth adults during the dry season? Do they die or become inactive? Or do *S. poliophaea* pupae undergo a form of obligatory diapause until the rains return? Might the adults move elsewhere to where moister conditions prevail, or possibly even seek out a second food source, possibly higher in the forest canopy?

Where do all the females come from in the early wet season to pounce, in unison, upon that first pulse of fresh mahogany leaves in the forest understory? It may be that *S. poliophaea* adults are active year-round, not unlike the highly host-specific *Eulepidotis* moth species — also in the family Noctuidae — which will attack flushing new leaves of host trees whenever available in Costa Rican dry forest in the latter half of dry season and first half of the rainy season (Janzen 1993). These are all open questions for future research.

For insect herbivores, lifetime fitness crucially depends on distinguishing among plants to find suitable hosts and evaluating their individual quality for oviposition (Visser 1986, Thompson and Pellmyr 1991, Bernays 2001, Schoonhoven et al. 2005). Clearly, *S. poliophaea* adult females are equipped with highly accurate identification capacity to be such exceptionally effective foragers for oviposition sites on mahogany seedlings and saplings. As mentioned above, attraction to unique host odors — which should increase in strength with host patch density — may be one strategy to enhance oviposition success and thus attack rates (Visser 1986, Thompson and Pellmyr 1991). While some Lepidoptera can make use of leaf optical properties (shape, color or reflectance) in choosing habitat or which plant to alight upon (Schoonhoven et al. 2005), highly specialized nocturnal foragers likely rely primarily upon strong, wind-borne chemical olfactory cues emanating from host plants to locate them across long distances (Visser 1986, Scoble 1992, Cardé and Willis 2008). These olfactory cues might even include host-specific volatile compounds associated with leaf chemical defenses against the more general herbivore community, and chemical deterrents from non-host plants (Renwick and Chew 1994, Bernays 2001).

In sum, because young leaves are a rare and ephemeral food resource, *S. poliophaea* must have a way to efficiently process sensory information. Following Bernays (2001), we expect this species to have evolved an efficient neural processing system that makes best use of high-contrast signals in the forest vegetation matrix, which further minimizes predation risk. This refined detection system may extend to adult mahogany trees too: we hypothesize that *S. poliophaea* might also use adult trees as resource markers in the forest. After all, seeds, and hence future seedlings, are concentrated near parent trees. In addition to mahogany's distinctive bark pattern, it is possible that *S. poliophaea* detects secondary compounds specific to mahogany, of which many have been described and to which this tree's timber at least in part owes its great durability. Moreover, such towering 'odor-plumes' would steer the moths' search for host plants primarily downwind, where the wind-dispersed mahogany seeds land and are more likely to be found (Grogan and Galvão 2006, Norghauer et al. 2011a).

3. Impact on Mahogany Juveniles and Populations in South America

That herbivory by invertebrates can negatively impact plant fitness, including that of trees, has been generally accepted for some time (Kulman 1971, Hawkes and Sullivan 2001). However, the impacts of insect herbivores on long-lived trees in tropical forests — in contrast to plantations (Gray 1972) — remains poorly understood (Swaine 1996, Turner 2001, Marquis 2005). Two early studies in Costa Rica demonstrated that heavy defoliation of mature plants can reduce seed production in six small-statured (5–20 m tall) tropical tree species (Rockwood 1973, see also Wong et al. 1990), and both individual growth and seed set in an understory *Piper* shrub (Marquis 1984). More recently, there is evidence that higher leaf damage can occur in gaps among conspecific tree seedlings of *Shorea leprosula* (Massey et al. 2006), and that rates of damage on mature (as opposed to newly flushed) leaves are associated with higher seedling mortality risk in the following year (Eichhorn et al. 2010).

Yet little empirical study has actually examined patterns of herbivory and their impact on tree juveniles in the context of the Janzen-Connell hypothesis (Clark and Clark 1984, 1985, Barone 1996, Blundell and Peart 1998, Sullivan 2003, Massey et al. 2005, 2006). The enemy-driven mechanism underpinning the J-C hypothesis presupposes a negative impact of herbivores on individual seedlings and saplings (seeds can also be affected in terms of mortality alone; Wilson and Janzen 1972). In the context of the *Steniscadia poliophaea–Swietenia macrophylla* interaction, a seed addition experiment at the logged site in southeast Amazonia, called Marajoara, found significantly higher seedling survival 50 m downwind from heavily fruiting mahogany trees than at 25 m and 10 m (Grogan and Galvão 2006). A similar positive distance effect in the forest understory for seedling establishment was reported in a field experiment at the unlogged Pinkaití forest (Norghauer et al. 2006a). In both studies, *S. poliophaea* was thought responsible for accelerated seedling mortality near mahogany adults. At Marajoara, up to 80% of new seedling regeneration was attacked by *S. poliophaea*.

In a separate seed addition experiment at Pinkaití, we reported clear evidence linking *S. poliophaea* attacks to higher seedling death near parent trees, in support of the J-C hypothesis (Norghauer et al. 2010). Percent damage and cases of 100% defoliation were disproportionately higher near parent trees, declining with distance downwind (Figure 3a). In turn, seedling survival after 10 months was significantly higher beyond 30 m and especially beyond 50 m from adult trees (Figure 3b). Based on more than 15 years of fieldwork, we are confident that attacks by the specialist *S. poliophaea* on newly germinating seedlings are an annual source of mortality for mahogany in this region.

In the early wet season following mahogany seed dispersal, population-wide seedling emergence is largely synchronized following moisture imbibition by seeds (Morris et al. 2000). These emerging seedlings represent a large resource-pulse of food for *S. poliophaea* which dissipates with distance from tree crowns. We suggest that adult trees are used by *S. poliophaea* as a cue for locating these new seedling cohorts. At a single heavily fruiting tree where all seeds were removed by hand except for those in three 1-m-wide downwind transects, *S. poliophaea* caterpillar defoliations peaked near the parent tree (Norghauer, unpublished data). Rarer seeds further away with fewer nearby siblings were in a better position to escape severe attack.

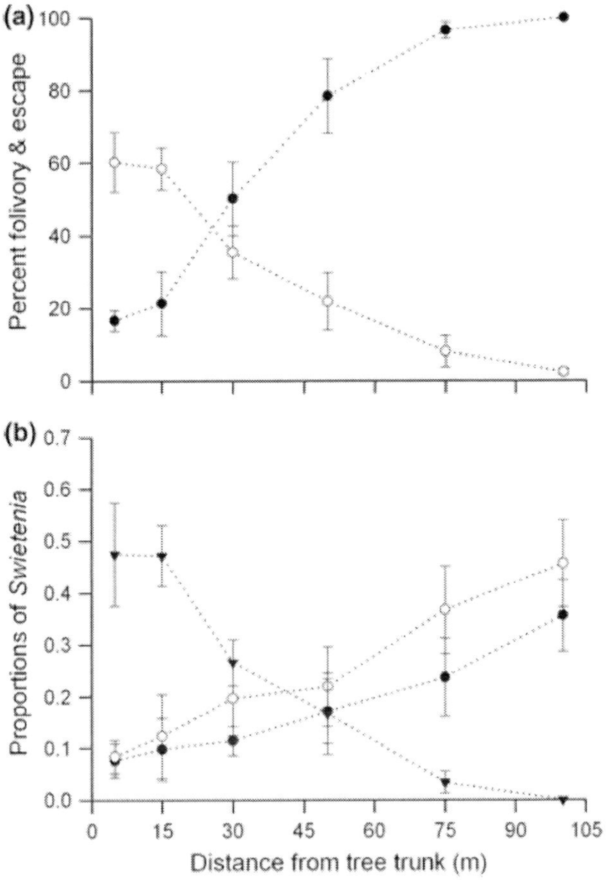

Figure 3. Impact of *Steniscadia poliophaea* caterpillars on new big-leaf mahogany seedlings at Pinkaití. Shown in (a) are percentages of leaf area damaged by the specialist (open circles) and new seedlings escaping attacks altogether (closed circles); in (b) proportions of 100% defoliation (triangles) and seedlings surviving 8 months after rainy season onset (closed circles = of germinating seedlings, open circles = all sown seeds). The experiment was conducted at seven reproductive trees in 2003 with a total sample size of 42 quadrats, each sown with 20 seeds. All symbols are means (± SE) and all linear regressions were highly significant ($P < 0.0001$). Reproduced with permission from Norghauer et al. 2010 in the journal Oecologia published by Springer-Verlag.

Newly germinating seedlings in higher light conditions of canopy gaps may not be as susceptible to *S. poliophaea* regardless of distance to the parent tree, possibly because of delayed germination and rodents that predate more seeds in gaps than in the forest understory, reducing seedling density (Grogan and Galvão 2006, Norghauer et al. 2006a), and/or because more dense vegetation or natural enemies present in gaps interferes with host location (Brokaw 1985, Denslow 1987, Uhl et al. 1988, Grogan et al. 2005, Schoonhoven et al. 2005).

Canopy gaps merit special attention because mahogany is a 'non-pioneer light-demanding' tree. Like many plant species it germinates well in shaded conditions, but early in its development mahogany needs above-average light levels found in or near gaps for vigorous growth, otherwise most seedlings will die within a few years (Lamb 1966). At Marajoara, a long-term experiment (80 months) found that *S. poliophaea* — along with other mortality factors, namely pathogens and drought stress — reduced overall seedling survivorship in the understory around 8 adult trees to just 1–2% (Grogan et al. 2005). Here mahogany seedlings lack the energy needed for leaf production. By contrast, canopy gaps can promote vigorous plant growth in terms of height and leaf production (Augspurger 1984b, Brokaw 1985, Uhl et al. 1988, Blundell and Peart 2001). For this reason, many tropical forest tree species are thought to require repeated local disturbances to reach the canopy and attain maturity (Hartshorn 1978, Denslow 1987, Richards 1996, Turner 2001) — probably in pulses, separated by years of suppression under low light following gap closure (Baker and Bunyavejchewin 2006). In contrast to the more shaded forest matrix, which, per unit area, is more food-limited in terms of leaf production (Richards and Coley 2007, 2008), it follows that herbivory of new plant material should be greater in gap microhabitats (Harrison 1987, Price 1991, Richards and Coley 2007, 2008).

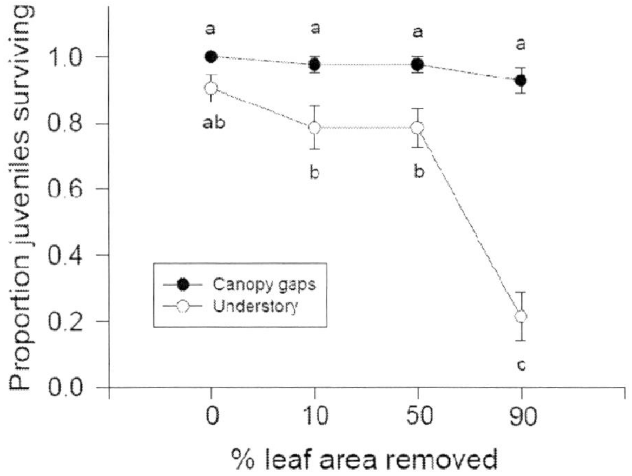

Figure 4. Survival over an 8-month period (April–December 2003) of big-leaf mahogany seedlings transplanted into canopy gaps and paired understory locations at Pinkaití under different levels of simulated herbivory. Nursery-grown seedlings had their leaves artificially clipped to simulate early specialist leaf damage to new germinants, but were left exposed to *Steniscadia poliophaea* herbivory during the experiment. Lower-case letters indicate significantly different means (± SE) from nested ANOVAs. Reproduced with permission from Norghauer et al. 2008a in the Journal of Ecology, published by Wiley-Blackwell on behalf of the British Ecological Society.

The impact of *S. poliophaea* on mahogany fitness in gaps is, we believe, primarily one of suppression, not outright death, weakening seedling competitive ability for diminishing light with heterospecific neighbors. At Pinkaití, nursery-grown mahogany seedlings were planted into 14 naturally formed canopy gaps paired with shaded sites in the forest understory, and their response to artificial defoliation and caterpillar damage to new leaves monitored. We found that light-rich gaps promoted both mahogany growth and tolerance to defoliation. Remarkably few individuals died in gaps despite removal of almost all leaf area, whereas mortality was significantly higher in the forest understory (Figure 4). On average, 50% or more leaf removal in gaps was required before seedlings exhibited significantly lowered height and basal stem growth rates (Figure 5a,c), similar to findings by Gerhardt (1998) for mahogany, and those reported by Blundell and Peart (2001) for another tropical tree species. In the forest understory seedling growth was negligible (Figure 5), but in gaps, although leaf production was not impaired during the 8-month study period (Figure 5b), the potential size of new leaves produced was reduced by heavy prior damage (Figure 5d). Sub-optimal growth will leave a mahogany seedling smaller and at greater risk for mortality following gap closure, requiring more gap events and thus longer time to reach maturity.

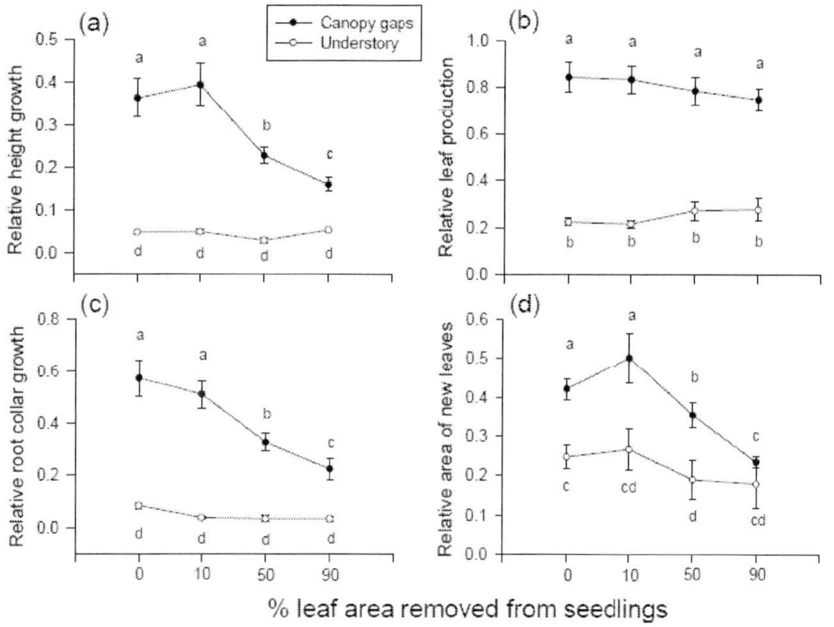

Figure 5. Proportional growth of big-leaf mahogany seedlings transplanted into canopy gaps and understory locations across different levels of simulated herbivory at Pinkaití. Lower-case letters indicate different means (± SE) from four separate nested ANOVAs that were all highly significant for the clipping treatment except for leaf production. Reproduced with permission from Norghauer et al. 2008a in the Journal of Ecology, published by Wiley-Blackwell on behalf of the British Ecological Society.

The impact of a specialist herbivore able to circumvent the constitutive defenses of its main host plant should further depend on the proximity to conspecific parent trees and local abundance of adults (or conspecific hosts) at larger spatial scales (Janzen 1970, Blundell and Peart 2004). At Pinkaití, first attacks in the shaded understory on seedlings established during the previous year occurred only very near adult mahogany trees (typically within 20–30 m), but the benefit of escape during synchronous germination and leaf flushing by emerging and established mahogany seedlings diminished where adults were aggregated (Norghauer, *unpublished manuscript*, Norghauer 2006b). At Pinkaití for example, specialist attack and damage on mahogany seedlings already established in the forest understory increased with local adult basal area, suggestion no satiation at the population level (Figure 6). By contrast, seedlings in gaps are easily discovered by *S. poliophaea*, consistent with the Plant Vigor Hypothesis (*sensu* Price 1991) and earlier studies. For example, Williams (1983) found that *Euphydras chalcedona*, a nearctic butterfly, also sought out sunlit habitats for oviposition on two preferred host plants. Likewise, in a Panamanian forest, the better quality of foliage in gaps vs. forest understory explained the higher levels of herbivory in the former by a specialist moth caterpillar, *Zunacetha annulata*, on its host shrub *Hybanthus prunifolius* (Harrison 1987).

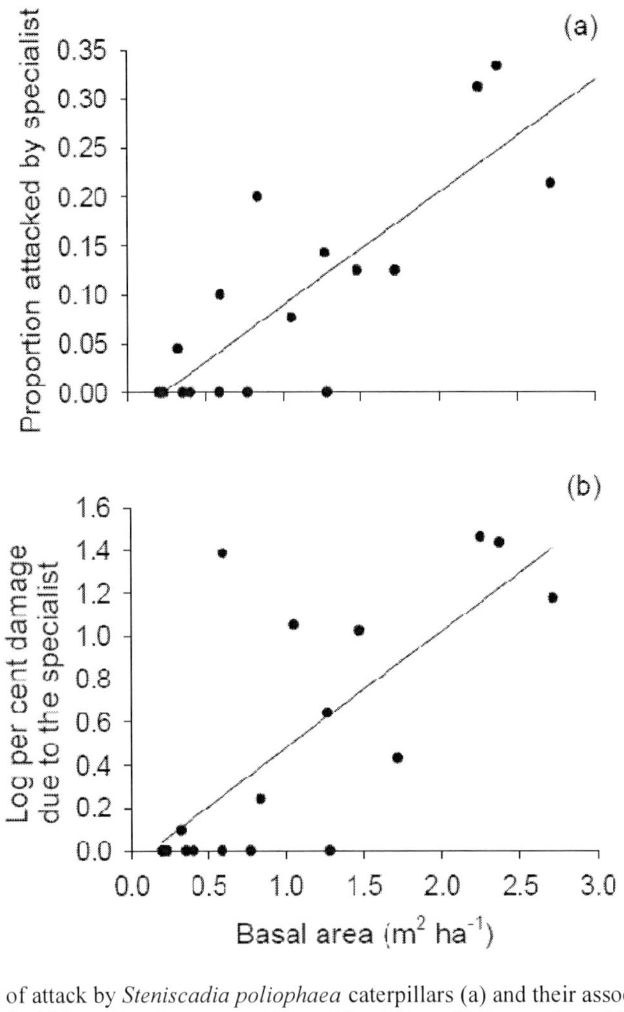

Figure 6. Incidence of attack by *Steniscadia poliophaea* caterpillars (a) and their associated leaf damage on naturally established big-leaf mahogany seedlings (b) within a 56.5-m radius of 18 adult mahogany trees at Pinkaití. The increases in specialist activity with increased adult tree basal area were significant in both regressions ($P < 0.0001$, $r^2 = 0.75$ and 0.68, respectively). Symbols are means of the averaged leaf damage for individual seedlings at each parent tree. Reproduced with permission from Norghauer et al. 2006a in the Journal of Tropical Ecology published by Cambridge University Press.

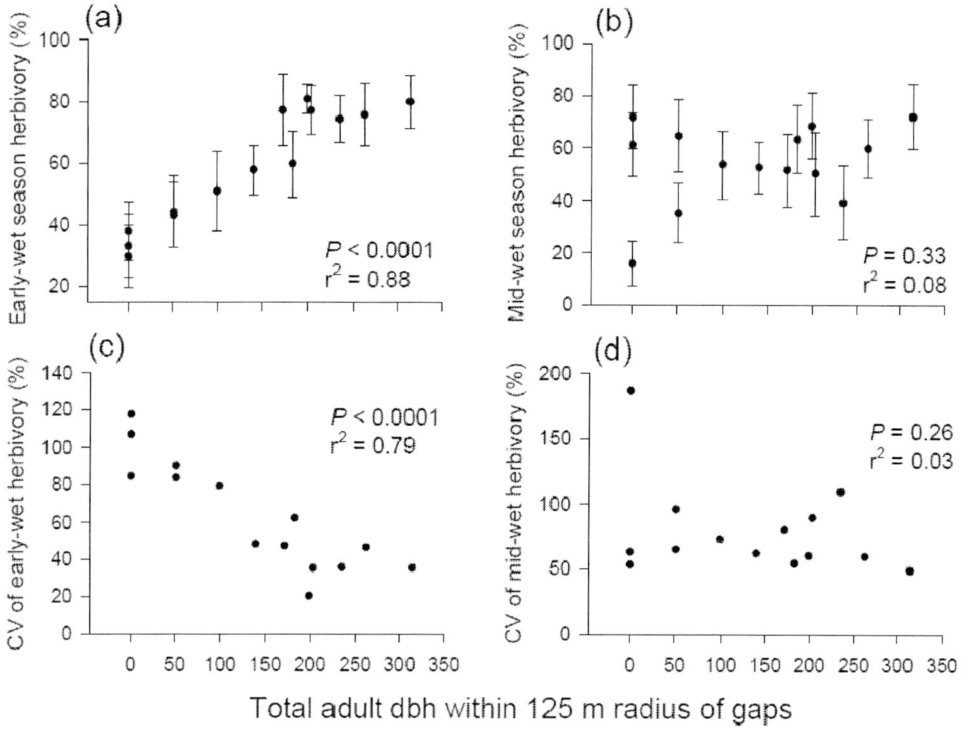

Figure 7. *Steniscadia poliophaea* damage to young leaves of mahogany seedlings in the early wet season (September–October) and mid wet season (November–December) in canopy gaps at Pinkaití in 2003. Damage is expressed as the mean (± SE) of 12 seedlings per gap, and as a function of mahogany adult tree size and density (i.e., total summed diameters) surrounding each canopy gap location. In generating these means, all new leaves were measured to the nearest 0.01 cm^2 and 0.25 cm^2 for damage and potential leaf area respectively, then averaged first at the individual seedling unit. The two lower panels show changes across these gaps in the coefficient of variation (CV) in specialist herbivory. The r^2 values for the linear regressions have been adjusted for the low sample size (n = 14 gaps). Reproduced with permission from Norghauer et al. 2008a in the Journal of Ecology, published by Wiley-Blackwell on behalf of the British Ecological Society.

In the early wet season (September–October) at our field sites, prior to germination, *S. poliophaea* caterpillar damage was also density-dependent on a larger spatial scale, that is, at the population level of mahogany: gaps with flushing mahogany seedlings that were further from one or more adult trees had the best chance to escape attacks at this time (Figure 7a; Norghauer et al. 2008a). During the next major leaf flush in November–December, the density-dependent pattern of attack was dampened as moths found these more remote seedlings in gaps (Figure 7b). The female moth, being active only at night, may be highly sensitive to specific odors emitted by mahogany seedlings flushing new leaves in gaps. In short, the moth engages in non-random host searches, orienting itself to odor plumes, as would be expected in more specialized insect herbivores (Visser 1986, Scoble 1992, Cardé and Willis 2008), and probably within a given 'discovered' gap it then searches specifically for mahogany hosts and compares their quality for oviposition via contact evaluation (Figure 7c; Renwick and Chew 1994, Bernays 2001, Schoonhoven et al. 2005). This ability to locate mahogany individuals in gaps moving downwind from parent trees should diminish as the density of host seedlings falls with distance.

Further evidence for a negative impact of *S. poliophaea* at the population level comes from a recent study of big-leaf mahogany introduced to the island of Dominica, in the Lesser Antilles, where the moth is absent. Here, first and second generation saplings and pole-sized trees are growing at high density beneath conspecific crowns — something never observed in forests within mahogany's natural range — and have little to no leaf area damaged (Norghauer et al. 2011b). Apart from expected self-thinning through larger size classes, the only limitation to growth and abundance of mahogany at this site seems to be light required by new seedlings. The 'Enemy-Release Hypothesis' (reviewed by Keane and Crawley 2002) — that plant species introduced to habitats outside their natural range are freed from top-down regulation, and thereby gain a competitive advantage versus native flora in the new community — can be seen as a biogeographical extension of Janzen-Connell 'escape' from native enemies via anthropogenic dispersal at an extreme spatial scale (DeWalt et al. 2004).

While herbivory may have consequences for host individuals, it does not necessarily follow that host plant populations must also be affected (Harper 1977, Crawley 1989). In other words, the population dynamics of mahogany could be unaffected if *S. poliophaea* were somehow magically excluded from local communities. But the J-C hypothesis suggests that insect herbivores can regulate population dynamics by slowing and spacing adult trees to some minimal distance within a community (Clark and Clark 1984, Schupp 1992). Our observations, like those of many other mahogany field researchers, indicate that saplings rarely occur beneath or near parent trees. This outcome cannot be solely the product of fewer gap formations per year near parent trees than further away on a per unit area basis (Augspurger 1983, Becker et al. 1985, Norghauer et al. 2011b). Instead, this suggests an herbivore-driven decoupling of the seed shadow from the larger juvenile shadow for a given tree species cohort through time (Augspurger 1983, Clark and Clark 1984), and runs counter to the expected null scenario of dispersal limitation (Makana and Thomas 2004): that is, near fruiting adult trees, more seeds should yield more recruits.

Nevertheless, the occurrence of mahogany adult trees in local aggregations is common in South American forests. Some researchers mistakenly cite clumping of adult trees as evidence against the J-C hypothesis. But local-scale aggregations may still arise via the J-C mechanism (Clark and Clark 1984, Becker et al. 1985): over many years of dispersal it is plausible that patches of mahogany juveniles could accumulate downwind through one or more gap events. In this way, patches of adult trees should cycle in space to create a moving mosaic over successive generations. This could occur as well in other tree species whose seed propagules are dispersed over long distances, often in clusters, by birds or mammals (Howe and Westley 1988, Wenny 2000, Fragoso et al. 2003, Russo and Augspurger 2004). Another interesting possibility not yet explored is that this density-dependent decoupling between key early stages in tree development fails to work soon after reproductive onset in mahogany trees (20-30 cm stem dbh). This is because the moth has not yet learned of that adult's presence, thus missing the first few years of seedling cohorts, especially if seed production rates are low, as expected from small-statured reproductive trees (Snook et al. 2005, Grogan and Galvão 2006, Norghauer et al. 2011a).

Taken together, the evidence presented here points to a strong Janzen-Connell effect in the Brazilian Amazonian forests of southeast Pará, driven by the Noctuid moth *S. poliophaea*, limiting recruitment of a fast-growing, potentially dominant tree species, big-leaf mahogany.

Herbivores have been increasingly shown to stabilize population dynamics of short-lived herbs and shrubs from temperate zones (Maron and Crone 2006). But to conclusively show that *S. poliophaea* is limiting the local abundance of mahogany adults would require long-term monitoring of very many individual juveniles. The long lifespan of the mahogany host species — and trees in general — makes this a daunting task. Other approaches are possible however. These include building population matrix projection and/or spatially explicit neighborhood models (Halpern and Underwood 2006) based on experimental field data from both Pinkaití and Marajoara, and observational data on gap disturbance regimes at these sites.

4. Ecological Significance of Steniscadia Poliophaea Vs. Hypsipyla Grandella

If big-leaf mahogany is indeed the only, or rather the locally most preferred, host plant of *S. poliophaea*, then the distance- and density-dependent impacts described above might be expected to extend throughout the host plant's range. Mahogany's natural range coincides with seasonally dry semi-evergreen forests from Mexico through Central America and into South America's Amazon Basin as far south as Bolivia (Figure 8; Lamb 1966, Grogan et al. 2002, 2010).

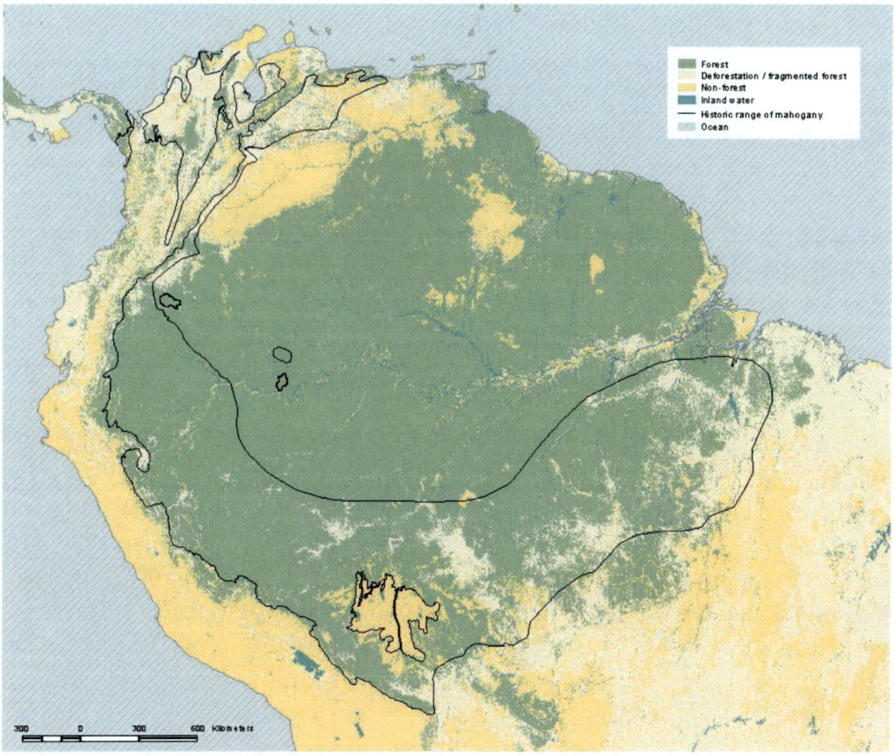

Figure 8. Historic range of big-leaf mahogany (*Swietenia macrophylla*) in South America overlaid on forest cover based on 2001 satellite data. Reproduced with permission from Grogan et al. 2010 in the journal Conservation Letters published by Wiley-Blackwell.

However, as far as we currently know, *S. poliophaea* is restricted to southern Amazonia, with confirmed sitings from our study region in southeast Pará to Acre in Brazil's far west. Outside Brazil, researchers in southeast Peru have documented caterpillar infestations on new mahogany seedlings in November in an area of undisturbed forest (G. Barrios, *pers. comm.*). For this reason, we suspect that *S. poliophaea*'s range extends to the forests of Peru and Bolivia, and possibly Ecuador. No sitings have been confirmed from Central America or Mexico.

A second and far more widely distributed mahogany 'pest' is the shoot-borer *Hypsipyla grandella* Zeller (Pyralidae), which occurs throughout big-leaf mahogany's neotropical range. *Hypsipyla grandella* is a nocturnal moth whose larval caterpillars bore into and consume expanding apical meristems on vigorously growing mahogany saplings (Lamb 1966). The flushing leader is hollowed out and frequently collapses, slowing vertical growth; many lateral shoots typically form from the damaged top, ruining stem form (Newton et al. 1993). In order for larval caterpillars to find sufficient food resources to mature through 5 to 6 instars, mahogany saplings generally must be at least 1 m tall and growing vigorously to serve as host plants. *H. grandella* attacks saplings and larger juveniles up to pole size and can also infest adult trees, especially injured or otherwise weakened individuals (Grogan 2001). They commonly infest woody fruit capsules in adult mahogany crowns (*S. poliophaea* has not been seen in adult mahogany crowns). A great deal of research has been conducted on the shoot-borer's biology and control due to its economic impact in plantations. Like *S. poliophaea*, *H. grandella* is most active when potential hosts are flushing new leaves, mainly in the wet season (Yamazaki et al. 1992, Newton et al. 1998, Taveras et al. 2004); females are attracted to flushing meristematic tissues via chemoreception, likely of essential oils (Soares et al. 2003); and they target vigorously growing juveniles, with the number of larvae increasing with shoot size. Might the shoot-borer, like *S. poliophaea*, generate a Janzen-Connell effect in natural and secondary forests? And if so, what interaction between these two insect herbivores might we expect where their natural ranges overlap?

Surprisingly, the ecological significance of *H. grandella* in the context of the J-C hypothesis is relatively unknown. No published study explicitly tests distance- or density-dependent shoot-borer attack on mahogany in natural forests (but see Yamazaki et al. 1992 for line plantings in Peruvian Amazon forest). At the unlogged Pinkaití forest, the shoot-borer's effects on mahogany saplings were not observed in naturally formed canopy gaps where nursery-grown seedlings were outplanted. But we did see signs of the shoot-borer in saplings > 2 m tall growing in very open conditions at the Pinkaití campsite. In thinned forest stands in Costa Rica, < 1% of 3-yr-old mahogany seedlings were attacked by *H. grandella*, and none in plantings made in natural forest in Peru (Yamazaki et al. 1990). At the selectively logged Marajoara forest, signs of the shoot-borer were more common, and incidence increased with canopy opening, host density and exposure of juvenile crowns (Grogan 2001 Appendix B.2, Grogan et al. 2005). In one outplanting experiment, the shoot-borer was not associated with any mahogany seedling deaths in 16 artificial gaps 200–400 m^2 in size (Grogan et al. 2003a Table 3). Because the minimum host size is larger than required for *S. poliophaea*, *H. grandella* will rarely kill mahogany saplings directly in well-lit conditions (Lamb 1966, Grogan et al. 2003a), since sufficient light allows saplings to continue growth after resprouting (i.e., tolerate attacks). Only repeated severe attacks over several years can cause juvenile mortality, all else being equal (Newton et al. 1993, Mayhew and Newton 1998, Nair 2007).

The ecological significance of *S. poliophaea* herbivory may extend to interactions with *H. grandella*, and possibly other insect herbivores as well. In one sense both predators are competing for the same host, but interspecific competition between the moths would best be described as asymmetric. Indeed, competition among multiple phytophagous insects sharing a host species is usually asymmetric, rarely satisfying the critical assumption of symmetry which can lead to competitive exclusion (Denno 1995, Kaplan and Denno 2007). Instead, herbivores might partition resources in time and space, so that specialized feeding on different plant parts can promote co-existence on the same host. In the present case there is partial temporal segregation in host use driven by plant ontogeny. Seedlings smaller than ~1 m height are only prone to attacks by *S. poliophaea*. But once they exceed this minimum size threshold and occur in gaps large enough to sustain vigorous shoot growth, *H. grandella* and *S. poliophaea* may overlap in host use, which may lead to direct interactions between them in the form of exploitative competition (Damman 1993). At Marajoara, attacks from both enemies have been observed on the same sapling(s), though whether this occurs on the same flushing event has not been confirmed. This seems unlikely, because *H. grandella*'s impact on the flushing apical leader is generally such that new leaves cannot survive long enough for *S. poliophaea* to grow to final instar size.

We propose that, by thinning and slowing growth of mahogany at early seedling and sapling stages, *S. poliophaea* functions in the herbivore community as a limiting factor on host numbers available later for the shoot-borer. This physically mediated resource limitation apparently imposed by *S. poliophaea* on *H. grandella* — coupled to very low density of individuals growing far downwind under natural dispersal — may in part explain the latter's near absence at Pinkaití. And because *S. poliophaea* is a highly effective distance- and density-dependent herbivore, we might expect a negative association between the two herbivores in their activity: *S. poliophaea*-attacked seedlings and saplings should appear less suitable to foraging *H. grandella* for being smaller and less vigorous than seedlings and saplings that escaped *S. poliophaea* attacks. We do not know, however, if *S. poliophaea* further negatively affects mahogany host availability and quality for foraging adult *H. grandella* via damage-induced chemical responses in attacked saplings, which might lead to complex resource-mediated interactions between these two Lepidoptera specialist herbivores (Damman 1993). These preliminary insights suggest that one potent specialized J-C enemy (*S. poliophaea*) might be able to disrupt the predicted J-C pattern (increased herbivory of hosts closer to parent trees) expected from another enemy (*H. grandella*) that also feeds on new mahogany plant tissue. We might also expect that in mahogany's northern range where *S. poliophaea* has not been reported, enemy compensation may occur such that *H. grandella* attacks in gaps are negatively correlated with proximity to adult mahogany trees, as predicted by the J-C hypothesis.

Resource limitation imposed by *S. poliophaea* does not necessarily translate into limited *H. grandella* abundance or densities because, unlike *S. poliophaea*, the shoot-borer can shift hosts to other Meliaceae genera such as *Carapa*, *Cedrela*, and *Guarea.* If among the three *Swietenia* species, which closely resemble *Khaya* spp. in Central Africa, *S. macrophylla* in South America is the oldest (i.e., evolved first) and spread north into Central America (Lamb 1966 p. 130), it is plausible that *S. poliophaea* herbivory may have exerted selection pressure on *H. grandella* leading to the latter's broader diet.

5. Forest Management Implications of *Steniscadia Poliophaea*

Steniscadia poliophaea's apparent obligate enemy relationship to *Swietenia macrophylla* is important for reasons extending beyond theoretical considerations of the Janzen-Connell hypothesis. Big-leaf mahogany is the world's most valuable widely traded tropical timber species, with a cubic meter of sawn timber fetching upwards of US$ 1700 (ITTO 2011). After centuries of overexploitation in neotropical forests, mahogany's conservation status had deteriorated by the late 1990s to the point where it was listed on Appendix II of the Convention on International Trade in Endangered Species of Wild Fauna and Flora (CITES) in late 2002. The CITES Appendix II listing requires cooperation between producer and consumer nations to verify that internationally traded volumes of sawn timber are harvested legally, and that harvests are non-detrimental to mahogany's role in ecosystems where it naturally occurs (Grogan and Barreto 2005). Mahogany's Appendix II listing has slowed the rate of predatory logging since 2003, especially in the three principal South American range nations of Brazil, Bolivia, and Peru, and contributed to improved forest management regulatory frameworks and practices (Mejía et al. 2008).

However, sustained yield management of mahogany from natural forests remains an elusive and unachieved goal for several reasons. Legal logging intensities remain too high, and minimum diameter cutting limits too low, for adequate population recovery between harvests on 25–30 year cutting cycles. Pre-logging densities of advance seedling regeneration and juvenile stems are typically too low for eventual replacement of logged trees. Growth rates by pole-sized and sub-adult trees, while relatively high for tropical forest species, are generally too low for production purposes on fixed-length cutting cycles. Silvicultural treatments reducing mortality and accelerating growth rates by individual trees are rarely implemented at industrial scales. And the mahogany shoot-borer, *Hypsipyla grandella*, plagues vigorously growing saplings and poles in logging gaps and plantations, slowing vertical growth rates and ruining bole form (Grogan et al. 2011). Now add *S. poliophaea* to these obstacles to sustainable management in South America: a specialist herbivore that targets seedlings and saplings, the most vulnerable phases of mahogany's life cycle.

Sustained yield management of mahogany requires balancing timber production — the extraction of commercial-sized adult mahogany trees — with reproduction, regeneration, recruitment and eventual replacement of logged stems with new adult trees. Management responses to *S. poliophaea* must therefore mitigate the moth's impacts on critical early life phases to reduce seedling mortality rates and accelerate growth. These efforts will be most successful if management practices are designed to coincide — or not — with *S. poliophaea*'s spatial and temporal appearance within the forest matrix.

The moth's distant- and density-dependent relationship with fruiting mahogany trees means that, all things being equal, seedlings experience declining early mortality rates moving away from parent trees. Though we have found that *S. poliophaea* is capable of locating flushing seedlings beyond the immediate perimeter of parent tree crowns, this ability necessarily declines with distance. Forest managers attempting to 'release' advance regeneration through targeted overhead canopy openings (Grogan et al. 2005) should therefore focus their efforts on seedlings or seedling patches occurring far (> 40–50 m) from rather than near retained seed trees or logged stems. However, as common-sensical as this recommendation may be, the fact remains that dispersal limitation of mahogany's wind-borne seeds means that few natural seedling clusters will be found in natural forests beyond 100 m of fruiting adults (Grogan and Galvão 2006, Norghauer et al. 2011a).

For seedlings and saplings growing in canopy gaps, the best defense against *S. poliophaea* will be excellent offense, that is, vigorous height growth so that juvenile crowns rapidly attain and exceed 5 m. For unknown reasons, the moth does not target mahogany saplings taller than this height. Robust mahogany seedlings are capable of growing up to 3 m height per year, though 1–1.5 m is more typical under optimal growing conditions (full sun exposure, in nutrient-rich and well-drained soil; Lopes et al. 2008). For forest managers, tending operations at the seedling and sapling stages must focus on encouraging height growth through removal of vines and directly overtopping competing vegetation (Snook and Negreros-Castillo 2004). Dense surrounding secondary vegetation probably shields seedlings and saplings from *S. poliophaea* attack to some degree, similar to the mahogany shoot-borer (Newton et al. 1993, Mayhew and Newton 1998), so surrounding vegetation should be retained for protection and 'training' vertical seedling growth.

Widely reported mahogany regeneration failure after logging combined with low background levels of advance regeneration mean that enrichment planting of nursery-grown seedlings into treefall gaps will be necessary for sustained yield production (Snook 1996; Gullison et al. 1996; Grogan et al. 2003b, 2005, 2008). *S. poliophaea* has proven a scourge in nurseries supplying mahogany seedlings for plantations in southeast Pará and Acre in western Brazil, especially where nurseries are located inside managed forests. Protection against *S. poliophaea* must be provided in these settings, either by mesh enclosure to prevent access to flushing seedlings by female moths or through periodic application of insecticides targeting larval instars.

S. poliophaea's density-dependent behavior indicates that mahogany seedlings should be outplanted into logging gaps at low densities, spaced 5 or more meters apart. Similar to released natural regeneration discussed above, tending operations should be restricted to vines and directly competing vegetation (overhead shading), leaving seedlings to grow in a dense matrix of secondary vegetation. As in nursery settings, periodic insecticide application may be warranted, but our experience has been that wet season rains may shorten the effective treatment period to a matter of days, requiring repeated application. As well, as seedlings gain height, insecticide application may become hazardous to management crews. The various costs of such treatment must be weighed against potential benefits.

Selection of enrichment planting gaps should consider the locations of both logged and retained mahogany seed trees as well as sub-commercial stems. All things being equal, *S. poliophaea*'s impacts are likely to be mitigated by selecting enrichment sites up- rather than downwind of adult trees, and far rather than close. However, locating enrichment gaps on portions of the landscape where adult mahoganies do not naturally occur, or where it occurs at extremely low densities, is unlikely to yield optimal seedling growth rates due to soil nutrient deficiencies or poor drainage (Grogan et al. 2003a, Norghauer et al. 2008b).

From a biodiversity monitoring and conservation perspective, the impacts on *S. poliophaea* and other insects of selective logging and forest management must also be considered. If *S. poliophaea* is truly monophagous, then local mahogany extirpations will mean the same for its specialized herbivore. However, short of forest conversion to other land uses such as pasture or agriculture, complete elimination of mahogany from forests once containing it will be a rare occurrence due to persistence of unlogged juveniles and natural regeneration. We have seen no evidence that extremely low density of mahogany at landscape scales reduces *S. poliophaea* occurrence correspondingly. For example, attack rates on seedlings in Acre, where 12 mahogany trees > 20 cm diameter occur per 100 ha, appear equivalent to attack rates on seedlings at Marajoara in southeast Pará, where 65 trees > 20 cm diameter occur per 100 ha (Grogan et al. 2008). For this reason we expect that *S. poliophaea* populations will adjust their size to match new post-logging mahogany population densities (as expected for a resource-limited insect herbivore, Dempster and Pollard 1981).

However, other aspects of logged environments could have unforeseen consequences for *S. poliophaea*. For example, the drier, more open environment that characterizes selectively logged forests may influence the moth's population dynamics by increasing food availability (Basset et al. 2001) and female moth detection of mahogany host plant odors against the contrasting backdrop of non-host odors, though this might be offset by greater interference from vegetation re-growth of early-succesional species. By contrast, drier forest conditions could also lead to reduced larval feeding rates and faster maturation rates by mahogany leaves, and to higher risks of larval desiccation. As well, changes in forest structure could lead to changes in natural enemy pressure on *S. poliophaea* larvae and adults. The possibility also exists that changes in microclimate associated with logging may exacerbate asynchrony in mahogany leaf phenology and/or reduce host plant quality, either of which might impact *S. poliophaea* population dynamics because of its restricted diet and limited larval mobility. Skid trails may create directional 'highways' allowing *S. poliophaea* to more easily find hosts, not unlike that suggested to have resulted from line plantings in Peruvian forests (Ikeda et al. cited in Mayhew and Newton 1998).

For these many reasons, research into the moth's seasonal dynamics and spatial patterns of herbivory in both logged and unlogged forests across years is vital. Depending on what this research reveals, the close association between *S. poliophaea* and mahogany invites the interesting possibility of using the former as a barometer of forest change and recovery after logging in Amazonia landscapes.

6. Are All Herbivores Equal? Moth Herbivores and the Janzen-Connell Hypothesis

At the heart of the J-C hypothesis are host plant–natural enemy interactions. In this context, these interactions are presumably widespread and involve co-evolution because of the negative impact on host fitness by herbivores (Ehrlich and Raven 1964, Howe and Westley 1988, Thompson 1999, Weiblen et al. 2006, Futuyma and Agrawal 2009). In this chapter we have reviewed and synthesized compelling evidence for one example of this antagonistic interaction, that between a Noctuid moth caterpillar and an emergent canopy tree, which generates a powerful J-C effect limiting the potential dominance of a widely distributed tree species in Amazonian lowland tropical forest. Top-down regulation is not a new idea in ecology, but it is necessarily spatially explicit for plants because of their mostly sessile life habit (Harper 1977). The J-C hypothesis proposes that the negative effect of parent trees on the probability of a given seed-seedling becoming a sapling, and later potentially an adult, decreases with dispersal distance (Figure 1) — as offspring escape high densities of conspecific siblings and predator attacks — and is species-specific. While distance from close relatives clearly matters (Howe and Smallwood 1982, Clark and Clark 1984, Swamy et al. 2011), the latter consideration is also crucial because it enhances intraspecific effects relative to interspecific effects of tree species, thereby promoting diversity by stabilizing species' populations (Chesson 2000).

The simplest explanation for strong species-specific J-C effects on tree seedling and sapling mortality and recruitment processes is not intraspecific competition for scarce nutrients (Wright 2002, Paine et al. 2008, Svenning et al. 2008), but rather insects and pathogens specialized to one or a few closely related host plants, particularly at the genus level (Novotny et al. 2002a, Leigh 2004, Kricher 2010, Swamy and Terborgh 2010). Whether insect herbivores can attack or nibble on leaves of other plants in laboratory choice tests is largely irrelevant; rather, it is their preference for and/or restricted diet in terms of what is locally available that matters in the J-C context. For example, many genus-specialized insect herbivores might effectively (*de facto*) be monophagous in part because of the lack of congener hosts in the local forest community. At the other extreme, a rodent species may function as a *facultative specialist* enemy (Janzen 1970) if it predates a single tree species' seeds during the dry season when these represent the best or only food available in the community. Nevertheless, it is well-established, both theoretically and empirically, that insect herbivores are more diverse, abundant, and better able than mammals to increase population growth rates in response to increasing food resource densities and thus avoid satiation (Janzen 1970, 1974, Dempster and Pollard 1981, Howe and Westley 1988, Hammond and Brown 1998, Muller-Landau et al. 2003, Nathan and Casagrandi 2004, Schoonhoven et al. 2005, Nair 2007, Speight et al. 2008).

Moth Lepidoptera, especially the species-rich Noctuoidae and Pyralidae super-families, may harbor many such specialized enemies, and thus we argue they are the best candidates to generate J-C effects on established woody seedlings and saplings in tropical forests. From rigorous, large-scale insect collecting field studies, a recurring theme is the narrow host-specificity of larval Lepidoptera (Janzen 1988, Novotny et al. 2002a, 2006, Dyer et al. 2007). For example, a randomly selected caterpillar from New Guinean forest vegetation will likely ($\geq$ 50% probability) have nearly all of its population ($\geq$ 90%) concentrated on a single host species (Novotny et al. 2004). In the same forest, Lepidoptera dominated the genus-level specialists, especially in terms of total biomass — which should correlate with impact on host plants — whereas the species-rich Coleoptera were mostly generalists and much lower in biomass (Novotny et al. 2002b). A recent phylogenetic study revealed that many more rain forest insect herbivores fed on closely related plants rather than on divergent hosts than expected by chance alone, and that Lepidoptera was the most specialized herbivore group (Weiblen et al. 2006).

Further favoring a key J-C role for moth Lepidoptera is plant leaf age. Along with tree ontogeny, plant leaf age represents a crucial yet neglected factor influencing the host-specificity of leaf-chewing insects (Novotny and Basset 2005). From the plant's perspective, young leaves are more valuable than mature leaves (Harper 1989). From an insect herbivore's perspective, mature and young leaves are very different kinds of food (Feeny 1970, Strong et al. 1984). Mature leaves are abundant and ever-present, but well-defended and tough to chew; young leaves are easier to chew, richer in nutrients, but ephemeral and often poisonous. In many forests most (> 70–80%) of a given leaf's lifetime damage occurs during the brief period of expansion, and this damage occurs irrespective of how fast young leaves expand or how toxic they are (Coley 1983, Coley and Aide 1991, Coley and Barone 1996, Kursar and

Coley 2003, Kursar et al. 2006, Kursar et al. 2009). It stands to reason then that whatever is eating new leaves is highly specialized to locate and use them as nourishment, whereas whatever is eating mature leaves is more likely to have a broader diet. Such a trend was recently seen in stick insect herbivores as well (Phasmida; Bluthgen and Metzner 2007). The narrow host-specificity of Lepidoptera suggests that many species target new leaves that are otherwise less edible to the general herbivore community (Coley et al. 2006). For example, of insects causing damage to young expanding sapling leaves in a Panamanian forest, 38 of 46 herbivore species were Lepidoptera, of which almost half were specialized to genus or species levels (Barone 1998). Similarly, 97% of caterpillars on 11 species of *Inga* tree saplings occurred on young leaves (Kursar et al. 2006). In their recent report of an exhaustive field study, Novotny et al. (2010) concluded that the herbivore guilds most likely to cause density-dependent effects on plant fitness — i.e., J-C type enemies — were those found to be most highly specialized in diet: larval leaf chewers (Lepidoptera), leaf-miners (Lepidoptera, Coleoptera, Diptera), and leaf-suckers (Hemiptera).

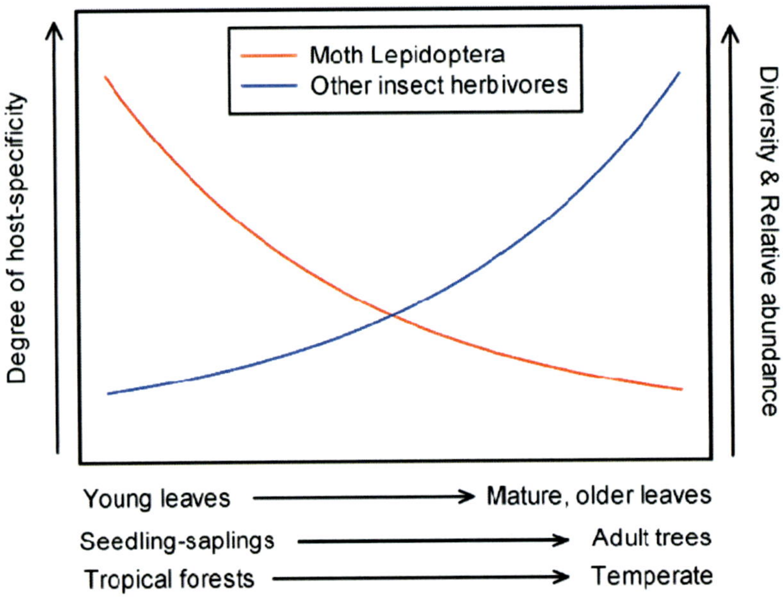

Figure 9. Conceptual diagram showing the hypothesized relevance of Lepidoptera larvae as key herbivores in the Janzen-Connell context in contrast to other insect herbivores. In this generalized scheme, the caterpillars' degree of local host-specificity is higher on younger plant parts and on younger or smaller-sized host plants or trees. Conversely, both species richness and relative abundance are higher moving further along the axes of host plant ontogeny. Finally, both trends should be less pronounced in temperate than in tropical broad-leafed forests when host ontogeny is taken into consideration.

We may refine this argument further: it is Lepidoptera feeding *on tree seedlings and saplings* that matter most in the context of the J-C hypothesis. Thus, sampling insects on large juvenile (> 5–10 cm diameter) and adult trees to gauge herbivore host-specificity yields little insight into the J-C hypothesis because its mechanistic effect has already transpired — these trees are now in an excellent position to recruit into the canopy. Further, as with *S. poliophaea*, the J-C hypothesis in its original formulation was never predicated on insect herbivores being present in parent tree crowns and from there descending upon conspecific seedlings and saplings, though this may happen (Janzen 1970). Most post-dispersal seed-eating small mammals and insects likely do not dwell in parent crowns, and can still generate J-C type effects (reviewed by Turner 2001, Carson et al. 2008). If anything, differences in the vertical stratification — and relative host impact — of herbivore assemblages in tropical forests are expected, perhaps more so than in temperate forests, which could reflect differences in food palatability or abundance of both trees and lianas; microclimatic changes and constraints (temperature, wind) on insect development and behavior; or changes in susceptibility to faunal predators, ants and birds in particular (Heinrich 1993, Basset et al. 2003). So, in addition to leaf age, we anticipate that insect herbivore host-specificity is likely very different between host plant species' ontogenetic stages (summarized in Figure 9) in part of because differences in host abundance and changes in plant resource allocation to defense and increased tolerance with tree age/size (Boege and Marquis 2005). That is, insect herbivores feeding on adult trees do not likely also feed on conspecific seedlings and saplings (Basset 2001, Barrios 2003).

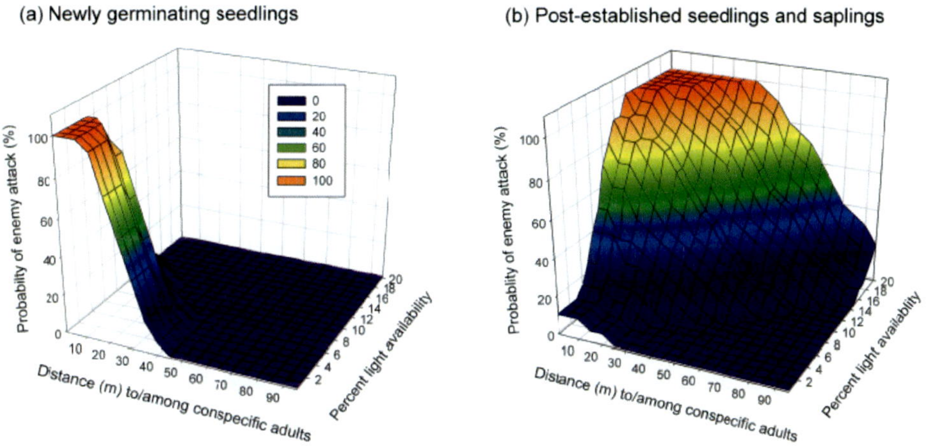

Figure 10. Three-dimensional graphics showing the proposed augmentation in the spatial scale of foraging and attacks by the specialist moth *Steniscadia poliophaea* on its host tree, big-leaf mahogany in the latter's South American range. Two factors, increased local light availability, mainly in canopy gaps, and proximity to conspecific adults interact to shape mahogany susceptibility to the specialist moth in forests. Once a year, newly germinating seedlings emerge in the vicinity of fruiting mahogany trees, mostly in shaded forest understory conditions, and the probability of attacks to the modal seedling declines with distance from the parent tree (a). At other times in the wet season, food for *S. poliophaea* is very limited in the understory because mahogany leaf production requires above-average light availability, and the probability of attacks to the modal established seedling or sapling increases both with flushing vigor in canopy gaps and proximity to nearby conspecific adults (b). The net result for this antagonistic interaction is that the Janzen-Connell effect generated by this moth on its host population is likely strengthened, operating over a larger area, thereby increasing the spacing apart of adult recruitment events (i.e., shifting the PRC further to the right in Figure1.).

As demonstrated by the *S. poliophaea* case study, an herbivore diet restricted to new leaves might increase the spatial scale of host searches. This could extend the J-C effect well beyond the parent tree crown to new forest canopy gaps where seedling growth is promoted by increased light levels (Figure 10b). In these microhabitats, depending on host plant density, flushing vigor, and proximity to conspecific adults, insect predators might exert a 'brake' on host recruitment rates. Because herbivores restricted to young foliage are likely very resource limited (Dempster and Pollard 1981), attacks in distant canopy gaps are thus predictable, as these herbivores are effectively food-limited in the shaded forest understory (Richards and Coley 2007, 2008). Lepidoptera specialists thus may search for hosts at further distances than other insect herbivores, and colonize them faster than polyphagous Lepidoptera, as reported for the monophagous butterfly *Heliconius hewitsoni* which preys on the neotropical vine *Passiflora pittieri* (Thomas 1990 Figure 1). Consequently the J-C effect on a host population is strengthened because the spatial repulsion between parent and offspring is enhanced. Considering that light availability is generally the most limiting resource for new seedlings and saplings (Chazdon and Fetcher 1984, Richards 1996, Turner 2001), the effects of light levels on their susceptibility to different enemy guilds (insects vs. pathogens vs. mammals) may prove useful to detect J-C effects, and provide a deeper understanding of them — and this should further benefit from consideration of different spatial scales for conspecific density-dependent effects (Angulo-Sandoval and Aide 2000, Sullivan 2003).

Little doubt remains that negative density dependence in tree dynamics is a widespread phenomenon, capable of promoting species coexistence in hyperdiverse tropical forest communities (Wright 2002, Carson et al. 2008, Zimmerman et al. 2008, Kricher 2010). Pests are the most plausible explanation for these patterns (Leigh 2004), as envisaged in the J-C hypothesis 40 years ago. Much attention has recently been paid to fungal pathogens, which can cause severe density-dependent mortality to cohorts of new seedlings (Bell et al. 2006, Bagchi et al. 2010, Mangan et al. 2010) and enhance species differences in shade tolerance (McCarthy-Neumann and Kobe 2008). But pathogen host-specificity is likely weak (McCarthy-Neumann and Kobe 2010) and their population-level impacts are probably limited to beneath tree crowns only (Figure 9a; Burdon and Chilvers 1982, Augspurger 1984b). However, as causal agents underpinning the J-C hypothesis, pathogens and moth Lepidoptera need not be mutually exclusive. As proposed here, their respective roles and impact may wax and wane depending on changing light availability to seedlings and saplings in forests and on the host density of conspecifics and possibly congeneric species (Figure 10). In this way, the presence and action of both enemy guilds can occur on a single or many canopy tree species, and in combination exert as a powerful force to help maintain community diversity.

References

Aide TM, Londono EC (1989) The effects of rapid leaf expansion on the growth and survivorship of a lepidopteran herbivore. *Oikos* 55: 66-70.

Angulo-Sandoval P, Aide TM (2000) Effect of plant density and light availability on leaf damage in *Manilkara bidentata* (Sapotaceae). *Journal of Tropical Ecology* 16: 447-464.

Augspurger CK (1983) Offspring recruitment around tropical trees: changes in cohort distance with time. *Oikos* 40: 189-196.

Augspurger CK (1984a) Seedling survival of tropical tree species: interactions of dispersal distance, light gaps, and pathogens. *Ecology* 65: 1705-1712.

Augspurger CK (1984b) Light requirements of neotropical tree seedlings: a comparative study of growth and survival. *Journal of Ecology* 72: 777-795.

Badenas-Perez FR, Johnson MT (2008) Biology, herbivory and host specificity of *Antiblemma leucocyma* (Lepidoptera: Noctuidae) on *Miconia calvescnes* DC. (Melastomataceae) in Brazil. *Biocontrol Science and Technology* 18: 183-192.

Bagchi R, Swinfield T, Gallery RE, Lewis OT, Gripenberg S, *et al.* (2010) Testing the Janzen-Connell mechanism: pathogens cause overcompensating density dependence in a tropical tree. *Ecology Letters* 13: 1262-1269.

Baker PJ, Bunyavejchewin S (2006) Suppression, release and canopy recruitment in five tree species from a seasonal tropical forest in western Thailand. *Journal of Tropical Ecology* 22: 521-529.

Barbosa P (1993) Lepidopteran foraging on plants in agrosystems: constraints and consequences. Pages 523-566 in *Caterpillars: Ecological and Evolutionary Constraints on Foraging*. Edited by Stamp NE, Casey TE. London: Chapman and Hall.

Barlow HS, Woiwood IP (1989) Moth diversity of a tropical rain forest in peninsular Malaysia. *Journal of Tropical Ecology* 5: 37-50.

Barone JA (1996) *Herbivory and tropical tree diversity: a test of the Janzen-Connell model.* PhD thesis. University of Utah, UT, USA.

Barone JA (1998) Host-specificity of folivorous insects in a moist tropical forest. *Journal of Animal Ecology* 67: 400-409.

Barrios H (2003) Insect herbivores feeding on conspecific seedlings and trees. Pages 282-290 in *Arthropods of Tropical Forests: Spatio-Temporal Dynamics and Resource Use in the Canopy*. Edited by Basset Y, Novotny V, Miller SE, Kitching RL. Cambridge: Cambridge University Press.

Basset Y (2001) Communities of insect herbivores foraging on saplings versus mature trees of *Pourouma bicolor* (Cecropiaceae) in Panama. *Oecologia* 129: 253-260.

Basset Y, Charles E, Hammond DS, Brown VK (2001) Short-term effects of canopy openness on insect herbivores in a rain forest in Guyana. *Journal of Applied Ecology* 38: 1045-1058.

Basset Y, Hammond PM, Barrios H, Holloway JD, Miller SE (2003) Vertical stratification of arthropod assemblages. Pages 17-27 in *Arthropods of Tropical Forests: Spatio-Temporal Dynamics and Resource Use in the Canopy*. Edited by Basset Y, Novotny V, Miller SE, Kitching RL. Cambridge: Cambridge University Press.

Becker P, Lee LW, Rothman ED, Hamilton WD (1985) Seed predation and coexistence of tree species: Hubbell's models revisited. *Oikos* 44: 382-390.

Bell T, Freckleton RP, Lewis OT (2006) Plant pathogens drive density dependent mortality in a tropical tree. *Ecology Letters* 9: 569-574.

Bernays EA (2001) Neural limitations in phytophagous insects: implications for diet breadth and host-affiliation. *Annual Review of Entomology* 46: 703-727.

Bernays EA, Graham (1988) On the evolution of host-specificity in phytophagous arthropods. *Ecology* 69: 886-892.

Bernays EA, Janzen DH (1988) Saturniid and sphingid caterpillars: two ways to eat leaves. Ecology 69: 1153-1160.

Blundell AG, Peart DR (1998) Distance-dependence in herbivory and foliar condition for juvenile *Shorea* trees in Bornean dipterocarp rain forest. *Oecologia* 117: 151-160.

Blundell AG, Peart DR (2001) Growth strategies of a shade-tolerant tropical tree: the interactive effects of canopy gaps and simulated herbivory. *Journal of Ecology* 89: 608-615.

Blundell AG, Peart DR (2004) Density-dependent dynamics of a dominant rain forest canopy tree. *Ecology* 85: 704-715.

Bluthgen N, Metzner A (2007) Contrasting leaf age preferences of specialist and generalist stick insects (Phasmida). *Oikos* 116: 1853-1862.

Boege K, Marquis RJ (2005) Facing herbivory as you grow up: the ontongeny of resistance in plants. *Trends in Ecology and Evolution* 20: 441-448.

Brokaw, NLV (1985) Gap-pahse regeneration in a tropical forest. *Ecology* 66: 682-687.

Burdon JJ, Chilvers GA (1982) Host density as a factor in plant disease ecology. *Annual Review of Phytopathol*ogy 20: 143-166.

Cardé MA, Willis RT (2008) Navigational strategies used by insects to find distant, wind-borne sources of odor. *Journal of Chemical Ecology* 34: 854-866.

Carson WP, Root RB (2000) Herbivory and plant species coexistence: community regulation by an outbreaking phytophagous insect. *Ecological Monographs* 70: 73-99.

Carson WP, Anderson Jr JT, Leigh EG, Schnitzer SA (2008) Challenges associated with testing and falsifying the Janzen-Connell Hypothesis: A review and critique. Pages 210–241 in *Tropical Forest Community Ecology*. Edited by Carson WP, Schnitzer SA. Oxford: Wiley-Blackwell.

Chazdon RL, Fetcher N (1984) Photosynthetic light environments in a lowland tropical rain-forest in Costa Rica. *Journal of Ecology* 72: 553-564.

Chesson P (2000) Mechanisms of maintenance of species diversity. *Annual Review of Ecology and Systematics* 31: 343-366.

Choong MF (1996) What makes a leaf tough and how this affects the pattern of Castanopsis fissa leaf consumption by caterpillars. *Functional Ecology* 10: 668-674.

Clark DA, Clark DB (1984) Spacing dynamics of a tropical rain forest tree: evaluation of the Janzen-Connell model. *American Naturalist* 124: 769-788.

Clark DB, Clark DA (1985) Seedling dynamics of a tropical tree – impacts of herbivory and meristem damage. *Ecology* 66: 1884-1892.

Coley PD (1983) Herbivory and defensive characteristics of tree species in a lowland tropical forest. *Ecological Monographs* 53: 209-233.

Coley PD, Bryant JP, Chapin FS (1985) Resource availability and plant antiherbivore defense. *Science* 230: 895-899.

Coley, PD, Aide TM (1991) Comparison of herbivory and plant defenses in temperate and tropical broad-leaved forests. Pages 25-49 in *Plant-Animal Interactions: Evolutionary Ecology in Tropical and Temperate Regions*. Edited by Price PW, Lewinsohn TM, Fernandes GW, Benson WW. New York: Wiley and Sons.

Coley PD, Barone JA (1996) Herbivory and plant defenses in tropical forests. *Annual Review of Ecology and Syste*matics 27: 305-335.

Coley PD, Kursar TA (1996) Ant-herbivore defenses of young tropical leaves: physiological constraints and ecological tradeoffs. Pages 305-336 in *Tropical Forest Plant Ecophysiology*. Edited by Mulkey SS, Chazdon RL, Smith AP. New York: Chapman and Hall.

Coley PD, Bateman ML, Kursar TA (2006) The effects of plant quality on caterpillar growth and defense against natural enemies. *Oikos* 115: 219-228.

Comita LS, Muller-Landau HC, Alguilar S, Hubbell SP (2010) Asymmetric density dependence shapes species abundances in a tropical tree community. *Science* 329: 330-332.

Connell JH (1971) On the role of natural enemies in preventing competitive exclusion in some marine animals and in rain forest trees. Pages 298-312 in *Dynamics of Populations*. Centre for Agricultural Publication and Documentation.

Crawley MJ (1989) Insect herbivores and plant population dynamics. *Annual Review of Ecology and Systematics* 34: 531-564.

Cyr H, Pace ML (1993) Magnitude and patterns of herbivory in aquatic and terrestrial ecosystems. *Nature* 361: 148-150.

Damman H (1993) Patterns of interaction among herbivore species. Pages 132-169 in *Caterpillars: Ecological and Evolutionary Constraints on Foraging*. Edited by Stamp NE, Casey TE. London: Chapman and Hall.

Dempster JP, Pollard E (1981) Fluctuations in resource availability and insect populations. *Oecologia* 50: 412-416.

Denno RF (1995) Interspecific interactions in phytophagous insects – competition reexamined and resurrected. *Annual Review of Entomology* 40: 297-331.

Denslow JS (1987) Tropical rainforest gaps and tree species diversity. *Annual Review of Ecology and Systematics* 18: 431-451.

Dethier VG (1954) Evolution of feeding preferences in phytophagous insects. *Evolution* 8: 32-54.

DeWalt SJ, Denslow JS, Ickes K (2004) Natural-enemy release facilitates habitat expansion of the invasive tropical shrub *Clidemia hirta*. *Ecology* 85: 471-483.

Dicke M (2000) Chemical ecology of host-plant selection by herbivorous arthropods: a multitrophic perspective. *Biochemical Systematics and Ecology* 28: 601-617.

Dussourd DE (1993) Foraging with finesse: caterpillar adaptations for circumventing plant defenses. Pages 92-131 in *Caterpillars: Ecological and Evolutionary Constraints on Foraging*. Edited by Stamp NE, Casey TE. London: Chapman and Hall.

Dyer LA (1995) Tasty generalists and nasty specialists? Antipredator mechanisms in tropical Lepidopteran larvae. *Ecology* 76: 1483-1496.

Dyer LA, Letorneau DK, Dodson CD, Tobler MA, Stireman III JO, *et al.* (2004) Ecological causes and consequences of variation in defensive chemistry of a neotropical shrub. *Ecology* 85 : 2795-2803.

Dyer LA, Singer, MS, Lill JT, Stireman JO, Gentry GL, *et al.* (2007) Host-specificity of Lepidoptera in temperate and tropical forests. *Nature* 448: 696-700.

Ehrlich PR, Raven PH (1964) Butterflies and plants: a study in coevolution. *Evolution* 18: 586-608.

Eichhorn MP, Nilus R, Compton SG, Hartley SE, Burslem DFRP (2010) Herbivory of tropical rain forest tree seedlings correlates with future mortality. *Ecology* 91: 1092-1101.

Feeny P (1970) Seasonal changes in oak leaf tannins and nutrients as a cause of spring feeding by winter moth caterpillars. *Ecology* 51: 565-581.

Fine PVA, Mesones I, Coley PD (2004) Herbivores promote habitat specialization by trees in Amazonian forests. *Science* 305: 663-665.

Fraenkel GS (1959) The raison d'etre of secondary plant substances. *Science* 129: 1466-1470.

Fragoso JMV, Silvius KV, Correa JA (2003) Long-distance dispersal by tapirs increases seed survival and aggregates trees. *Ecology* 84: 1998-2006.

Futuyma DJ, Agrawal AA (2009) Macroevolution and the biological diversity of plants and herbivores. Proceedings of the National Academy of Sciences 106: 18054-18061.

Gerhardt K (1998) Leaf defoliation of tropical dry forest tree seedlings - implications for survival and growth. *Trees-Structure and Function* 13: 88-95.

Gray B (1972) Economic tropical forest entomology. *Annual Review of Entomology* 17:313-352.

Grogan JE (2001). *Bigleaf mahogany (*Swietenia macrophylla *King) in southeast Pará, Brazil: a life history study with management guidelines for sustained production from natural forests.* PhD thesis, Yale University, New Haven, CT.

Grogan J, Barreto P, Veríssimo A (2002) Mogno na Amazônia Brasileira: Ecologia e Perspectivas de Manejo. Imazon: Belém, Pará, Brazil. 58 p.

Grogan J, Ashton MS, Galvão J (2003a) Big-leaf mahogany (*Swietenia macrophylla*) seedling survival and growth across a topographic gradient in southeast Para, Brazil. *Forest Ecology and Management* 186: 311-326.

Grogan J, Galvão J, Simões L, Veríssimo A (2003b) Regeneration of big-leaf mahogany in closed and logged forests of southeastern Pará, Brazil. Pages 193-208 in *Big-Leaf Mahogany: Genetics, Ecology, and Management.* Edited by Lugo A, Figueroa Colón JC, Alayón M. Springer-Verlag: New York.

Grogan J, Barreto P (2005) Big-leaf mahogany on CITES Appendix II: big challenge, big opportunity. *Conservation Biology* 19: 973-976.

Grogan J, Landis RM, Ashton MS, Galvão J (2005) Growth response by big-leaf mahogany (*Swietenia macrophylla*) advance seedling regeneration to overhead canopy release in southeast Pará, Brazil. *Forest Ecology and Management* 204: 399-412.

Grogan J, Galvão J (2006) Factors limiting post-logging seedling regeneration by big-leaf mahogany (*Swietenia macrophylla*) in southeastern Amazonia, Brazil, and implications for sustainable management. *Biotropica* 38: 219-228.

Grogan J, Jennings S, Landis RM, Schulze M, Baima AMV, *et al.* (2008) What loggers leave behind: big-leaf mahogany (*Swietenia macrophylla*) population structure and potential for post-logging recovery in the Brazilian Amazon. *Forest Ecology Management* 255: 269-281.

Grogan J, Blundell AG, Landis RM, Youatt A, Gullison RE, *et al.* (2010) Over-harvesting driven by consumer demand leads to population decline: big-leaf mahogany in South America. *Conservation Letters* 3: 12-20.

Grogan J, Peña-Claros M, Günter S (2011) Managing natural populations of big-leaf mahogany. Pages xx-xx in *Silviculture in the Tropics.* Edited by Günter S, Stimm B, Weber M, Mosandl R. Institute of Silviculture, Technische Universität München.

Gullison RE, Panfil SN, Strouse JJ, Hubbell SP (1996) Ecology and management of mahogany (*Swietenia macrophylla* King) in the Chimanes Forest, Beni, Bolivia. *Botanical Journal of the Linnean Society* 122: 9-34.

Halpern SL, Underwood N (2006) Approaches for testing herbivore effects on plant population dynamics. *Journal of Applied Ecology* 43: 922-929.

Hammond DS, Brown VK (1998) Disturbance, phenology, and life-history characteristics: factors influencing distant/density-dependent attack on tropical seeds and seedlings. Pages 51–79 in *Dynamics of Tropical Communities.* Edited by Newberry DM, Prins HTT, Brown ND. Cambridge: Blackwell Science.

Hampson GF (1912) Catalogue of the Lepidoptera Phalaenae in the British Museum: Vol.11, Catalogue of the Noctuidae. London: British Museum (Natural History). Department of Zoology.

Hanley M (1998) Seedling herbivory, community composition and plant life history traits. *Perspectives in Plant Ecology, Evolution and Systematics* 1/2: 191-205.

Harms KE, Wright SJ, Calderón O, Hernández A, Herre EA (2000) Pervasive density dependent recruitment enhances seedling diversity in a tropical forest. *Nature* 404: 493-495.

Harper JL (1977) *Population Biology of Plants*. London: Academic Press. 892 p.

Harper JL (1989) The value of a leaf. *Oecologia* 80: 53-58.

Harrison S (1987) Treefall gaps versus forest understory as environments for a defoliating moth on a tropical forest shrub. *Oecologia* 72: 65-68.

Hartshorn GS (1978) Tree falls and tropical forest dynamics. Pages 617-638 in *Tropical Trees as Living Systems*. Edited by Tomlinson PB, Zimmermann MH. Cambridge: Cambridge University Press.

Hawkes CV, Sullivan JJ (2001) The impact of herbivory on plants in different resource conditions. *Ecology* 82: 2045-2058.

Hebert PDN, Penton Eh, Burns JM, Janzen DH, Hollwachs W (2004) Ten species in one: DNA barcoding reveals cryptic species in the neotropical skipper butterfly *Astrapes fulgerator*. Proceedings of the National Academy of Sciences 101: 14812-14817.

Heinrich (1993) How avian predators constrain caterpillar foraging. Pages 224-247 in *Caterpillars: Ecological and Evolutionary Constraints on Foraging*. Edited by Stamp NE, Casey TE. London: Chapman and Hall.

Howe HF, Smallwood J (1982) Ecology of seed dispersal. *Annual Review of Ecology and Systematics* 13: 201-228.

Howe HF, Westley LC (1988) *Ecological Relationships of Plants and Animals*. Oxford: Oxford University Press, 273 p.

Huntly N (1991) Herbivores and the dynamics of communities and ecosystems. *Annual Review of Ecology and Systematics* 22: 477-503.

ITTO (International Tropical Timber Organization) (2011) *Tropical Timber Market* (TTM) Report, 16 – 30 April, Vol. 16 (8): 8.

Janzen DH (1970) Herbivores and the number of tree species in tropical forests. American *Naturalist* 104: 501-528.

Janzen DH (1971) Escape of juvenile *Dioclea megacarpa* (Leguminosae) vines from predators in a deciduous tropical forest. *American Naturalist* 105: 97-112.

Janzen DH (1974) Tropical blackwater rivers, animals, and mast fruiting by the Dipterocarpaceae. *Biotropica* 6: 69-103.

Janzen DH (1988) Ecological characterization of a Costa Rican dry forest caterpillar fauna. *Biotropica* 20: 120-135.

Janzen DH (1993) Caterpillar seasonality in a Costa-Rican dry forest. Pages 448-477 in *Caterpillars: Ecological and Evolutionary Constraints on Foraging*. Edited by Stamp NE, Casey TE. London: Chapman and Hall.

Janzen DH (2003) How polyphagous are Costa Rican dry forest saturniid caterpillars? Pages 369-379 in *Arthropods of Tropical Forests: Spatio-Temporal Dynamics and Resource Use in the Canopy*. Edited by Basset Y, Novotny V, Miller SE, Kitching RL. Cambridge: Cambridge University Press.

Kaplan I, Denno RF (2007) Interspecific interactions in phytophagous insects revisited: a quantitative assessment of competition theory. *Ecology Letters* 10: 977-994.

Keane RM, Crawley MJ (2002) Exotic plant invasions and the enemy release hypothesis. *Trends in Ecology and Evolution* 17:164-170.

Kometter RF, Martinez M, Blundell AG, Gullison RE, Steininger MK, et al. (2004) Impacts of unsustainable mahogany logging in Bolivia and Peru. *Ecology and Society* 9: 12. http://www.ecologyandsociety.org/vol9/iss1/art12.

Kricher JC (2010) Tropical Ecology. Princeton: Princeton University Press. 683 p.

Kristensen NP, Scoble MJ, Karsholt O (2007) Lepidoptera phylogeny and systematics: the state of inventorying moth and butterfly diversity. *Zootaxa* 1668: 699-747.

Kulman HM (1971) Effects of insect defoliation on growth and mortality of trees. *Annual Review of Entomology* 16: 289-324.

Kursar TA, Coley PD (2003) Convergence in defense syndromes of young leaves in tropical rainforests. Biochemical Systematics and Ecology 31: 929-949.

Kursar TA, Wolfe BT, Epps MJ, Coley PD (2006) Food quality, competition, and parasitism influence feeding preference in a neotropical Lepidopteran. Ecology 87: 3058-3069.

Kursar TA, Dexter KG, Lockvam J, Pennington RT, Richardson JE, *et al.* (2009) The evolution of antiherbivore defenses and their contribution to species coexistence in the tropical tree genus *Inga*. Proceedings of National Academy of Sciences 106: 18073-18078.

Lamb FB (1966) *Mahogany of Tropical America: Its Ecology and Management*. Ann Arbor: University of Michigan Press. 220 p.

Laurance WF (1999) Reflections on the tropical deforestation crisis. *Biological Conservation* 91: 109-117.

Leigh EG (2004) The diversity of tropical trees: the role of pest pressure. *Pages* 315-319 in *Tropical Forest Diversity and Dynamism: Findings from a Large-Scale Plot Network*. Edited by Losos EC, Leigh EG. Chicago: University of Chicago Press.

Letourneau DK, Armbrecht I, River BS, Lerma JM, Carmona EJ, et al. (2011) Does plant diversity benefit agroecosystems ? A synthetic review. *Ecological Applications* 21: 9-21.

Lopes JCA, Jennings SB, Matni NM (2008) Planting mahogany in canopy gaps created by commercial harvesting. *Forest Ecology and Management* 255: 300-307.

Louda SM, Potvin MA (1995) Effects of inflorescence feeding insects on the demography and lifetime fitness of a native plant. *Ecology* 76: 229-245.

Makana JR, Thomas SC (2004) Dispersal limits natural recruitment of African mahoganies. *Oikos* 106: 67-72.

Mangan SA, Schnitzer SA, Herre EA, Mack KML, Valencia MC, *et al.* (2010) Negative plant-soil feedback predicts tree-species relative abundances in a tropical forest. *Nature* 466: 752-755.

Maron JL, Crone E (2006) Herbivory: effects on plant abundance, distribution and population growth. *Proceedings of the Royal Society B* 273: 2575-2584.

Marquis RJ (1984) Leaf herbivores decrease fitness of a tropical plant. Science 226: 537-539.

Marquis RJ (2005) Pages 328-346 in *Biotic Interactions in the Tropics: Their Role in the Maintenance of Species Diversity*. Edited by Burslem DFRP, Pinard MA, Hartley SE. Cambridge: Cambridge University Press.

Massey FP, Press MC, Hartley SE (2005) Have the impacts of insect herbivores on the growth of tropical tree seedlings been underestimated? Pages 347-365 in *Biotic*

Interactions in the Tropics: Their Role in the Maintenance of Species Diversity. Edited by Burslem DFRP, Pinard MA, Hartley SE. Cambridge: Cambridge University Press.

Massey FP, Massey K, Press MC, Hartley SE (2006). Neighbourhood composition determines growth, architecture and herbivory in tropical rain forest tree seedlings. *Journal of Ecology* 94: 646-655.

McCarthy-Neumann S, Kobe R (2008) Tolerance of soil pathogens co-varies with shade tolerance across species of tropical tree seedlings. *Ecology* 89: 1883–1892.

McCarthy-Neumann S, Kobe R (2010) Conspecific plant–soil feedbacks reduce survivorship and growth of tropical tree seedlings. *Journal of Ecology* 98: 396-407.

Mejía E, Buitrón X, Peña-Claros M, Grogan J (2008) Bigleaf mahogany (*Swietenia macrophylla*) in Peru, Bolivia and Brazil. Case Study for the International Expert Workshop on CITES Non-Detriment Findings, 17-22 November 2008, Cancún, Mexico.

Metz MR, Sousa WP, Valencia R (2010) Widespread density-dependent seedling mortality promotes species coexistence in a highly diverse Amazonian rain forest. *Ecology* 91: 3675-3685.

Montllor CB, Bernays EA (1993) Invertebrate predators and caterpillar foraging. Pages 170-202 in *Caterpillars: Ecological and Evolutionary Constraints on Foraging*. Edited by Stamp NE, Casey TE. London: Chapman and Hall.

Morris MH, Negreros-Castillo P, Mize C (2000) Sowing date, shade, and irrigation affect big-leaf mahogany (*Swietenia macrophylla* King). *Forest Ecology and Management* 132: 173-181.

Muller-Landau HC, Levin SA, Keymer JE (2003) Theoretical perspectives on the evolution of long-distance dispersal and the example of specialized pests. *Ecology* 84: 1957-1967.

Nair KSS (2007) *Tropical Forest Insect Pests: Ecology, Impact and Management*. Cambridge: Cambridge University Press. 422 p.

Nascimento MT, Proctor J (1994) Insect defoliation of a monodominant Amazonian rainforest. *Journal of Tropical Ecology* 10: 633-636.

Nathan R, Casagrandi R (2004) A simple mechanistic model of seed dispersal, predation and plant establishment: Janzen-Connell and beyond. *Journal of Ecology* 92: 732-746.

Newton AC, Baker P, Ramnarine S, Mesén JF, Leakey RRB (1993) The mahogany shoot borer: prospects for control. *Forest Ecology and Management* 57: 301-328.

Newton AC, Cornelius JP, Mesen JF, Corea EA, Watt AD (1998) Variation in attack by the mahogany shoot borer, *Hypsipyla grandella* (Lepidoptera : Pyralidae), in relation to host growth and phenology. *Bulletin of Entomological Research* 88: 319-326.

Norghauer JM, Malcolm JR, Zimmerman BL (2006a) An experimental test of density- and distant-dependent recruitment of mahogany (*Swietenia macrophylla*) in southeastern Amazonia. *Oecologia* 148: 437-446.

Norghauer JM, Malcolm JR, Zimmerman BL (2006b) Juvenile mortality and attacks by a specialist herbivore increase with conspecific adult basal area of Amazonian *Swietenia macrophylla* (Meliaceae). *Journal of Tropical Ecology* 22: 451-460.

Norghauer JM, Malcolm JR, Zimmerman BL (2008a) Canopy cover mediates interactions between a specialist caterpillar and seedlings of a neotropical tree. *Journal of Ecology* 96: 103-113.

Norghauer JM, Malcolm JR, Zimmerman BL, Felfili JM (2008b) Experimental establishment of big-leaf mahogany (*Swietenia macrophylla* King) seedlings on two soil types in native forest of Pará, Brazil. *Forest Ecology and Management* 255: 282-291.

Norghauer JM, Grogan J, Malcolm JR, Felfili JM (2010) Long-distance dispersal helps germinating mahogany seedlings escape defoliation by a specialist caterpillar. *Oecologia* 162: 405-412.

Norghauer JM, Nock CA, Grogan J (2011a). The importance of tree size and fecundity for wind dispersal of big-leaf mahogany. *PLoS ONE* 6: e17488.

Norghauer JM, Martin AR, Mycroft EE, James A, Thomas SC (2011b) Island invasion by a threatened tree species: evidence for natural enemy release of mahogany (*Swietenia macrophylla*) on Dominica, Lesser Antilles. *PLoS ONE* 6: e18790.

Novotny V, Basset Y, Miller SE, Weiblen GD, Bremer B, et al. (2002a) Low host specificity of herbivorous insects in a tropical forest. *Nature* 416: 841-844.

Novotny V, Basset Y, Miller SE, Drozd P, Cizek L (2002b) Host specialization of leaf-chewing insects in a New Guinea rainforest. *Journal of Animal Ecology* 71: 400-412.

Novotny V, Miller SE, Leps J, Basset Y, Bito D, *et al.* (2004) No tree an island: the plant–caterpillar food web of a secondary rain forest in New Guinea. *Ecology Letters* 7: 1090-1100.

Novotny V, Basset Y (2005) Host specificity of insect herbivores in tropical forests. *Philosophical Transactions of the Royal Society* B 272: 1083-1090.

Novotny V, Drozd P, Miller SE, Kuflan M, Janda M, *et al.* (2006) Why are there so many species of herbivorous insects in tropical rainforests? *Science* 313: 1115-1118.

Novotny V, Miller SE, Baje L, Balagawi S, Basset Y, *et al.* (2010) Guild-specific patterns of species richness and host specialization in plant–herbivore food webs from a tropical forest. *Journal of Ecology* 79: 1193-1203.

Paine CET, Harms KE, Schnitzer SA, Carson WP (2008) Weak competition among tropical tree seedlings: implications for species coexistence. *Biotropica* 40: 432-440.

Price P (1991) The plant vigour hypothesis and herbivore attack. Oikos 62: 244-251.

Price PW, Diniz IR, Morais HC, Marques ESA (1995) The abundance of insect herbivore species in the tropics: the high local richness of rare species. *Biotropica* 27: 468-478.

Price PW (2002) Species interactions and the evolution of biodiversity. Pages 3-25 in *Plant-Animal Interactions: An Evolutionary Approach*. Edited by Herrera CM, Pellmyr O. Oxford: Blackwell Publishing.

Queenborough SA, Burslem DFRP, Garwood NC, Valencia R (2007) Neighborhood and community interactions determine the spatial pattern of tropical tree seedling survival. *Ecology* 88: 2248-2258.

Reavey D, Gaston KJ (1991) The importance of leaf structure in oviposition by leaf-mining Microlepidoptera. *Oikos* 61: 19-28.

Renwick JAA, Chew FA (1994) Oviposition behaviour in Lepidoptera. *Annual Review of Entomology* 39: 377-400.

Richards LA, Coley PD (2007) Seasonal and habitat differences affect the impact of food and predation on herbivores: a comparison between gaps and understory of a tropical forest. *Oikos* 116: 34-40.

Richards LA, Coley PD (2008) Combined effects of host plant quality and predation on a tropical Lepidopteran: a comparison between treefall gaps and the understory in Panama. *Biotropica* 40: 736-741.

Richards PW (1996) *The Tropical Rain Forest* (2nd Ed.). Cambridge: Cambridge University Press. 575 p.

Richardson ML, Mitchell RF, Reagel PF, Hanks LM (2010) Causes and consequences of cannibalism in noncarnivourous insects. *Annual Review of Entomology* 55: 39-53.

Rockwood LL (1973) The effect of defoliation on seed production of six Costa Rican tree species. *Ecology* 54: 1363-1369.

Root RB (1973) Organization of a plant-arthropod association in simple and diverse habitats: the fauna of collards (*Brassica oleracea*). *Ecological Monographs* 43: 95–124.

Russo SE, Augspurger CK (2004) Aggregated seed dispersal by spider monkeys limits recruitment to clumped patterns in *Virola calophylla. Ecology Letters* 7: 1058-1067.

Schoonhoven LM, van Loon JJA, Dicke M (2005) *Insect-Plant Biology* (2nd Ed.). Oxford: Oxford University Press, 421 p.

Schupp EW (1992) The Janzen-Connell model for tropical tree diversity – population implications and the importance of spatial scale. *American Naturalist* 140: 526-530.

Scoble MJ (1992) *The Lepidoptera: Form, Function and Diversity.* Oxford: Oxford University Press. 404 p.

Snook LK (1996) Catastrophic disturbance, logging and the ecology of mahogany (*Swietenia macrophylla* King): grounds for listing a major tropical timber species in CITES. *Botanical Journal of the Linnean Society* 122: 35-46.

Snook LK, Negreros-Castillo P (2004) Regenerating mahogany (*Swietenia macrophylla* King) on clearings in Mexico's Maya forest: the effects of clearing method and cleaning on seedling survival and growth. *Forest Ecology and Management* 189: 143-160.

Snook LK, Camara-Cabrales L, Kelty MJ (2005) Six years of fruit production by mahogany trees (*Swietenia macrophylla* King): patterns of variation and implications for sustainability. *Forest Ecology and Management* 206: 221-235.

Soares MG, Batista-Pereira LG, Fernandes JB, Correa AG, da Silva MFGF, *et al.* (2003) Electrophysiological responses of female and male *Hypsipyla grandella* (ZELLER) to *Swietenia macrophylla* essential oils. *Journal of Chemical Ecology* 29: 2143-2151.

Speight MR, Hunter MD, Watt AD (2008) *Ecology of Insects: Concepts and Applications.* West Sussex: Wiley-Blackwell Publishing, 628 p.

Strong DR, Lawton JH, Southwood TRE (1984) *Insects on Plants, Community Patterns and Mechanisms.* Oxford: Blackwell Scientific Publications, 313 p.

Sullivan JJ (2003) Density-dependent shoot-borer herbivory increases the age of first reproduction and mortality of neotropical seedlings. *Oecologia* 136: 96-106.

Summerville KS, Crist TO (2002) Effects of timber harvest on forest Lepidoptera: community, guild, and species responses. *Ecological Applications* 12: 820-835.

Svenning JC, Fabbro T, Wright SJ (2008) Seedling interactions in a tropical forest in Panama. *Oecologia* 155: 143-150

Swaine MD (1996) Foreword to *The Ecology of Tropical Forest Tree Seedlings.* UNESCO-Parthenon. 340 p.

Swamy V, Terborgh JW (2010) Distance-responsive natural enemies strongly influence seedling establishment patterns of multiple species in an Amazonian rain forest. *Journal of Ecology* 98: 1096-1107.

Swamy V, Terborgh JW, Dexter KG, Best BD, Alvarez P, *et al.* (2011) Are all seeds equal? Spatially explicit comparisons of seed fall and sapling recruitment in a tropical forest. *Ecology Letters* 14: 195-201.

Taveras R, Hilje L, Hansen P, Mexzón R, Carballo M, *et al.* (2004) Population trends and damage patterns of *Hypsipyla grandella* (Lepidoptera: Pyralidae) in a mahogany stand, in Turrialba, Costa Rica. *Agricultural and Forest Entomology* 6: 89-98.

Thomas CD (1990) Herbivore diets, herbivore colonization, and the escape hypothesis. *Ecology* 71: 610-615.

Thompson JN, Pellmyr O (1991) Evolution of oviposition and host preference in Lepidoptera. *Annual Review of Entomology* 36: 56-89.

Thompson, JN (1999) What we know and do not know about coevolution: insect herbivores and plants as a test case. Pages 7-30 in *Herbivores: Between Plants and Predators*. Edited by Olff H, Brown VK, Drent RH. Oxford: Blackwell Science Publishing.

Turner IM (2001) *The Ecology of Trees in the Tropical Rain Forest*. Cambridge: Cambridge University Press. 298 p.

Uhl C, Clark K, Dezzeo N, Maquirno P (1988) Vegetation dynamics in Amazonian treefall gaps. *Ecology* 69: 751-763.

Visser JH (1986) Host odur perception in phytophagous insects. *Annual Review of Entomology* 31: 121-144.

Weiblen GD, Webb CO, Novotny V, Basset Y, Miller SE (2006) Phylogenetic dispersion in host use by tropical insect herbivore community. *Ecology* 87: S62-S75.

Wenny DG (2000) Seed dispersal, seed predation, and seedling recruitment of a neotropical montane tree. *Ecological Monographs* 70: 331-351.

Williams (1983) The coevolution of *Euphydryasc halcedona* butterflies and their larval host plants III. Oviposition behavior and host plant quality. *Oecologia* 56: 336-340.

Wilson DE, Janzen DH (1972) Predation on *Scheelea* palm seeds by bruchid beetles: seed density and distance from the parent palm. *Ecology* 53: 954-959.

Wong M, Wright SJ, Hubbell SP, Foster RB (1990) The spatial pattern and reproductive consequences of outbreak defoliation in *Quararibea asterolepis*. *Journal of Ecology* 78: 579-588.

Wright SJ (2002) Plant diversity in tropical forests: a review of mechanisms of species coexistence. *Oecologia* 130: 1-14.

Yamazaki S, Taketani A, Fujita K, Vasques-Pacheco C, Ikeda T (1990) Ecology of *Hypsipyla grandella* and its seasonal changes in population density in Peruvian Amazon forest. *Japan Agricultural Research Quarterly* 24: 149-155.

Yamazaki S, Ikeda T, Taketani A, Vasques-Pacheco C, Sato T (1992) Attack by the mahogany shoot borer, *Hypsipyla grandella* (Lepidoptera: Pyralidae), on the Meliaceous trees in the Peruvian Amazon. *Applied Entomological Zoology* 27: 31-38.

Zahiri R, Kitching IJ, Lafontaine D, Mutanen M, Kaila L, Holloway JD, Wahlberg N (2010) A new molecular phylogeny offers hope for a stable family level classification of the Noctuoidae (Lepidoptera). *Zoologica Scripta* 40: 158-173.

Zhang BC (1994) *Index of Economically Important Lepidoptera*. Wallingford: CABI International. 591 p.

Zimmerman JK, Thompson J, Brokaw N (2008) Large tropical forest dynamics plots: testing explanations for the maintenance of species diversity. Pages 89-117 in *Tropical Forest Community Ecology*. Edited by Carson WP, Schnitzer SA. Oxford: Wiley-Blackwell.

In: Moths: Types, Ecological Significance and Control... ISBN: 978-1-61470-626-7
Editor: Luis Cauterruccio, pp. 75-113

Chapter 3

MICROLEPIDOPTERA OF ECONOMIC SIGNIFICANCE IN FRUIT PRODUCTION: CHALLENGES, CONSTRAINS AND FUTURE PERSPECTIVES FOR INTEGRATED PEST MANAGEMENT

Petros T. Damos and Matilda Savopoulou-Soultani
Laboratory of Applied Zoology and Parasitology, Department of Plant Protection, School of Agriculture, Aristotle University of Thessaloniki.

ABSTRACT

One of the leading concerns of pest control in modern fruit production, and for both fruit quality assurance and environmental preservation, has been how conventional control methods affects biodiversity and how they can be altered to mitigate pesticide side effects in all aspects. This chapter discusses the significance of economically important microlepidoptera-moth species in fruit production and is mostly focused on their Integrated Pest Management (IPM). Microlepidoptera is a cluster of moth families commonly known as the 'smaller moths'. Since the group is characterized by polyphyletic diversity this is not, from a taxonomical standpoint, a restrict definition albeit commonly used to group small moth species which in most cases display similar life cycles and habits that are not found in larger Lepidoptera (i.e. butterflies). An overview of the current status of representative codling moths, tortrix, Gelechiidae and leaf-roller moths including: *Cydia pomonella*, *Grapholitha molesta*, *Anarsia lineatella* and *Adoxophyes orana* are presented. The detailed habits and bionomics are documented from prior studies and compared to older and latest references. The work proceeds by the description of numerous control methods and tactics that are currently used in IPM and as part of the wider framework of Integrated Fruit Production (IFP). The development of forecasting models based on degree-days, as well as the development of Economic Injury levels and Thresholds as decision tools to determine the optimal treatment time for biorational insecticides and insect growth regulators is presented. Efforts are also made to discuss and weight constrains of the 'Economic Injury Level concept' to be applicable on a realistic basis in fruit orchards. The major properties of bio-rational chemical compounds and biological control agents (i.e. bacteria and parasitic nematodes) and possible side effects on beneficial species are short reviewed. Novel control methods such as matting disruption, the attract and kill and push and pull strategies are briefly outlined with the view to be developed and incorporated in future IPM programs on a regular basis to control fruit moths. Finally, actual facts and challenges such as pesticides resistance and restrictions due to the implementation of the latest European Union council directives for pesticides are also discussed.

1. Introduction

Lepidoptera is one of the largest orders in the class Insecta, with more than 105.000 known species. Although Lepidoptera can be divided into four suborders: Zeugloptera, Aglossata, Heterobathmiina and Glossata it is quite difficult to further divide the order into sound, manageable groups, since several classification properties are subjectively defined and most divisions are based more or less on the type of study rather than on a scientific restrict taxonomic standpoint [6][109].

The division of Lepidoptera to butterflies (Rhopalocera) and moths (Heterocera) and/or to macro and micro moths are two popular approaches that have had wide use, but are without scientific validity. Actually, the distinction between micro and macro Lepidoptera is artificial, concerning that the latter group includes species large enough to be pinned conveniently into collector's cabinet, while the former are all too small Lepidoptera for the collector to bother with [109]. Hence, the division to Rhopalocera and Heterocera is made to distinguish between larger butterflies and small moths species respectively. As a rule, butterflies are diurnal, brightly colored, and have knobs or hooks at the tip of the antennae. The wings of butterflies at rest, are held vertically over the body, while in contrast, most (but not all) moths are nocturnal and at rest their wings are held horizontally against the substrate, folded flat over the back, or curled around the body. Additionally, moth appearance is typically drab, and individuals have thread-like, spindle-like or comb-like characteristic antennae [109].

From a phylogenetic standpoint, butterflies as a group are probably very compact but to obtain generally equivalent grouping of the moth families would require several categories.

Nevertheless, in this chapter we will adopt the more general term of 'micro lepidoptera', or 'micro moths' to group small species which in most cases display similar life cycles and habits which are not found in larger Lepidoptera or Butterflies. Additionally, we emphasize to those species that cause significant economic damage in fruit production and are recognized among Agricultural Entomologists as 'key pests'.

The definition of key pests is essential to be made in Agro Ecosystems since it constitutes the basis for the development of Integrated Pest Management (IPM) strategies and the development of reliable decision tools for Integrated Fruit Production (IFP).

Micro Lepidoptera of Economic Significance in Fruit Production

Apple, peach and pear are all deciduous fruit trees of major economic importance in Southern Europe and all these crops share common problems in pest management and IFP

[6][24][27][91][92]. Lepidopteran larvae in fruit orchards are the most important pests, followed by aphids and mites. Among the most serious lepidopteran species are the codling moth *Cydia pomonella* Linnaeus, the oriental fruit moth *Grapholitha molesta* Busck (Lepidoptera: Tortricidae), the summer fruit totrix moth *Adoxophyes orana* (Fisher von Röslerstamm) (Lepidoptera: Tortricidae) and the peach twig borer *Anasria lineatella* Zeller (Lepidoptera: Gelechiidae).

The codling moth, *C. pomonella*, is probably regarded as the most serious moth of apple worldwide, and is becoming an increasing problem also in walnuts, prunes, and a few varieties of plums [24] *C. pomonella* lays its eggs on leaves or fruits and neonate larvae bore into the fruit causing significant fruit damage and yield loss. Mature larvae (5th instar) leave the fruit in search of rough bark and cryptic habitats in which they spin their cocoons and pupate, and the 3rd generation ones for overwintering.

The oriental fruit moth *G. molesta is* considered a regular problem in stone fruits as well as in apples when there are nearby stone fruit orchards. Apart of peaches (*Prunus persica*) and nectarines, *G. molesta* attacks a great variety of hosts including apricots (*Prunus armeniaca*), almonds (*Prunus amygdalus*), quince (*Cydonia oblonga*), pears (*Pyrus sp.),* plums (*Prunus domestica*) and cherries (*Prunus sp.*) as well as woody ornamental plants. The Oriental fruit moth has three full generations and occasionally a partial 4th and 5th generation in Southern Europe [55]. Flight patterns however are very confusing and generations are difficult to be distinguished. The moths overwinter as full-grown larvae in cocoons, settled in various sites, mostly tree bark crevices and weed stems, trash on the ground, fruit containers and packing sheds.

Like *C. pomonela*, *G.molesta* has a marked daily flight period in the evening. On peach orchards, the species lays the majority of its eggs on the leaf surfaces and first generation larvae usually damage two or three shoots during the first generation. Larvae of subsequent generations damage both shoots and fruits. Larvae of these generations are the major cause of wormy fruit production at harvest, often with little or no external sign of injury. Additionally, about half of the injury in late ripening peach varieties is characterized by no visible entrance (concealed injury). The oriental fruit moth *G. molesta* attacks also apples, but is considered a regular pest of stone fruits.

The summer fruit totrix *Adoxophyes orana* (Fisher von Röslerstamm) (Lepidoptera: Tortricidae) attacks a wide variety of plants with a preference for Rosaceous plants, especially apple, pear and peach. The species however, is reported to feed and develop on more than 50 plant species of multiple families including fruits, forest trees, and ornamentals [160].

The peach twig borer, *Anarsia lineatella* Zeller (Lepidoptera: Gelechiidae), is one of the major economic pests of stone fruits in central and southern Europe [50][51][53]. It is referred to be oligophagous, preferring mainly peaches, apricots, and almonds [144]. In southern Europe (i.e. Greece), *A. lineatella* has three or usually four generations per year depending on prevailing temperatures [50][51][52][53][54][55][56]. The species overwinters in bark crevices as second or third instars, forming hibernacula. Larvae become active in spring and are able to cause early season injury burrowing into new twigs. Later during summer, newly hatched larvae, originating from next generations, feed mainly on fruit, causing significant damage. In most cases, the above species appear simultaneously during a growth season and therefore species-specific detailed phenology is essential for management success [52][54][56].

Integrated Pest Management and Integrated Fruit Production

The traditional use of non-selective insecticides is associated to a variety of problems including: environmental effects, insecticide resistance, negative impacts on natural enemies, and safety for pesticide applicators and the food supply [8][9][30][63][65][90][100][107].

Concerns about these consequences have increased the interest in the development of alternative means for pest control that have little or no impact on humans, beneficial organisms and sensitive ecosystems [171][180][188][189].

Integrated Pest Management (IPM) is a decision-based process, involving coordinated use of multiple tactics for optimizing the control of all classes of pests (insects, pathogens, vertebrates and weeds) in an ecologically and economically sound manner [48][49][51][70]. Traditionally, IPM programs use current, comprehensive information on the life cycles of pests and their interaction with the environment.

Moreover, IPM is an essential component of Integrated Fruit Production (IFP). IFP[1] provides an economical and high quality fruit production framework, giving priority to ecologically safer methods, minimizing the undesirable side effects and use of agrochemicals, to enhance the safeguards to the environment and human health [48].

Lately, the major IPM principles[2] have been outlined by the European Commission and the European Parliament (adopted in the second reading). Eight general principles for Integrated Pest Management (IPM) are currently identified related to the following topics [1][67][68][97]:

(1) Measures for prevention and/or suppression of harmful organisms
(2) Tools for monitoring
(3) Threshold values as basis for decision-making
(4) Non-chemical methods to be preferred
(5) Target-specificity and minimization of side effects
(6) Reduction of use to necessary levels
(7) Application of anti-resistance strategies
(8) Records, monitoring, documentation and check of success

Current IPM programs are based on specific knowledge of the species life cycles in relation to its environment which is combined with the application of those control methods that are selected among several available by economical, social (least possible hazard to people) and environmental friendly means [70][97][106][170].

---

[1]IFP originated in Europe in the 1950s as extension to first IPM efforts, but did not experience much growth until the late 1980s [66]. Marked demand on IFP is steadily increasing as a result of latest European directives and during the last years almost fifty percent (790,000 acres) of the apple and pear acreage in Western Europe is managed under an IFP program. However, although independent organizations such as IOBC have recently established IFP guidelines [81][82], regulations used by each country or organization vary and are not always consistent with the Guidelines.

[2]The above principles are also part of the IPM guidelines as defined by the International Organization of Biological Control (IOBC) [47][48]. IOBC formed specific working groups that serve as a liaison between basic research and practical application, with members drawn from academic and government research institutions, plant protection industry and extension services.

Adoption of peach IPM programmes in practical terms includes reducing pest status and accepting the presence of a tolerable pest density. This approach relies on the development and application of accurate monitoring techniques and action thresholds. Additionally the development and incorporation of decision tools such as phenology models in IPM programs is a prerequisite. The successful application of alternative and novel ingredients, with no or very low negative effects on natural enemies, is strongly related to forecasting models since they are characterized by specificity of action to specific developmental stages.

2. Insect Forecasting Models

Insect phenology is the seasonal occurrence of stages, including the temporal pattern of adult emergence and can be characterized by events such as the onset of emergence, its duration, the synchrony of individuals, any skew from normality, and several variations including irregularities and modalities [41][57][59][62][133][181].

A typical life cycle of an insect describes the phenological patterns of a species with normally distributed emergence that takes place over a growth season. In a typical life cycle of a moth, adult phenology evolves during the growth season, in which more than one generation are observable. The latter induces diapause and overwinters at a particular stage of development.

Phenology models serve to predict the exact time of the phenological development of pest populations. However, several approaches existing in modeling insect development and related phenology, at a basic level, all implicit that insect poikilotherm development is directly related to ambient temperature and time [123][132][124]. Additionally, observable manifestations on phenology are further related to the basic assumption that most[3] enzyme-catalyzed reactions are the rate of limiting for growth and development.

One possible approach is that of temperature driven time-varying distributed delay models, based on an Erlang density function to generate the frequency distribution of the individual developmental times [159]. The function is parameterized with the species and stage specific thermal constant and its variance. The process in total (i.e. life cycle stage specific phenology) can be further modeled by the application of simple algorithms, while the underlying relationships between temperature and developmental rates are prior calculated by linear and/or non-linear models which are established by thorough laboratory studies for each stage of the life cycle.

Hence, one prerequisite in application of all temperature driven models is the definition of species-specific temperature thresholds (especially lower) and thermal constants [19][61][87][128][172].

A principal method for calculating the lower developmental threshold (LTT) is the x-intercept method after growth rate fitting to a simple linear equation and then extrapolated to zero: $y=a+bx$, where y is the developmental rate (1/developmental time) at temperature x and a and b estimates using the least square equations [50] [119][120][131].

The lower theoretical temperature threshold is derived from the linear function as: $T_o=-a/b$ and is used to define a temperature based BioFix for starting temperature summations.

[3]Since enzyme concentrations may be also regulated hormonally, as affected by abiotic factors photoperiod or genetics, deviations in poikilothermic development based on heat accumulations should not be excluded [86].

Table _1 lists important developmental events of the moth *A. lineatella* in relation to degree-days heat accumulations and related management actions for its control.

Probably the simplest case of phenology model is that which is based on a non-linear regression function that correlates accumulated moth catch to heat accumulations. These models are most useful in predicting the date of the emergence of the first flight or the peak emergences of successive generations and are actually empirical regression models of the general form:

$$y = f(x, \vartheta) + \varepsilon_i$$

in which y is an (n-1) vector of observations of the response variable, X is an (n-p) design matrix determined by the predictors, and θ a (p-1) vector of unknown parameters to be estimated.

Thus, f is any function of x and θ and ε_i is an (n-1) vector of the independent, identically distributed random disturbances, with zero mean and constant variance across observations $\varepsilon_i \approx N(0, \sigma^2)$.

Most regression lines are empirically fitted curves that represent the best mathematical fit and do not possess any clear inherent meaning. However, they provide useful insights into the modelled phenomenon, if they adequately describe the relations between the variables. Regression equations are also important in defining relationships between pest population and infestations and/or damage (see also economic thresholds). On the other hand it should be noted that one of the most serious defects of regression models is that they give predictions within the range of data under which the relationships are originally derived [63]. The Richards' function or the Generalised Logistic Curves are s-shaped mathematical functions which can be twisted around to fit most conceivable variations of its basic shape and Figure 1, for instance describes cumulative moth phenology of *A. lineatella*, *G. molesta* and *A.orana* in Northern Greece orchards as generated by a unique logistic model [52][54][56]:

$$f(x, \vartheta) = d + \frac{a}{(\frac{x}{c})^b}$$

Cumulative moth catch is correlated to accumulated degree-days by using the above regression function [56] and predicting the dates of the emergence of the first flights or the peak emergences of successive generations of three peach moth species. In the above function f represents the cumulative percentage of moth catches, x corresponds to accumulated degree-days, and α, b, c and d are constant parameters (which calibrate the final shape of generated function).

Actually, the above forecasting model is an s-shaped type of a cumulative distribution function, in a parameterization stable for numerical work.

Since all the above microlepidopteran species occur simultaneously, the model in Figure 1 is generated for each species and flight respectively, but based on the same lower temperature threshold of the reference species *A. orana* which was calculated after a relative degree-day adjustment [54][58].

Table 1. Major developmental events of *A. lineatella* in peach orchards of Northern Greece and relative management actions in IPM programs

Major developmental events of A. lineatella and relative IPM action				
Generation	DD	Developmetnal Stage		Management Actions
		Moth emergence (%)	Egg hatch (%)	
Overwintering	0-100	0	0	Degree-day Summation
	150-200	1-5	0	Trap placement and/or Mating disruption diffusers placement
1st	50-75	5-9	0	Shoot flaging monitoring Model for start of Egg hatch
	100-200	15-40	0	Application of Biorational 1 (i.e. ovividal IGR)
	220-250	45-50	1-3	DD Model generation for max Egg hatch Biorational 2 (depending on residual activity of comp.1)
	340-640	67-98	12-80	Critical period of management succes Inspect moth levels and EIL, DD model for moth phenology Appliaction of Biorational 2 if nececary (first Rotation, i.e. B.t.)
	640-920	100	99	
2nd	1000-1050	5-8	0	DD Model for second generation egg hatch Application of biorational 2(second Rotation, IGR 2)
	1100	13	1	Inpect moth levels and EIL, DD model for 2nd moth flight Appliaction of Biorational Compount if nececary (second Rotation, i.e. B.t.)
	1320-1720	46-93	11-71	Inpect moth levels and EIL, DD model for 2nd moth flight Biorational 3, if nececary (depending on prior application)
	2100	100	99	
3rd	2160	1	15	Model for third generation egg hatch Critical only for late ripening varieties

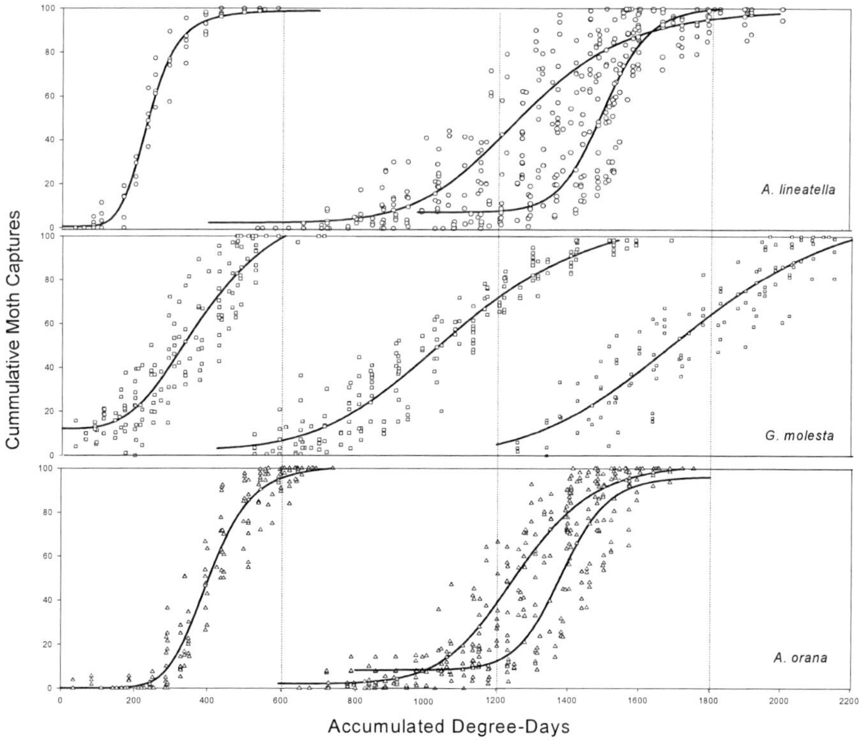

Figure 1. Moth phenology and generated unique logistic model for *A. lineatella*, *G. molesta* and *A. orana* in Northern Greece (Temperature Threshold: 7,3°C, BioFix: 1st Mars).

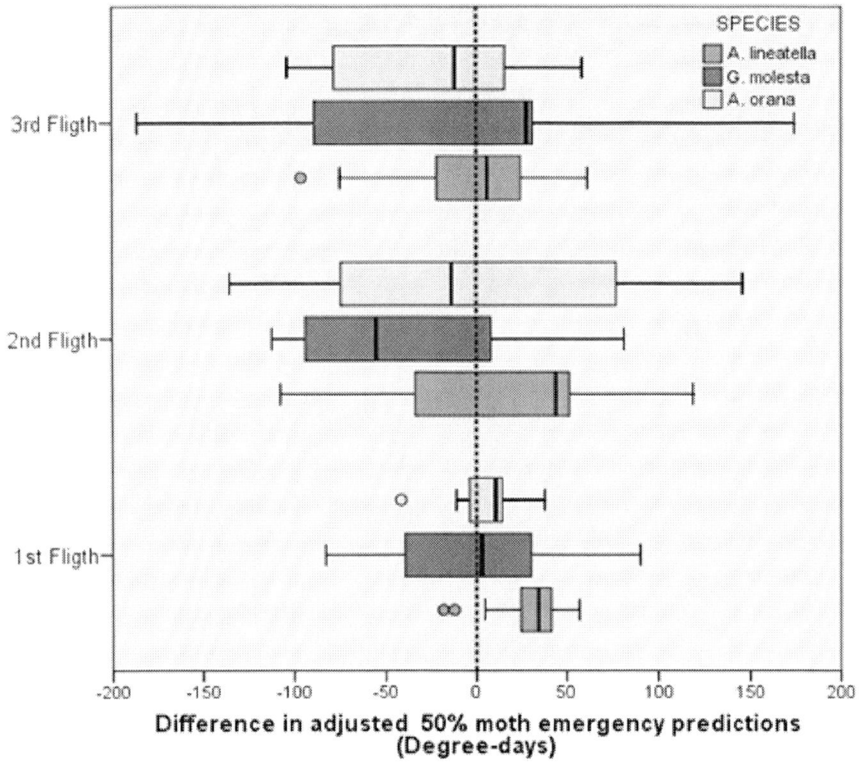

Figure 2. Box plot of 50% moth capture prediction, according to the unique Logistic model in describing of *A. lineatella*, *G. molesta* and *A. orana* moth phenology for three successive flight periods.

Accuracy evaluation of the forecasting model success in predicting moth emergence, can be made by using particular data sets (in most cases new ones) and examining the observed versus predicted data appearance as plotted in Figure 2.

Additionally, slight deviations can be indicated by using several statistical criteria such as ordinary root mean square error (RMSE). In the cases in which we are interested in selecting the model that best performs among several available, having different parameters, it is convenient to apply the adjusted coefficient of determination (adj.R2) and informational statistical measures (i.e. Akaikes and Bayes-Schwarts information criteria) [50][52][54][56][57].

3. Damage Functions and Economic Injury Levels

Developing a relationship between pest abundance and damage to crops is essential for the calculation of economic injury levels (EILs) and to anticipate informed management decisions [34][35][36][55][93][121][147].

Additionally, a subjective evaluation of the control benefits during the growth season requires a good understanding of the relationships between insect abundance and damage

[140][139]. Several approaches can be followed in order to assess crop response to pest abundance [34][35][36].

The mathematical relation between insect pest injury and yield loss is called damage function or damage curve [140] and consists the most challenging function to be plugged on an EIL model to be used by farmers as a decision tool for management, while the EIL is the lowest population density that will cause sufficient economic damage to justify the cost of corrective control measures.

Generally several models (damage function types) can be applied on data, although the model that best fits to data depends upon several factors and in most cases a linear regression is preferred for simplicity reasons [55].

Unfortunately, several factors such as temperature, predation, chemical properties and release of pheromone compounds, sex ratio, mating behavior and age dependent response to pheromones can influence normality of moth flights as observed by pheromone traps and make this approach not always successful [55][147] [103]. Moreover, if physical fluctuations on the population phenology are not apparent, artificial infestations of different population levels should be made and then correlated to observed damage. This however, is quite difficult for perennial crops in large field trials in contrast to annuals or protected crops [55].

Therefore, regression and yield loss functions based on absolute measures (i.e. shoot flagging and larvae on fruits) are more accurate when compared to indirect measures (moth catches) [34].

Respective yield-loss linear functions between the number of *A. lineatella* moth larvae damaging peach fruits during cultivation season and observed yield loss are given in [54]. Since final yield loss due to larval activity is constant regardless of its instar and species, it should be noted that the above concept can be extended to the construction of yield loss functions for more than one species. The slope of the above functions is probably the most essential and challenging, as well as the only biological variable of the Economic Injury Model that follows and EIL is actually the slope of the regression equation that defines the damage function. For more details on the concept of EIL refer to [36], for application on peach orchards refer to [54][55].

The concept of the EIL integrates biology and economics and uses pesticides (or generally control actions) only when economic loss is anticipated), while the Econonic Threshold (ET) is an operational criterion rather than a theoretical one and represents the population density at which control measures should be initiated to prevent an increasing pest population from reaching the EIL [110][139] [140].

The EIL is based on the relation among five primary variables and can be estimated according to the following [55][86]:

$$EIL = \frac{C}{V \cdot I \cdot D \cdot K}$$

in which:

-(C) represents the cost of management tactic per production unit

-(V) is the price of commodity

-(I) is the injury units per pest

-(D) is the damage per unit of insect injury and

-(K) is the proportionate reduction of injury averted by the application of a tactic.

The variables *I* and *D* are related to each other and consist of the parameter *D'* which represents the yield loss in association per insect[4]. The above model can additionally be improved by incorporating the environmental cost, apart from the management cost. However, indirect effects are neither simple to delineate nor available for most compounds and it is not always very clear how to quantify and include the environmental costs [86]. To illustrate the concept of an economic injury level which was developed for *A. lineatella*, the above model was generated in Figure 3 by using original data published in [54] and [55]. This model was developed to provide objective guidelines for decision making and especially pesticide use [54][55][110]. It is feasible that for a given species, specific damage function (i.e. parameter *D'*) and pesticide efficacy (*K*) the EIL is strongly governed by the cost of management (*C*) and the commodity value (*V*). In other words, different combinations among the four variables provide the respective EIL expressed as the level of crop injury and measured as number of moth larvae that will result in yield loss equal to costs associated to the management of the pest. On the other hand the ET is the operational criterion and a direct function of the EIL. Although in several cases the ET represents a fixed and lower value of the EIL, it varies with logistical considerations associated with time delays that may vary from one situation to another and thereby, implementation of the concepts of the EIL through the ET has been a difficult and rather inexact process [140]. Additionally, since the requirement of considerable work should be carried out to establish the relationships between the levels of injury and yield loss, the EIL and ET are not available for most fruit moths.

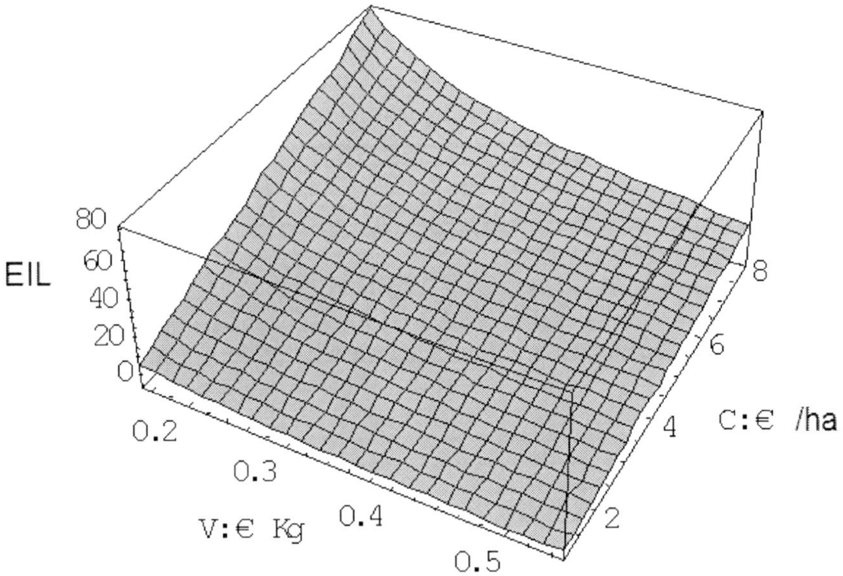

Figure 3. Extrapolation of the Economic Injury Level (EIL) model for larvae of *A. lineatella* (Lepidoptera: Gelechiidae) and effects of commodity value (V) and cost of management tactic (*C*) on the EIL in respect to a standard insect–yield function (*D'*) and proportionate reduction of injury averted be the application of a tactic (*K*) [54].

[4]Parameter *D'* is obtained from the slope of the yield or damage function (Y = a + bx), where Y = yield; b = yield loss/insect; a = 0 and x = number of insects on fruit per sampling unit [54][55]. The coefficient b represents the loss per insect, which is equal to *I*D* or *D'* and therefore *D'=I*D* [86].

Moreover, one additional goal in IPM is the ability to make management decisions for more than one pest. The development of multispecies EILs can be proved quite difficult and time consuming since a large amount of time and research is required to construct species specific damage functions. Nevertheless, EIL are an important component of a successful long term pest management programs in fruit orchards if it is possible to define them.

4. Biorational Insecticides

Since IPM uses all available means to maintain pest populations below levels that would cause economic loss while minimally impacting the environment, the replacement of current synthetic insecticides with bio-rational is now a fact in most of the EU members without dispute. Actually, the increasing use of bio-rational products in managing Lepidoptera in fruit orchards is the result of numerous side effects that have been raised due to the extensive use of conventional synthetic insecticides. Although several environmental implications, such as groundwater pollution and cases of human and animal poisoning as well as address affects resulting to insect resistance and beneficial extinctions have been long time ago documented, most IPM programs still relay heavily upon the use of conventional insecticides to control fruit pests. Table 2, for instance, presents a list of insecticide products that are available for the control of codling moth in the USA as well as in the EU (Greece in particular after the latest EU directives). Besides, insecticide resistance in key pests will continue to be a major impetus for adopting novel insecticides [3][33][47]. Recently the term "biorational" has been proposed to describe those insecticides that are efficacious against the target pest but are less detrimental to natural enemies. This term at times has been used to describe only those products derived from natural sources, i.e. plant extracts, insect pathogens, etc. Nerveless, in the following short section we adopt the terminology proposed in [160][162] and define a biorational pesticide as "any type of insecticide active against pest populations, but relatively innocuous to non-target organisms, and, therefore, non-disruptive to biological control" [160][162][167][168][195].

Hence, the term biorational[5] is used under the sense of a set of novel products which are increasingly used in IPM fruit programs and are distinguishable to most known neurotoxic agents due to unique and specific modes of actions that are characterized by high target selectivity and very low side effects to humans.

Traditionally, soaps/detergents, oils and botanical products have been termed biorational; however, in the present discussion, systemic insecticides, insect growth regulators and biological products containing *Bacillus thuringiensis* or its components, nematodes and viruses will be also included, although they are mostly microorganisms and do not consist a group of synthetic chemical compounds.

Table 3 is a presentation of 'biorational' products that are available for the control of codling moth in the USA as well as in the EU (Greece in particular) and which are widely used in IPM programs for the control of *C. pomonella.*

---

[5]The term biorational is somehow contradictory since it is addressed to substances of microbial or plant origin and by displaying certain physical or contact routes of entry, however most of them are fully interchangeable with organic pesticides, while others define biorationals as pesticides of natural origin or that that are synthesized as analogues to natural occurring plant or insect chemicals. For more details refer to [160][162][167][168].

Table 2. List of representative compounds that have registration for the control of *C. pomonella* in the US, in the EU and Greece in particular (in italics)

Synthetic Insectides		
Category	Ingredient	Commercial
Neonikitinoids	*Acetamiprid*	*(Assail)*
	Clothianidin	(Clutch)
	Thiacloprid	*(Calypso)*
IGR (are strongly suggested for IPM programs)	azadirachtin	(Aza-Direct)
	methoxyfenozide	*(Interprid)*
	Novaluron	(Rimon)
	pyriproxyfen	(Esteem)
	fenoxycarb (25%)	(Insegar 25G)
	buprofezin (25%)	(Aplaud WP)
	tebufenozide	*(Confirm)*
3rd generation	Abemectin 1.8%	Vertimec 18EC
	Spinosad	*Laser 480 SC*
Organophosphates* (*are used in IPM programs but under consideration)	azinphosmethyl	(Guthion)
	chloropyriphos	*(Losban)*
	diazinon	(Diazinon)
	dimethoate	(Dimethoate)
	malathion	(Malathion)
	phosmet	*(Imidan)*
Carbamates*	Carbaryl	(Sevin)
	indoxacarb	(Avount)
Pyrethroids (are not suggested for use in IPM programs)	esfenvalerate	(Asana)
	fenopropathrin	(Danitol)
	cypermethrin	*(Delear, Assist 10EC)*

Table 3. List of representative biorational compounds that have registration for the control of *C. pomonella* in the US, in the EU and Greece in particular (in italics)

Biorational Insecticides		
Oils [1]	*Parafin Oil*	*(Summer Oil)*
IGR	azadirachtin	(Aza-Direct, Azatin)
	methoxyfenozide	*(Interprid)*
	novaluron	(Rimon)
	pyriproxyfen	(Esteem)
	fenoxycarb (25%)	(Insegar 25G)
	buprofezin (25%)	(Aplaud WP)
	tebufenozide	*(Confirm)*
Microbial'	Abemectin 1.8%	Vertimec 18EC
	Spinosad	*Laser 480 SC*
Bio-insecticdes	*Bacillus thuregiensis*	*(Dipel)*
		(Bactospeine 32 WG)
		(Agrec)
	Granulovirus (CpGV)	*Cyd-X*
Plant derived	rotenone	
	pyrethrum	
	ryana	
Facing compounds	Kaolin clay	Surround

[1] References for negative impacts on bloom and fruit ripening

[2] Management success for low moth populations and questionable side effects for beneficial species.

Insect Growth Regulators (IGR's)

IGR's are products or materials that inhibit or generally interrupt the life cycle of a pest. According to their mode of action they can be classified as hormonal IGR's which inhibit or mimic the juvenile hormone (JH) and ecdysone, chitin synthesis inhibitors (Figure 4) which prevent chitin formation, and anti-juvenile hormone agents which block the physiological process of juvenile hormone production [134][145][148][166]. The JH and related compounds affecting the developmental changes associated with embryogenesis, morphogenesis, and reproduction and thereby act as insect growth regulators. Generally, under normal physiological conditions immature stages (larvae or nymphs) express high JH levels which decrease prior to pupation (or adult eclosion).

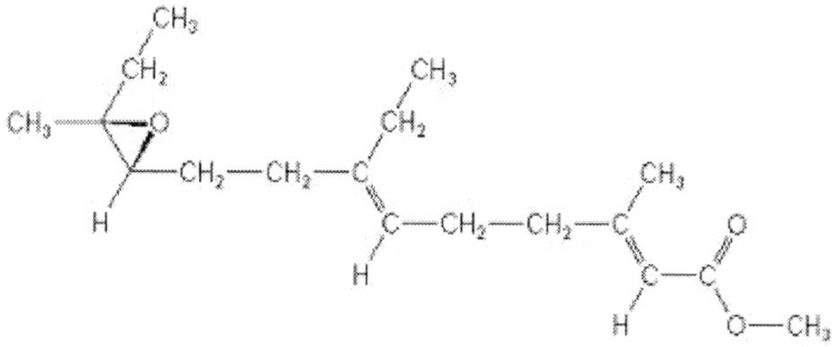

Figure 4. Juvenile hormone (JH) methyl (2E,6E,10Z)-10,11-epoxy-7-ethyl-3,11-dimethyl-2,6-tridecadienoate.

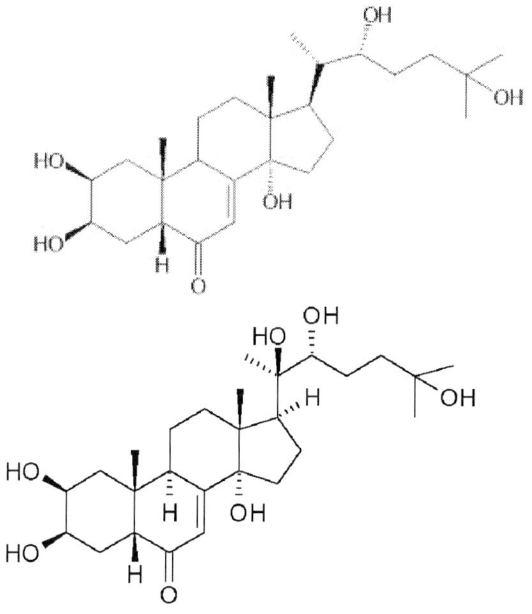

Figure 5. Ecdysone and 20-hydroxyecdysone belong to the family of insect ecdysteroids and simulates the molting process by mimicking the action of molting hormone by initiating ecdysis during the immature stages of development. It is apparent that their chemical structure is similar to that of human reproductive hormones (estrogen, progesterone, and testosterone).

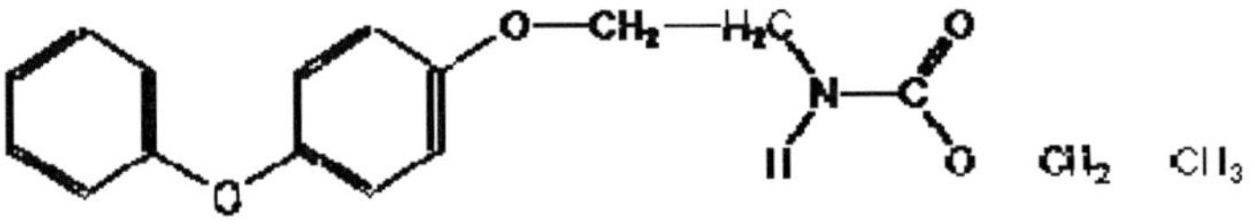

Figure 6. Chemical structure of fenoxycarb: ethyl 2-(4-phenoxyphenoxy)ethylcarbamate (IUPAC[6]).

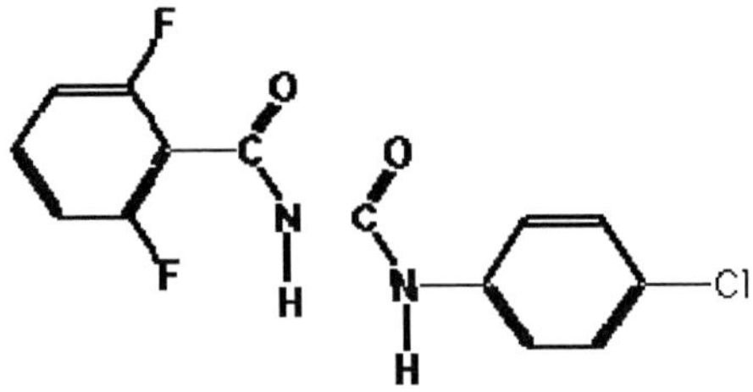

Figure 7. Chemical structure of diflubenzuron: 1-(4-chlorophenyl)-3-(2,6-difluorobenzoyl)urea.

Fenoxycarb

IGR's and ecdysteroid agonists (Figure 5) are usually compatible with natural enemies [60], because the structure and biochemical properties of receptors differ among insect species [39]. Fenoxycarb was discovered in 1981 and is probably the most representative compound that displays IGR properties (Figure 6) [33][175]. Fenoxycarb is a non-neurotoxic carbamate insect growth regulator used to control a wide variety of insect pests including moths [79][171]. Fenoxycarb is practically non toxic to honeybees but is considered moderately to highly toxic to fish, highly toxic to the aquatic invertebrate and affects growth and reproduction after chronic exposures. However, several synthetic substances with novel mode of action have activity primarily affecting lepidopteran pests [75] and compounds such as diflubenzuron, indoxacarb, pyriproxyfen, tebufenozide (Confirm), methoxyfenozide (Interprid) and emamectin benzoate (Denim, Proclaim) are already registered for crops in the US (California) and in most of the EU state members.

Dilfubenzuron

This chemical is blocking the activity of chitin synthetase and inhibits the molting process (apolysis) (Figure 7). This mode of action is highly specific for arthropods and disrupts any process that involves construction of new cuticle (e.g. hatching, molting, pupation) [11]. However, it is most effective against immature stages (early instars). Probably one of the disadvantages is that they act rather slowly and therefore the first effects are observable within two to five days. Diflubenzuron, currently registered under the trade name Dimilin, is used for controlling gypsy moth, boll weevils and various other pests.

Indoxacarb

Indoxacarb is an oxadiazine insecticide that blocks the sodium channels in insect nerve cells (Figure 8). As a result, lepidopteran larvae stop feeding within four hours and die within

[6]Onomatology – chemical name according to the International Union of Pure and Applied Chemistry (http://iupac.org/).

two to five days [75][129]. Indoxacarb expresses high selectivity by acting primarily against lepidopteran larvae.

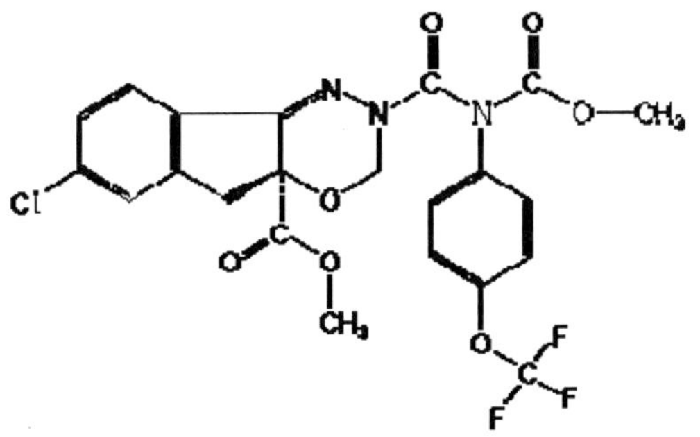

Figure 8. Chemical structure of Indoxacarb: methyl (*S*)-*N*-[7-chloro-2,3,4a,5-tetrahydro-4a-(methoxycarbonyl)indeno[1,2-*e*][1,3,4]oxadiazin-2-ylcarbonyl]-4-(trifluoromethoxy)carbanilate (IUPAC).

Tebufenozide

Tebufenozide is a dialcylhydrazine that has IGR properties and blocks the completion of the physiological molting process acting as a molting hormone analogue (Figure 9) [39]. Tebufenozide is a dibenzoylhydrazine stomach poison specifically for Lepidoptera and the insect stops feeding within a few hours. Most of the IGRs' application time is critical for success, because the compound is more active on first larval instars [185].

Tebufenozide is a slow acting compound with residual activity of 14 to 21 days and must be ingested to take effect. In contrast to other IGRs, tebufenozide is non toxic to honeybees and expresses high selectivity by preserving most of the natural enemies [64]. Apart from pome and stone fruits, in which tebufenozide is currently registered in the US, it is also registered for several crops including cotton, grapes, lettuce, and tomatoes [129].

Methoxyfenozide

Methoxiphenozide is a dibenzoylhydrazine IGR (Figure 10), similar to tebufenozide in its mode of action. It induces a lethal molt and its specificity for Lepidoptera is very high [39] because it displays a higher binding connectivity to the lepidopteran receptors compared to the non lepidopteran species and thereby is characterized by high selectivity to non target species. Compared to tebufenozide methoxiphenozide expresses also a longer residual activity.

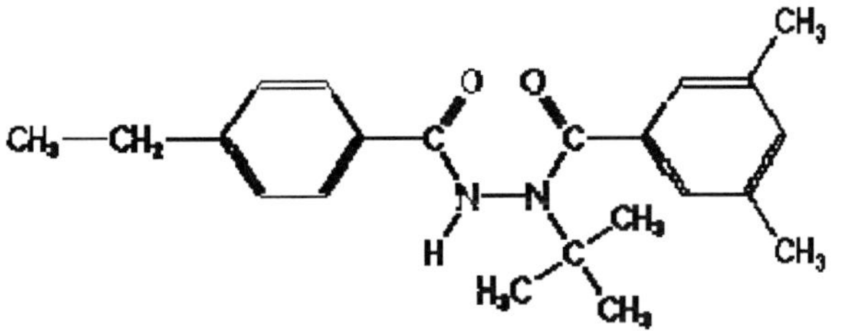

Figure 9. Chemical structure of Tebufenozide: *N*-*tert*-butyl-*N'*-(4-ethylbenzoyl)-3,5-dimethylbenzohy drazide (IUPAC).

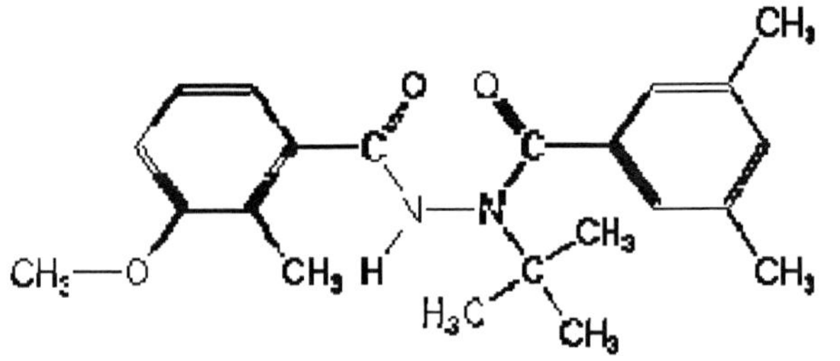

Figure 10. Chemical structure of methoxyfenozide: N-tert-butyl-N′-(3-methoxy-o-toluoyl)-3,5-xylohydrazide (IUPAC).

Emamectin- Benzoate

This compound is a second generation avermectin analog [$C_{56}H_{81}NO_{15}$ (emamectin B_{1a} benzoate) + $C_{55}H_{79}NO_{15}$ (emamectin B_{1b} benzoate)] (Figure 11) with exceptional activity against lepidopteran species. It causes decrease of the excitability of neurons. Emamectin-benzoate toxicity has a broader spectrum compared to methoxyphenozide, tebufenozide or indoxacarb and controls a wide variety of lepidopteran pests [129]. On the other hand, this broad spectrum activity increases the possibility of the lepidopterous prey to be toxic to more natural enemies.

TC: $C_{49}H_{75}NO_{13}$ (emamectin B1a) + $C_{48}H_{73}NO_{13}$ (emamectin B1b). Although this compound is photodegradable, approximately within 5 days, fragments of the maternal compound are persistent in the plant tissues, expressing laminar activity, and thus emamectin benzoate has longer residual activity. The exposed moth larvae, after contact or feeding, become irreversibly paralyzed, stop feeding and die within three to four days. The ingredient is not ovicidal, although it can penetrate into plant tissues.

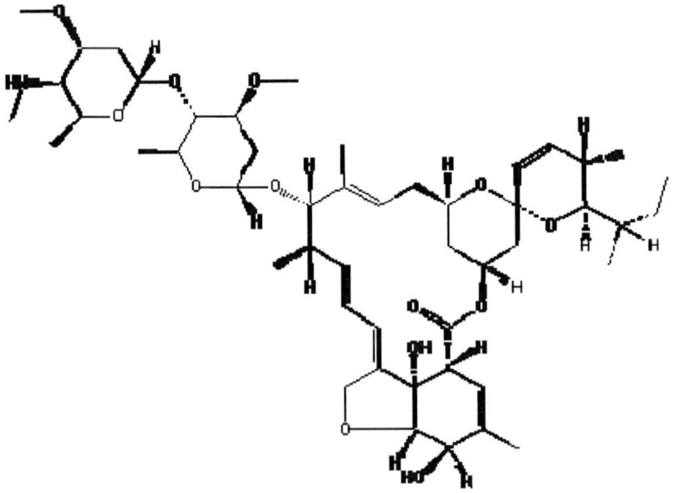

Figure 11. Chemical structure of Emamectin- benzoate: (4″R)-4″-deoxy-4″-(methylamino)avermectin B1 benzoate (1:1) (CAS[7]).

[7]Onomatology – chemical name according to the American Chemical Society (http://cas.org/index.html).

Spinosad

Spinosad has recently been registered for fruit orchards in Northern America for the control of tortricidae leafrollers as well as other Lepidoptera. Spinosad has a unique mode of action, causing rapid excitation of the insect nervous system [158]. The major ingredient of the product comprises a mixture of spinosyns (SPA A and SPA D) derived from the fermentation of the *Saccharopolyspora spinosa* (Actinobacterida) [108][130]. The structures of SPA and its many analogues have a characteristic perhydro-*as*-indacene core, which is glycosylated by a rhamnose at C-9 and a forosamine at C-17 (Figure 12). Both the aglycone (AGL 3) and the sugar appendages contribute to the observed activity of the spinosyns, among which SPA is most potent [108]. Although spinosad shows low toxicity when ingested by mammals and no adverse chronic exposure effects, it is highly toxic to bees, oyster and other marine mollusks, as well as Trichogrammatidae and Braconidae species.

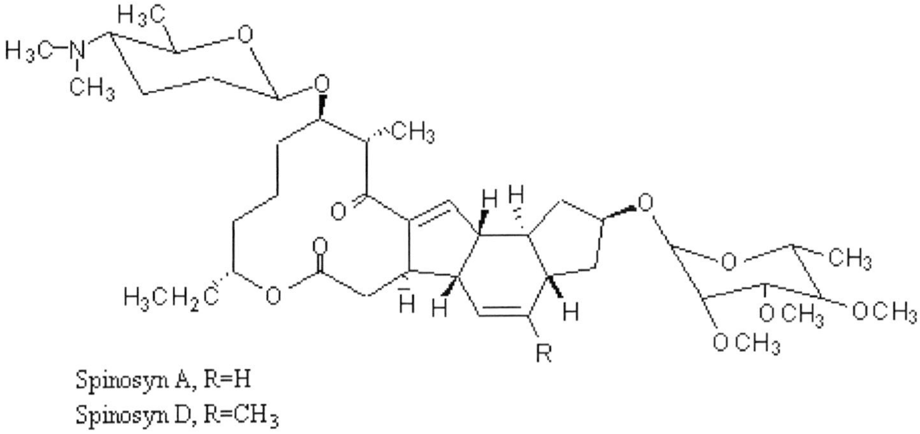

Figure 12. Chemical structure of Spinosad [$C_{41}H_{65}NO_{10}$ (spinosyn A) + $C_{42}H_{67}NO_{10}$ (spinosyn D)] Spinosyns are macrolides with a 21-carbon, tetracyclic lactone backbone to which the deoxysugars forosamine and tri-O-methylrhamnose are attached.

Azadirachtin

Azadirachtin is a major compound of neem oil, pressed from the fruits and seeds of *Azadirachta indica* (Indica Neem tree) (Figure 13) [137]. The insecticidal activity of the azadirachtin is rather complex, since it can act as an insect growth regulator, but also poses feeding and ovipositional deterrent activities [5][22][111][163].

Azadirachtin interferes with the neuroendocrinal control of metamorphosis, affecting both ecdosteroidal and juvenile titers and IGR effects are expressed through various post-embryonic, reproductive, and growth inhibitory effects, causing dose-dependent mortality [22].

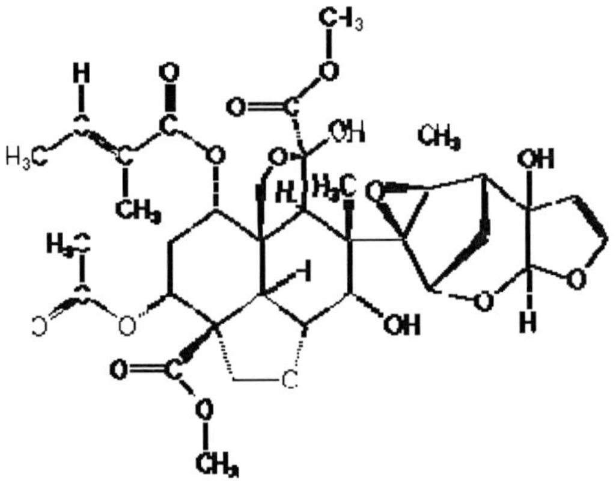

Figure 13. Chemical structure of Azadirachtin, botanical insecticide: $C_{35}H_{44}O_{16}$.

Bio-Insecticides

Bacillus thuringiensis

Bacillus thuringiensis is a gram+ bacterium that is pathogenic to larvae of several species and in particular to lepidopterous larvae. The infection of the larvae is caused by the endospore of the species [10][31]. *B. thuringiensis* produces during sporulation crystalline inclusions consisting of one or more insecticidal proteins known as d-endotoxins or Cry-proteins [10].

The main symptom of infected larvae is paralysis. The endotoxin causes lysing to the insect gut and as a result the larva stops feeding and mortality occurs due to starvation. Lately, this endotoxin has been widely used in IPM programs for the control of several moth species, such as the codling moth.

The first products for control of lepidopterous larvae were based upon two subspecies of *B. thuringiensis*, var: *kurstaki* (i.e. Dipel , Javelin) and var: *aizawai* (i.e. XenTari), or a combination of the two (i.e. Agree). Second generation products are based on the conjugation of the two subspecies. The third and fourth generation products are based on new B.t. strains, using recombinant DNA technology.

CpGV - Granulovirus of Cydia pomonella

The granulovirus of codling moth (CpGV) is an insect-specific granulovirus that offers new means of highly selective control of *C. pomonella* in fruit orchards [20][21][46][70][71][83][98]. CpGV was first isolated from infected larvae in Mexico. The virions are infecting several insect tissues including the epithelial cells of the gut, as well as the fat body. Although replication takes places 48h after the first infection, the first symptoms are observable approximately after 4-5 days and the larvae stops feeding after 7 days. Commercial products that are used in IPM programs contain the virus in an aqueous suspension and are sprayed during the egg hatch. Thus, the timing of CpGV application is critical in order for occlusion bodies to be ingested by neonate larvae before they bore into the fruits. CpGV is highly selective and the host range is actually limited to *C. pomonella* and a few other Tortricidae species [74][115]. CpGV thus offer new options for management of the codling moth in organic or 'soft' orchards, although their short residual activity may limit adoption in conventional insecticide programs, while the limited protection to fruit also makes CpGV less effective under high pest pressure [20]. Commercial products of CpGV are now produced in Europe and North America and used by farmers worldwide, while currently three insecticidal formulations of CpGV are registered for commercial use in North America [115][116] [117][118].

Entomopathogenic Nematodes

Steinernema and *Heterorhabditis* species (order Rhabditida) are insect specific parasitic nematodes. They are lethal to their host due to the transmission of bacteria [143][190]. This makes this group of nematodes more suitable for biological control of insects and has therefore become increasingly popular in IPM programs. Most of the research concerning the potential of entomopathogenic nematodes to control insect species has been conducted with *Steinernemia feltiae* and *Steinernema carpocapsae.* In specific the lethal action of the entomopathogenic nematodes is associated to symbiotic bacteria such as *Xanorhabbdis spp.* [115][116]. The symbiond is released after actively seeking out and penetrating the larval host by the nematode. Generally all steinernematid and heterorhabdid infective juveniles carry species-specific symbiotic bacteria. *Steinernema carpocapsae* was one of the first entomopathogenic nematodes that were isolated and commercially used for the control of *C. pomonella,* especially in controlling the overwintering larvae of the moth. This overwintering population actually represents the most of the codling moth population and therefore successful control can considerably reduce early season infestations, providing substantial protection during growth season. *Steinernema feltiae* is also used in IPM programs in controlling overwintering larvae of *C. pomonella.* Unlike *S. carpocapsae,* which expresses limited host search behavior, *S. feltiae* expresses higher search capacity and is considered an intermediate search strategist. The best application time is when moth larvae are known to be present. However, ideal environmental conditions are essential for success. One of the limitations of the commercial products of *S. caprocapsae* is that it must be applied under optimal temperature, not below 15°C, and high moisture conditions [77]. On the other hand, *Steinernema feltiae* is active at 10°C and lower temperatures [77][78][79].

5. Semiochemicals and Principal Strategies in IPM

A semiochemical is a chemical substance, or mixture, that carries a message within or between species and includes pheromones, allomones, kairomones, attractants and repellents. In general, chemical compounds that mediate interactions between organisms are called infochemicals or semiochemicals. Signals that are transmitted between individuals of different species (inter-specific) are called allelochemicals, while signals perceived by individuals of the same species (intraspecific) are termed pheromones [99][149][165][191][196].

The use of pheromones[8] to monitor insect phenology [103] is often incorporated in IPM programs, although now this knowledge is associated to change pest behavior or its enemies

[8]Several types of insect pheromones have been demonstrated [97], such as sex, aggregation, alarm and trail-pheromones. Since mate finding in moths is largely based on olfactory communication, via sex pheromones, they have attracted most attention by researchers.

as well [104][105][106][138][146][192]. Table 4 lists the major chemical compounds that are first cited and are mostly related to the pheromone blend of representative moths [18][150][151][152][153][154][155][156][174][176][179][187].

Additionally, semiochemicals play an important role in chemical ecology, and semiochemical-mediated communication affects several types of interactions in organisms in natural habitats, while it has been employed to manipulate the behavior of economically important pest species [12][13][16][17][32] [37][38][72][73][80][84][85][193].

Semiochemicals act at an ecological level by expressing the following action modes: *Habituation*, which is the progressive acustomization of individual males to the pheromone. Since pheromone is perceived as a natural part of the chemical environment, males do not notice female chemical signals. *Camouflage*, is the cataclysm of the environment (i.e. orchard) by pheromone camouflages and signals produced by the females are not distinguishable. *Completive attraction,* is exerted by the lures rather than the female signals. Males are actively sought to find the lures and reduce the chance to mate with females. *Trapping*, lures with pheromones are placed in trap devices and males are first trapped and then removed.

One of the major advantages of these technologies is that they do not exhibit adverse effects on non target pests, while insect pest resistance is unlikely to appear. Additionally, from an environmental perspective, the comprising properties allowed them to be characterized as non persistent displaying relatively low toxicity to non target species and are environmentally safe [40][142].

Mating Disruption

Although pheromone technology has traditionally been used to monitor pest activity and pest populations, aiming to precise timing of pesticide application, during the last years new strategies such as mating disruption have been developed and present promising solutions in environmentally friendly control of moth populations [23][25][26][28][29].

Mating disruption (MD) is a relatively new commercial strategy in the control of microlepidoptera of economic importance. In principal this technology is based on the use of semiochemicals. The pest management strategy is using synthetically produced pheromones through controlled release to confuse males in order to limit their ability to locate females for mating. The main objective is to reduce the likelihood of successful mating through the distribution of pheromone dispensers in the orchard [94][95][96][112][113].

Within this frame, methods based on the same principles are also cited as confusion methods and false-trail or disorientation methods. In practice, such a method consists in the setting up of several prevailing pheromone trails, released by an appropriate number of dispensers loaded with low pheromone dosage, able to compete with those of the female insect and thus disorientate males in their search for partners [135][136]. On a behavioural level, the effectiveness of the method is likely to depend on keeping males busy most of the time by visiting the artificial pheromone sources rather than the females which are actually present in the treated orchard.

Table 4. Major pheromone compounds of micro moths of economic significance

Moth species	Chemical name	Chemical structure	Reference
Adoxophyes orana (Summerfruit tortrix)	(Z)-11-Tetradecenyl acetate (Z)-9-Tetradecenyl acetate	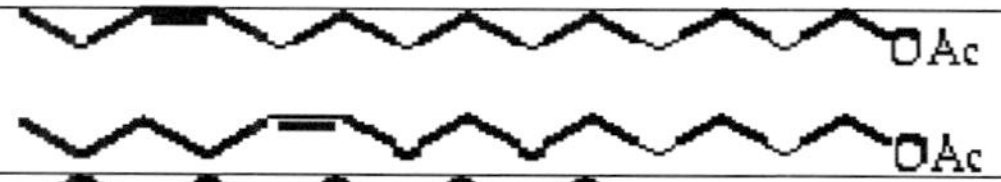	**Tamaki et al. (1971) [174][194]**
Anarsia lineatella (Peach twig borer)	(E)-5-Decenyl acetate (E)-5-Decen-1-ol	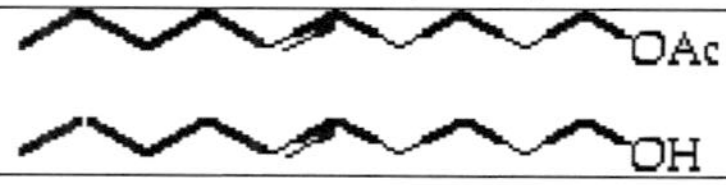	Roelofs (1975) [154]
Cydia pomonella (Codling moth)	(E,E)-8,10-Dodecadien-1-ol, Codlemone	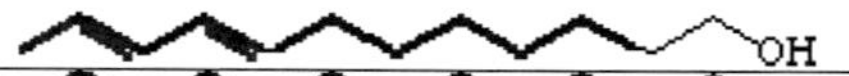	Roelofs (1971) [152]
Cydia splendana (Chestnut tortrix)	(E,E)-8,10-Dodecadienyl acetate, Codlemone acetate		Wall (1976) [187]
Eupoecilia ambiguella (European grape moth)	(Z)-9-Dodecenyl acetate	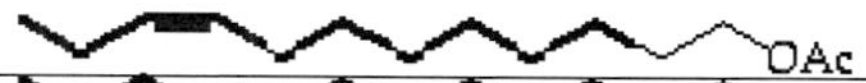	Roelofs (1971) **[194]**
Grapholitha funebrana (Plum fruit moth)	(Z)-8-Dodecenyl acetate		Roelofs and Comeau (1969) **[194]**
Grapholitha lobarzewskii (Small fruit tortrix)	(E)-8-Dodecenyl acetate (Z)-8-Dodecenyl acetate	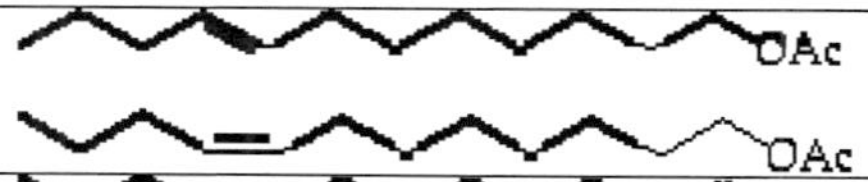	Biwer (1978) [29]
Grapholitha molesta (Oriental fruit moth)	(Z)-8-Dodecenyl acetate		Roelofs (1969) [150]
Lobesia botrana (Grapevine moth)	(E,Z)-7,9-Dodecadienyl acetate		Roelofs (1973) [153]
Pammene rhediella (Fruitlet mining tortrix)	(E,E)-8,10-Dodecadien-1-ol, Codlemone		Brakefield (1975) [32]
Pandemis heparana (Apple brown tortrix)	(Z)-11-Tetradecenyl acetate (Z)-11-Tetradecen-1-ol (Z)-9-Tetradecenyl acetate	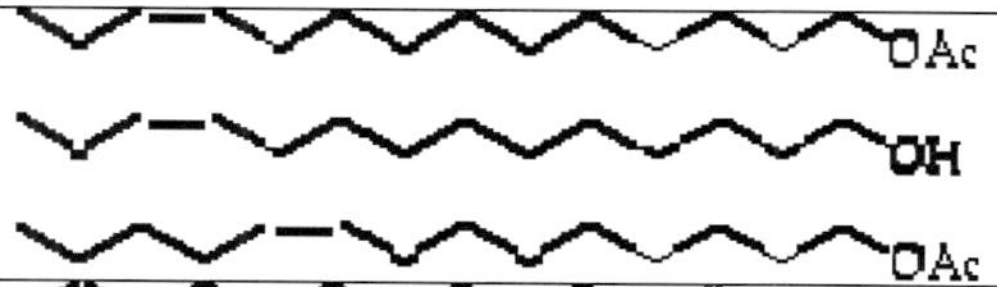	Frérot (1979) [194]
Sparganothis pilleriana (Long-palped totrix)	(E)-9-Dodecenyl acetate (E)-9-Dodecen-1-ol	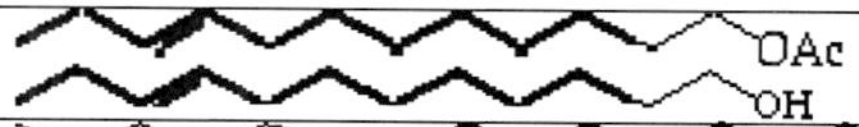	Roehrich (1977) [156]
Spilonota ocellana (Eyespotted bud moth)	Z)-8-Tetradecenyl acetate		Arn (1974) [16][18]

Table 4. (Continued).

Moth species	Chemical name	Chemical structure	Reference
Zeuzera pyrina (Leopard moth)	(E)-2-Octadecenyl acetate E,Z)-2,13-Octadecadienyl acetate (Z)-13-Octadecenyl acetate	  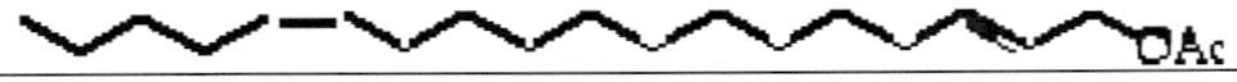	Castellari(1986) [40]

Since simultaneous presence of more than one moth species is very often, the possibility to combine two different pheromone-based techniques is also of particular interest, applied on diverse pests and at different times according to the real needs which further improve control methods and additionally reduce multispecies managing costs [54][56]. Figure 14 for instance presents the almost complete elimination of catches of *G. molesta* in the pheromone traps in fruit orchards in which mating disruption dispensers were installed [58].

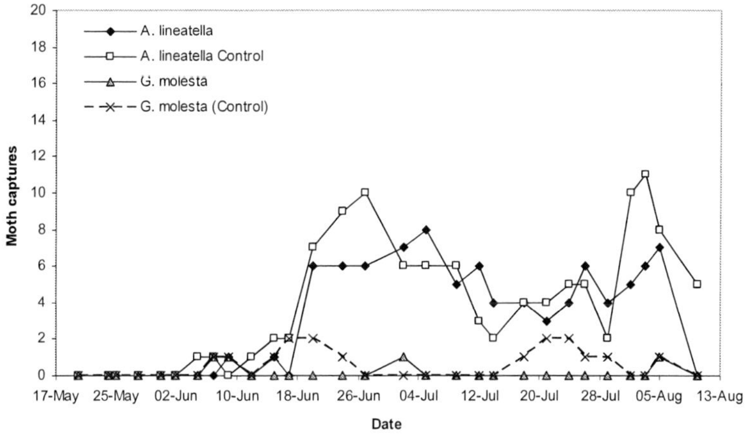

Figure 14. Effect of simultaneous mating disruption on *G. molesta* and *A. lineatella* in treated and untreated peach orchards (control) in Northern Greece. Field trials were carried out during 2010 in four plots (~2 hectares, industrial variety Loadel, no insecticides were applied during the trials) [58].

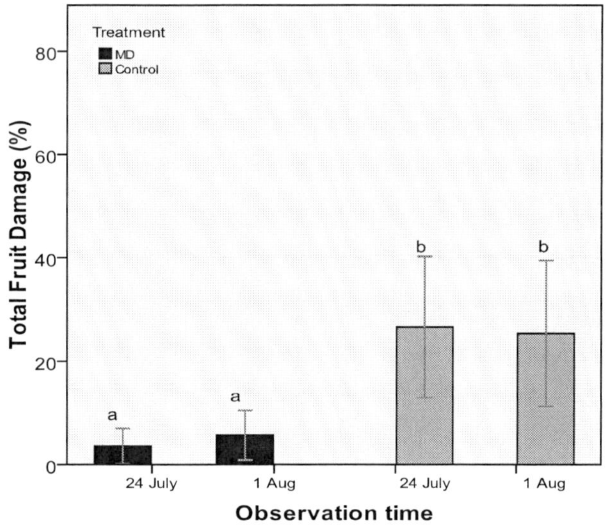

Figure 15. Mean percentage fruit damage in treated (MD) and untreated (Control), with mating disruption dispensers for the control of *G. molesta* and *A. lineatella* in Northern Greece.peach orchards (F=6,664, P<0.05).

In most cases, mating disruption involves the distribution of a high number of dispensers (300-1000 per ha, depending upon manufacturer and loaded pheromone concentrations) in the fruit orchard. Installation time of dispensers is critical for the success and therefore this strategy should be integrated with information derived by forecasting models. However, in general, dispensers should be installed a few days before the moth's first flight, while they remain effective through the second and third flights if population pressure is low [135][136]. Furthermore, new technologies involve long season mating disruption by the use of machine application of pheromone in wax drops [169].

Promising results of the mating disruption strategy in managing *G. molesta* have been reported in Greece as well as in several countries including South Africa [24], Italy [176][177] [135][136] and Slovenia [157], while successful pheromone formulations have already been developed also for the codling moth *C. pomonella* [113][178][182]. Additionally, mating disruption could be potentially combined with pesticides in IPM [127][183].

Attract and Kill

The attract and kill strategy is another semiochemical-based approach involving the combination of a semiochemical lure with an insecticide [7][42][43][44][102]. This principle has been successfully used to control the codling moth, although some critical parameters affecting its efficacy have been reported [14][15]. This approach is already successfully applied to the control of codling moth [88][89][125], although environmental factors can alter its performance [126][176][177].

A commercial product of this novel strategy has been developed recently by Ciba–Geigy (now Novartis Crop Protection) [44]. The principle enables the distribution of viscous paste, which contains a sex pheromone and insecticide[1], in the form of small droplets in the crop [44] [96].

The droplets attract males, which are stimulated to express mating behaviour and are killed by the insecticide. Commercial products have already been registered for the control of codling moth in Switzerland, South Africa and the United States [42][43][88][89].

The principal mechanism responsible for the success of the "attract and kill" method is the elimination of males by use of permethrin and not the interruption of sexual communication or confusion due to multiple pheromone sources [42]. This is a major difference of strategy compared to sexual confusion and mating disruption. Additionally, the combination of the pheromone should be made with a contact insecticide for an acceptable effectiveness of the "attract and kill" approach.

Push and Pull

The 'push-pull' strategy is a novel tool for IPM programs that uses a combination of behavior-modifying stimuli to manipulate the distribution and abundance of insect pests and/or natural enemies [4][45][101].

In brief, in this strategy the pests are repelled or deterred away from the main crop (push) by using stimuli that mask host appearance or are repelled, while they are simultaneously attracted (pull), using highly apparent and attractive stimuli, to other areas. The attractive

[1] The commercial product (Sirene CM) contains 0.16% codllemone E-8, E-10-dodecadien-1-ol (E8,E10-12:OH) and 6.0% permethrin.

stimuli can be either traps or crops or alternative hosts, which serve to concentrate the pest, thereby facilitating their control. The moth species can be pushed by volatile chemicals by intercropped plant which attracts natural enemies. On the other hand, volatile chemicals are produced by border trap plans and attract moths to lay their eggs.

CONCLUSIONS

Most IPM strategies which aim to control the major lepidopterous fruit pest should use all available knowledge and technology in order to be successful and according to the context of IFP [47]. However, this is not always possible due to lack of alternatives. For instance, in most cases, there are either no available EIL and ET levels for perennial crop, or they are very difficult to be developed. Additionally, although the calculation of EIL is quite simple, the development of the relative ET is somehow subjectively defined (based on the expert's experience), while a detailed calculation can be proved problematic due to numerous complex variables that are involved (e.g. time delays on insect development and related phenology, injury rates, host performance) [86]. Hence the simplest approach is to use fixed and empirical ETs, which corresponds to a predefined level of the EIL (i.e. 25-50%) and can be used where injury rates or dynamics of pest population are unknown.

One important task in IPM is also the simultaneous management of more than one species which appear during the growth season. This is very common situation in regions in which more than one hosts are closely cultivated in nearby areas and environmental factors allow high population levels. It is therefore essential to define the key pest and to detect similarities (if any) among the pest species, in order to design optimum management action that can control at once more than one species. This is an important challenge of current IPM in fruit production. In practical terms, this information simply helps plant protection managers and growers reduce time spent for scouting, decrease management costs and eliminate and mitigate side effects of pesticide applications of no necessity.

In general, most pest control tactics are species-specific and growers often focus in controlling the major key-pest, which is causing most of the damage, while take on additional measures in order to manage outbreaks of secondary pests. Management actions therefore occur at different time and space scales according to subjective criteria and have disputable success. Detailed models that are forecasting the phenology of each species are therefore essential, while unified models ameliorate significant forecasting calculations based on common base temperatures and biofix. For instance, a common prediction in time occurrence for peach moth complex allows the determination of a 'control windows' for each generation in which a simultaneous management is either possible or not [56]. Hence, initiation of control measures should ideally assess multi-pest population dynamics and take into account similarities or differences in phenology during the growth season.

In some cases, it is important to ensure that the various components of IPM are available and compatible with each other. Most of the biorational compounds that are prior listed are generally considered as environmentally and toxicologically reduced risk pesticides. Spinosad for instance, degrades quickly and generally shows little residual activity a few days after application and is therefore classified as a reduced risk pesticide by the US environmental protection agency. However, there are concerns about its compatibility with biological control since several harmful effects have been documented for beneficial species such as parasitic hymenoptera, as well as honeybees and other pollinators.

One additional problem in modern IPM is the progressive establishment of insect population that expresses high resistance also to biorational insecticides [173][197]. It should be noted that several moth species have developed resistance to some of the above novel compounds. For instance, as a result of cross resistance to older groups of insecticides, several moth species recently exhibited resistance also to indoxacarb [3]. Furthermore, low levels of resistance to tebufenozide and methoxyfenozide have been referred to for *C. pomonella* and other species without prior exposition to this insecticide [3].

Thus, one important constraint that should be taken into account on a long term establishment of IPM programs is the rotation of chemical compounds during the growth season. Actually, rotation of chemical compounds is suggested for all classes of insecticides as well as to major novel groups of insecticides. The major principle to avoid resistance is the application of products that have ingredients which express different modes of action. Rotation, for instance, of indoxacarb with other compounds such as emamectin and dibenzoylhydrazine IGRs can decrease the possibility of resistance development.

Additionally, most of these compounds are effective when applied only on particular developmental stages of the target pest. Phenology models are therefore once again an essential part of an IPM program and forecasting models provide manners to introduce control methods at the time in which they can express their maximal activity on the target pest and the lowest side effects to beneficial species. Therefore, to integrate biological control into IPM programs more extensively, it is essential to have a thorough knowledge of the life stages of the pest and its natural enemies that are present, in order to maximize effectiveness of these biorational products in their management [2][114].

On the other hand, it has been demonstrated that not all life stages of beneficials are affected at the same grade. In other words, the time of application in relation to the stage and abundance of pests and beneficial species define the biorational nature of a compound.

Unfortunately, little is known about the actual phenology of beneficial species compared to pest and how it is affected by different management actions [51]. Additionally, while for most novel compounds much is known regarding the activity on target pests, less in known concerning their potential side effects on beneficial species [161]. Thus, although most IGR regulators are thought to have minimal impact on natural enemies, this has been demonstrated only for a few species and compounds [122].

The use of semiochemicals in Europe is extensively increasing in IPM programs and IFP. Mating disruption, in particular, is used to control several economically important tortricid moths in orchards and vineyards on ca. 50 000ha. In particular, wide area applications of the mating disruption techniques in Europe has been successfully implemented for the management of the codling moth *C. pomonella* and the grape berry moths *Eupoecilia ambiguella* and *Lobesia botrana*, [146][164] [184], for *A. orana* [67] [186], *Sparganothis sulfureana* in cranberries [141], while promising results for simultaneous management of *A. lineatella* and *G. molesta* have been registered in peach orchards of low and moderate population moth densities in Greece [58].

However, the management of pests via manipulation of sexual communication, although very effective in low population pressures, must be still combined with other control measures in order to be effective in warmer regions (i.e. Southern Europe), where moth population levels in some cases can be extremely high. However, the facts that beneficials are not affected, as well as the lower pesticide dependency, are major advantages and driving forces for future research to this direction.

One other important constraint in development of IPM programs and successful establishment of IFP is also the particular legislation frame that determines the production, registration of ingredients and final availability of insecticides that have permission to be used only for certain pests and fruit crops.

Since the status of pesticide approvals in the EU is continually changing, as many of the older generation pesticides developed in the 1950s-1980s are either under review or completely banned[2], the registration of newer products such as Biorational and Bio insecticides as well as alternative control methods are extremely important for realistic implementation of IPM. The framework in which these products can be used in IPM programs is also rapidly changing [67][68].

In the European Union (EU) for instance, member states must adopt specific defined "National Action Plans" in order to reduce the "risks and impacts" of pesticide use on human health and the environment. These framework directives[3] and plans include timetables and targets for use reduction. It is noteworthy to state that a thematic strategy on the sustainable use of pesticides was launched in 2006 by the European Commission concerning plant protection products [67][68].

Additionally, a framework directive setting minimum rules of pesticide use aims to fill the legislative gap concerning the application phase of the available compounds among the EU state members. This is feasible through setting minimal rules concerning pesticides applications, while the framework also intends to include biocides as well. Moreover, several compounds that where allowed for use have been, or will be, completely banned, while several new and older compounds fail to gain registration according to the new legislation. These products must be accompanied by clear evidence that action has been taken to protect health and environment, while approximately 500 active ingredients have been withdrawn from the EU market either due to the review process or by the industry for market reasons.

It is commonly accepted that traditional insect management will change in the forthcoming years as a result of higher environmental concerns and pressures exhibited by all parts, including not only consumers and growers but also political authorities, industry and international organizations. Worldwide the growing demand for environmentally safe food production is expected to lead to an increased need for alternative control methods in fruit production.

The development of alternatives on a realistic basis that will reduce our dependency on conventional agrochemicals and traditional management actions requires a full understanding of all aspects of the underpinning science of Economic Entomology. Additionally, IPM is probably among the very few concepts that take into account concerns of most enabling parts (environment, society, governments, industry and incorporations) by providing a realistic

---

[2]The EU started to implement during the 80s a limited number of bans on specific compounds, mainly persistent organochlorines. DDT was banned in 1986, while in 1991 an EU directive (79/117/EEC) consolidated decisions on 20 active ingredients and prohibited their sale. One year later (1992), the EU embarked on a full review of about 1000 active ingredients. Since 1993 manufacturers applying to gain EU-wide approval for a compound have to submit new data sheets that prove that the substance can be used without unacceptable risks and meeting stricter environmental and health standards.

[3]The discussions on the Framework Directive on the sustainable use of pesticides started in the European Parliament and the Council in 2006. A second-reading agreement was found in December 2008 to be adopted on 2009.

management framework that can substitute conventional Agriculture. Continued work to provide reliable alternatives is essential to permit IPM to be feasible and successful in sustainable Agriculture in the future.

REFERENCES

[1] Agra CEAS Consulting (2002). Integrated Crop Management Systems in the EU Amended Final Report for European Commission, DG Environment. Brussels (B) / Wye (GB), May 2002.

[2] Akhurst, R. J. (1990). Safety to nontarget invertebrates of nematodes of economically important pests, pp. 233-240. *In*: M. Laird, L. A. Lacey and E. W. Davidson (eds.), Safety of Microbial Insecticides. CRC Press, Boca Raton, FL.

[3] Ahmad, M, R. M. Hollingworth, J.C. Wise (2002). Broad-spectrum insecticide resistance in obliquebanded leafroller *Choristoneura rosaceana* (Lepidoptera: Tortricidae) from Michigan. *Pest Manag. Sci* 58, 834–8.

[4] Ahmed, H., H. Herren, Z. R. Khan, J. A. Pickett, C. M. Woodcock (2007). Integrated pest management: the push–pull approach for controlling insect pests and weeds of cereals, and its potential for other agricultural systems including animal husbandry. *Phil. Trans. R. Soc. B: Biological Sciences*, 363, 611-621.

[5] Aldhous, P. (1992). Neem chemical: the pieces fall into place. *Science*, 258-893.

[6] Alford, D.V., (1984). A Color Atlas of Fruit Pests. Their recognition, biology and control. Wolfe Publishing Ltd.

[7] Alma, A., A. Arzone, A. Galliano, and F. Vittone (2001). "Attract and kill" a new IPM method in apple orchards against *Cydia pomonella* (L.). Bull. OILB-SROP. 24, 139-143.

[8] Altieri, M. A., Nicholls, C. I.(2000). Agroecology in action. Indigenous and modern approaches to IPM in Latin America. ESPM Division of Insect Biology, University of California,Berkeley,USA.http://www.cnr.berkeley.edu/~agroeco3/indigenous_and_modern_approaches.html

[9] Amano, H. (2001). Species structure and abundance of invertebrate natural enemies in sustainable Agroecosystems. In: *Structure and Function in Agroecosystem Design and Management,* Shiyomi M. andH. Koizumi (Eds.). CRC Press, New York, U.S.A., pp. 167-182.

[10] Andermatt, M., E. Mani, T. H. Wildbolz, P. Lüthy. (1988). Susceptibility of *Cydia pomonella* to *Bacillus thuringiensis* under laboratory and field conditions. *Entomol. Exp. Appl.*, 49, 291-295.

[11] Anderson, D. W., (1978). Diflubenzuron: control of the codling moth, *Laspeyresia pomonella* L., and the obliquebanded leafroller, *Choristoneura rosaceana* (Harris). MSc Thessis, Department of Plant Science, The University of British Columbia.

[12] Ando, T., S. Yoshida, N. Takahashi, (1977). Sex attractants for male Lepidoptera. *Agricultural and Biological Chemistry*, 41, 1485-1492.

[13] Angeli, G., C. Ioriatti, S. Finato, (2000). A new method for the control of codling moth. *Inf. Agrar.*, 56, 63-66.

[14] Angeli, G., Berti, M., Maines, R., and Ioriatti, C. (2003). Control of *Cydia pomonella* (L.) and *Cydia molesta* (Busck) by false trail following and attract and kill techniques. *Inf. Fitopatol.*, 53, 45-50.

[15] Angst, M., D. Hofer (1990). Attract and kill of *Cydia pomonella*. Abstract of OILB/SROP pheromone meeting "Pheromones in Mediterranean pest management"; Granada 10-15, Attract and kill in codling moth 9 September 1990.

[16] Arn, H., C. Schwarz, H. Limacher, E. Mani, (1974). Sex attractant inhibitors of the codling moth *Laspeyresia pomonella* L. *Experientia*, 30, 1142-1144.

[17] Arn, H., B. Delley, M. Baggiolini, P. Charmillot, (1976). Communication disruption with sex attractant for control of the plum fruit moth *Grapholitha funebrana*: a two year field study. *Ent. Exp. Appl.*, 19, 139-147.

[18] Arn, H., M. Tóth, e. Priesner (2000). List of sex pheromones of female Lepidoptera and related male attractants, Internet Edition: http://www-pherolist.slu.se

[19] Arnold, C. Y. (1959). The determination and significance of the base temperature in a linear heat unit system. *Proc. Am. Soc. Hortic. Sci.* 76, 430-445.

[20] Arthurs, L.A. Lacey, R. Fritts (2005). Optimizing the use of the codling moth granulovirus: effects of application rate and spraying frequency on control of codling moth larvae in Pacific Northwest apple orchards, *J. Econ. Entomol.*, 98, 1459–1468.

[21] Arthurs, S. P., L. A. Lacey, E. R. Miliczky (2007). Evaluation of the codling moth granulovirus and spinosad for codling moth control and impact on non-target species in pear orchards. *Bio. Control*, 41, 99–109.

[22] Ascher, K. R. S., Klein, M. and J. Meisner (1992). Azatin, a Neem formulation, acts on nymphs of the western flower thrips. *Phytoparasitica* 20: 305-306.

[23] Audemard, H., J. P. Gendrier, (1992). Control of the oriental fruit moth (*Cydia molesta* Busck) by the mating disruption technique and integrated plant protection in peach orchards. *Acta Phytopathol. Entomol. Hung.* 27, 59-64.

[24] Barnes M.M (1991). Tortricids in pome and stone fruits, codling moth occurrence, host race formation and damage. In: L.P.S. van der Geest and H.H. Evenhuis, Editors, *Tortricid Pests, Their Biology, Natural Enemies and Control*, Elsevier Science Publishers, Amsterdam, The Netherlands (1991), pp. 313–327.

[25] Barnes, M.M., Millar, J.G., Kirsch, P., and Hawks, D.C. (1992). Codling moth (Lepidoptera: Tortricidae) control by dissemination of synthetic female sex pheromone. *J. Econ. Entomol.* 85:1274-1277.

[26] Barnes, B. N., T. L. Blomefield (1997). Goading growers towards mating disruption: the South African experience with *Grapholita molesta* and *Cydia pomonella* (Lepidoptera,Tortricidae). *IOBC/wprs Bulletin* 20, 45-56.

[27] Barrett, _(2001). Integrated Pest Management. Insects and mite pests of Apples. In: *Plant Protection Programs*. University of Missouri-Columbia.

[28] Bartell, R. J. (1982). Mechanisms of communication disruption by pheromone in the control of Lepidoptera: a review. *Physiol. Entomol.* 7, 353-364.

[29] Biwer, G. (1976). Specificite de l'acetoxy-1-dodecene-8-Z (Z-8 DDA) vis 2 vis de.*G molesta* Busck, de *G. funebrana* et de quelques tordeuses des verges. Pub.1, INRA Centre de Recherches de Bordeaux, pp. 60-78.

[30] Boller, E. F., J. Avilla, E. Jörg, C. Malavolta, F. Wijnands, P. Esbjerg (2004). Integrated Production: Principles and Technical Guidelines, 3rd edition 2004. http://www.iobc.ch/iobc_bas.pdf

[31] Bosch, D., B. Schipper, H. van der Kleij, R.A. de Maagd, W. J. Stiekema (1994). Recombinant *Bacillus thuringiensis* crystal proteins with new properties: possibilities for resistance management, *Bio Technology*, 12, 915–918.

[32] Brakefield, J. J., D. H. Bartlett, (1975). The use of pheromones in trapping moth pests in U.K. top fruit orchards. *Proc. 8th British Insecticide Fungicide Conf.,* Vol. 1, British Crop Protection Council, London, 277-292.

[33] British Crop Protection Council (1991). *The Pesticide Manual: A World Compendium.* Ninth Edition. Ed. Worthing, C.R., Hance, R.J. Unwin Brothers Limited. Great Britain.

[34] Buntin G. D., (1994). Developing a primary sampling program. In: *Handbook of sampling methods for arthropods in agriculture*. Ed. by Pedigo LP, Buntin GD, CRC Press, Boca Raton, FL, 99–115.

[35] Buntin, G. D., L.P. Pedigo (1981). Dispersion and sequential sampling of green cloverworm eggs in soybeans. *Environ. Entomol.* 10, 980–985.

[36] Buntin G.D. (1996). Economic thresholds for insect management. In: Economic thresholds for integrated pest management. Ed. by Higley LG, Pedigo LP , University of Nebraska Press, Lincoln, NE, London, UK, 128–147.

[37] Burkholder, W. E. (1990). Practical use of pheromones and other attractants for stored-product insect. In Behavior-modifying chemicals for insect management. R. Ridgway R. Silverstein and M. Inscoe Ed. New York and Basel.

[38] Carde, R. T., A. K. Minks (1995).Control of Moth Pests by Mating Disruption: Successes and Constraints. *Ann. Rev. Entomol.*, 40, 559-585.

[39] Carlson G. R., (2000). Tebufenozide: a novel caterpillar control agent with unusually high target selectivity. In: Anastas, P.T., Heine, L.G., Williamson, T.C. (Eds.), Green chemical synthesis and processes. ACS Symposium Series 767. Washington, DC, pp. 8–17

[40] Castellari, P. L., (1986). *Zeuzera pyrina* L. (Lep. Cossidae): indagini biologische e prove in campo sull'attractività di miscele di component del feromone sessuale. *Boll. Inst. Ent., Univ. Bologna*, XL, 239-270.

[41] Charmillot, P. J., J. F. Brunner (1989). Summerfruit tortrix, *Adoxophyes orana*: lifecycle, warning system, and control. *Ent. Hellenica*, 7, 17-26.

[42] Charmillot P.J., D.Pasquier, A. Scalco, D. Hofer (1996). Essais de lutte contre le carpocapse *Cydia pomonella* L. par un procédé attratticide. *Mitt. schweiz. ent. Ges.*, 69, 431-439.

[43] Charmillot, P. J., D. Hofer (1997). Control of codling moth, *Cydia pomonella* L., by an attract and kill formulation. *IOBC/wprs Bulletin*, 20, 139-140.

[44] Charmillot, P.J. Hofer D. Pasquier D. (2000). Attract and kill: a new method for control of thecodling moth *Cydia pomonella. Entomol. Exp. Appl.*, 94, 211-216.

[45] Cook, S. M., Z. R. Khan and J. A.Pickett (2007). The use of push-pull strategies in Integrated Pest Management. *Ann. Rev. Entomol.*, 52, 375–400.

[46] Cossentine J.E., L. B. Jensen (2004). Persistence of a commercial codling moth granulovirus product on apple fruit and foliage, *J. Entomol. Soc. Brit. Columbia*, 101, 87–92.

[47] Cross, M. G., D. Solomon, P. Chandler, P. Jarrett, P.N. Richardson, D. Winstanley, H. Bathon, J. Huber, B. Keller, G.A. Langenbruch, G. Zimmermann (1991). Biocontrol of pests of apples and pears in northern and central Europe: 1. Microbial agents and nematodes, *Biocontrol Sci. Techn.*, 9, 125–149.

[48] Cross, J. V., E. Dickler, (1994). *Guidelines for Integrated Production of pome fruits in Europe: IOBC Technical Guideline III.* IOBC/wprs Bulletin 17, 1–8. International Organization for Biological and Integrated Control of Noxious Animals and Plants, Montfavet, France.

[49] Cross, J. V. (1996). A pheromone trap survey of tortricid moths (Lepidoptera: Tortricidae) in apple orchards in England subject to different insecticide management. *The Entomologist*, 115, 168-180.

[50] Damos, P., M. Savopoulou-Soultani, (2008a). Temperature dependent bionomics and modeling of *Anarsia lineatella* (Lepidoptera: Gelechiidae) in the laboratory. *J. Econ. Entomol.* 101, 1557–1567.

[51] Damos, P., m. Savopoulou-Soultani M, (2008b). About the presence of beneficials in overwintering sites of *Anarsia lineatella* Zeller (Lep., Gelechiidae) in Northern Greece peach orchards. IOBC/wprs Bull. 35, 44–50.

[52] Damos P, Savopoulou-Soultani M, (2008c). Development and validation of models in forecasting the seasonal emergency and population dynamics of the peach twig borer *Anarsia lineatella* (Lepidoptera: Gelechiidae) in northern Greece. In: Proceedings of XXIII international congress of entomology, 6–12 July, Ed. by ICE Org. Comm., Durban, South Africa, ICE section 1, 267, Abstract No. 414.

[53] Damos, P., M. Savopoulou-Soultani M, (2008d). Des notes sur la préférence de ponte par Anarsia lineatella (Lepidoptera: Gelechiidae) sur le pêcher et des autres hôtes enlaboratoire. IOBC/WPRS Bull. 37, 35–42.

[54] Damos, P. (2009). Bioecology of microlepidopterous pests of peach trees and their management according to the principles of integrated fruit production. PhD Thesis. Aristotle University of Thessaloniki, Greece.

[55] Damos, P. and Savopoulou-Soultani, M. (2010a). Population dynamics of *Anarsia lineatella* (Lep:Gelechiidae) in relation to crop damage and development of Economic Injury Levels. *J. Appl. Entomol.*, 134, 105-115.

[56] Damos, P. and Savopoulou-Soultani. (2010b). Development and statistical evaluation of models in forecasting major lepidopterous peach pest complex for integrated pest management programs. *Crop Prot.*, 29, 1190-1199.

[57] Damos, P., A. Rigas, M. Savopoulou-Soultani (2011a). Application of Markov Chains and Brownian motion models in Insect Ecology. In: *Brownian Motion: Theory, Modelling and Applications.* Editors: Robert C. Earnshaw and Elizabeth M. Riley© 2011 Nova Science Publishers, Inc. ISBN: 978-1-61209-537-0.

[58] Damos, P., K Goudis, M. Savopoulou-Soultani (2011b). Re-evaluation of the mating disruption technique in IPM peach orchards. (In preparation).

[59] Danks, H. V., (1987). Insect dormancy: An ecological perspective. *Biological Survey of Canada*, Ottawa, Canada.

[60] Darvas, B., L. A. Polgar (1998). Novel type insecticides: specificity and effects on non-target organisms. In: Ishaaya, I. and Degheele, D., Editors, 1998. *Insecticides with Novel Modes of Action*, Springer, Berlin, pp. 188–259.

[61] Davidson, J., (1942). On the speed of development of insect eggs at constant temperatures. *Aust. J. Exp. Biol. Med. Sci.*, 20, 233-239.

[62] Dennis, B., P. W. Kemp, C. Beckwith, (1986). Stochastic Model of Insect: Estimation and Testing. *Env. Entomol.*, 15, 540-546.

[63] Dent , D. (1994). *Integrated Pest Management.* Chapman and Hall, London.

[64] Dhadialla, T. S., G. R. Carlson,D.P. Le (1998). New insecticides with ecysteroidal and juvenile hormone activity. *Ann. Rev. Entomol.* 43, 545–69.

[65] Dickler, E. (1992). Current situation of integrated plant protection (IPP) in orchards. *Act.Phyt. Entomol. Hung.*, 27, 23-28.

[66] Dickler, E., S. Schäfermeyer (1991). General principles, guidelines and standards for integrated production of pome fruit in Europe and procedures for endorsement of national and regional guidelines and standards. OILB Bulletin SROP1991/XI/3.

[67] Doerr, M., J. Brunner, (2001). Update on codling moth mating disruption. Washington State University, Tree Fruit Research and Extension Center. http://entomology.tfrec.wsu.edu/ jfbhome/growerarticles/CMMD2001.pdf

[68] European Commission, (2009a). Directive of 21 October 2009 concerning the sustainable use of pesticides (2009/128/EC).

[69] European commission, (2009b). *Final Report Development of guidance for establishing Integrated pests management* (IPM) principles 2009 pp3.

[70] Ehler. L. E., (2006). Perspective Integrated pest management (IPM): definition, historical development and implementation, and the other IPM. *Pest Manag Sci*, 62,787–789.

[71] Falcon L.A., J. Huber (1991). Biological control of the codling moth. In: L.P.S. van der Geest and H.H. Evenhuis, Editors, *Tortricid Pests, Their Biology, Natural Enemies and Control*, Elsevier Science Publishers, Amsterdam, The Netherlands, pp. 355–369.

[72] Gallois F. M., J. Einhorn, (1979). La phéromone sexuelle produite par la famelle vièrge de *Pandemis heparana, C.r. hebd Séanc. Acad. Sci., Paris Ser. D*, 228, 1611–1640.

[73] Gadi, V. P., A. Guerrero (2010). New Pheromones and Insect Control Strategies. *Viatamins and Hormones,* 83, 493-519.

[74] Glen, D. M., C. W. Wiltshire, N. F. Milsom, P. Brain, (1984). Codling moth granulosis virus: effects of its use on some other orchard arthropods. *Ann. Appl. Biol.*, 104, 99–106.

[75] Grafton-Cardwell, E., E., L. D. Godfrey, W. E. Chaney, W. J. Bentley (2005). Varius novel insecticides are less toxic to humans, more specific to key pests. Review Article. *California Agriculture*, 59, 29-34.

[76] Grenier, S, A. Grenier (1993). Fenoxycarb, a Fairly new Insect Growth Regulator: a Review of its Effects on Insects. *Ann. Appl. Biol.*, 3, 476-481.

[77] Grewal, P. S., S. Selvan, R. Gaugler (1994). Thermal adaptation of entomopathogenic nematodes: Niche breadth for infection, establishment and reproduction. *J. Therm. Biol.*, 19, 245-253.

[78] Grewal, P. S., R. Gaugler, Y. Wang., (1996). Enhanced cold tolerance of the entomopathogenic nematode *Steinernema feltiae* through genetic selection. *Ann. Appl. Biol.*, 129, 335-341.

[79] Greval, P. S., R.-U. Ehler, D. I. Shapiro-Ilan (2005). Nematodes as Biological Control Agents. CABI Publishing, Wallingford, Oxfordshire, U.K.

[80] Guerin,P. M, H. Arn, H. R. Buser, P. J. Charmillot, (1986). Sex pheromone of *Adoxophyes orana*: additional components and variability in ratio of (Z)-9- and (Z)-11-tetradecenyl acetate. *J. Chem.. Ecol.*, 12, 763-772

[81] Guidelines for Integrated Production of Pome Fruits. (45 pp.) Technical Guideline III.3rd edition 2002 Edited by J.V. Cross, 10 BC IOBC WPRS Bull. Vol. 25 (8), 2002. ISBN 92-9067-145-4 (English, French,German, Italian, Spanish)

[82] Guidelines for Integrated Production of Stone Fruits. (52 pp.) Technical Guideline III. 2nd edition 2003.Edited by C. Malavolta, J.V. Cross, P. Cravedi and E. Jörg. IOBC WPRS Bull. 26 (7) 2003. ISBN 92-9067-155-4. (English, French, German, Italian, Spanish, Portuguese)

[83] Guillon, M., G. Biache, (1995). IPM strategies for control of codling moth (*Cydia pomonela* L.) (Lepidoptera: Olethreutidae) interest of CmGV for long biological control of pest complex in orchards. *Med. Fac. Landbouww. Univ. Gent.*, 60, 695-705.

[84] Gut, L.J., J.F. Brunner (1998). Pheromone-based management of codling moth (Lepidoptera: Tortricidae) in Washington apple orchards, *J. Agric. Entomol.*, 15, 387–405.

[85] Hall, D.R., B.F. Nesbitt, G.J. Marrs, A.S.J. Green, D.G. Campion, B.R. Critchley (1982). Development of microencapsulated pheromone formulations. In: *Insect pheromone technology: chemistry and applications. ACS Symp. Ser* 190, Washington DC. pp. 131-143.

[86] Higley, L. G., L. P., Pedigo and K. R. Ostlie (1986). Degday: A program for calculating degree-days, and assumptions behind the degree-day approach. *Environ. Entomol.* 15, 999-1005.

[87] Hilbert, D. W., J. A. Logan (1983). Empirical model of nymphal development for the migratory grasshopper, *Melanoplus saguinipes* (Orthoptera: Acrididae). *Environ. Entomol.* 12, 1-5.

[88] Hofer, D. and J. Brassel, (1992). 'Attract and kill' to control *Cydia pomonella* and *Pectinophora gossypiella*. *Bulletin OILB/srop* 15, 36–39.

[89] Hofer, D., P. J. Charmillot , M. Angst, (1996). Control of codling moth (*Cydia pomonella)* with SIRENE ® CM, a novel attract and kill formulation. Report of XX International Congress of Entomology, Firenze, Italy, August 25–31.

[90] Hollingsworth, C.S. (1995). Integrated Fruit Production (IFP): a status report. *Fruit Notes*. (Fall), 14–15.

[91] Howitt, A. (1993). Fruit IPM Fact Sheet, Common Tree Fruit Pests NCR 63, Michigan State University Extension.

[92] Hoyt, S. C., E. C. Burts (1974). Integrated Control of Fruit Pests. *Ann. Rev. Entomol.*, 19, 231-252.

[93] Ifoulis, A., Savopoulou-Soultani M, (2006). Developing the optimum sample size and multistage sampling plans for *Lobesia botrana* (Lepidoptera: Tortricidae) larval infestation and injury in northern Greece. *J. Econ. Entomol.* 99, 1890–1898.

[94] Ilichev, A. L., D. G. Williams, A. D. Milner, (2004). Mating disruption barriers in pome fruit for improved control of oriental fruit moth *Grapholitha molesta* Busck (Lep., Tortricidae) in stone fruit under mating disruption. *J. Appl. Ent.* 128, 126–132.

[95] Ioriatti, C., A. Dalla-Serra (1992). Five-year experience with mating disruption as a control method against codling moth and leafroller species in apple orchards of Trentino. *Bull. OILB-SROP*. 15:88-93.

[96] Ioriatti C, G.Angeli (2002). Control of codling moth by attract and kill. Use of pheromones and other semiochemicals in integrated production *IOBC wprs Bulletin*, 25, 1-9.

[97] IPM Europe (2000). For the harmonisation of European support to developing countries in the use of IPM to improve agricultural sustainability.

[98] Jaques, R., J. Hardman, J. Laing, R. Smith (1994). Orchard trials in Canada on control of *Cydia pomonella* (Lep: Tortricidae) by granulosis virus. *Entomophaga*, 39, 281–292.

[99] Jutsum, A.R., R.F.S. Gordon (1989). Pheromones: importance to insects and role in pest management. In: *Insect pheromones in plant protection*. Ed. A.R. Jutsum and R. F. S. Gordon. John Wilen and Sons Ltd. pp. 1-13.

[100] Kenawy E-R. (1998). Recent advances in controlled release of agrochemicals. *JMS-Rev Macromol Chem Phys* C38:365-390.

[101] Khan, Z.R. , J.A. Pickett (2004). The 'push-pull' strategy for stemborer management: a case study in exploiting biodiversity and chemical ecology. In: *Ecological engineering for pest management: advances in habitat manipulations for arthropods*. G. Gurr, S.D. Waratten, and M.A. Altieri (Eds.) CSIRO and CABI Publishing. p 155-164, 232.

[102] Klassen W.; R. L. Ridgway and M. Inscoe (1982). Chemical attractants in integrated pest management programs. In: *Insect suppression with controlled release pheromone systems*. Vol. I. A.F.Kydonieus and M. Berosa. CRC Press. pp. 65-71.

[103] Knight A.L., L.A. Hull, (1989). Predicting seasonal apple injury by tufted apple bud moth (Lepidoptera:Tortricidae) with early-season sex pheromone trap catches and brood fruit injury. *Environ. Entomol.*18, 939–944.

[104] Knight, A.L., J.F., Howell, L.M. McDonough, M. Weiss (1995). Mating disruption of codling moth (Lepidoptera: Tortricidae) with polyethylene tube dispensers: determining emission rates and the distribution of fruit injuries. *J. Agric. Entomol.*, 12, 85-10.

[105] Knight A.L., J.E. Turner, B. Brachula (1997). Predation on eggs of codling moth (Lepidoptera: Tortricidae) in mating disrupted and conventional orchards in Washington, *J. Entomol. Soc. Brit. Columbia*, 94, 67–74.

[106] Knight, A. L. (2003). Testing an attracticide hollow fibre formulation for control of codling moth, *Cydia pomonella* (Lepidoptera: Tortricidae). *J. Entomol. Soc.* B.C. 100, 71-78.

[107] Kidd, H., D. R. James (1991). The Agrochemicals Handbook, Third Edition. Royal Society of Chemistry Information Services, Cambridge, UK, 1991, pp. 3-11.

[108] Kim H. J., J.A. White-Phillip, Y. Ogasawara, N. Shin, E. A. Isiorho, H. W. Liu, (2010). Biosynhesis of spinosyn in *Saccharopolyspora spinosa*: Synthesis of permethylated rhamnose and Characterization of the function of SpnH, Spnl, and ApnK. *J. Am. Chem. Soc.*, 132, 2901–2903.

[109] Kristensen, N. P. (1984). Studies on the morphology and systematics of primitive Lepidoptera (Insecta). *Steenstrupia* 10, 141-191.

[110] Kogan M., (1998). Integrated pest management: historical perspectives and contemporary developments. *Ann. Rev Entomol.*,_ 243–270.

[111] Koul, O., M. B. Isman,C. M. Ketkar (1990). Properties and uses of Neem, Azadirachta indica. *Can. J. Bot.* 68: 1-11.

[112] Kovanci, O.B., C. Schal, J.F. Walgenbach, G.G. Kennedy (2005). Comparison of mating disruption with pesticides for management of Oriental fruit moth (Lepidoptera: Tortricidae) in NorthCarolina apple orchards. *J. Econ. Entomol.*, 98, 1248–1258.

[113] Kutinkova, H. J. Samietz, V. Dzhuvinor, V. Veronelli, A. Iodice. (2010). Control of oriental fruit moth *Cydia molesta* (Busck), BY Isomate OFM rosso dispensers in peach orachards in Bulgaria – preliminary results. IOBC/wprs, 54, 331-336.

[114] Lacey L. A., P. V. Vail, D. F. Hoffman (2002).Comparative activity of baculoviruses against the codling moth *Cydia pomonella* and three other tortricid pests of tree fruit. J. *Invert. Path.* 80, 64-68.

[115] Lacey, L.A., S. Arthurs, A. Knight, K. Becker and H. Headrick (2004a). Efficacy of codling moth granulovirus: effect of adjuvants on persistence of activity and comparison with other larvicides in a Pacific Northwest apple orchard, *J. Entomol. Sci.*, 39, 500–513.

[116] Lacey, L.A., S.P. Arthurs, D. Thomson, R.J. Fritts and D. Granatstein (2004b). Codling moth granulovirus and insect specific nematodes for control of codling moth in the Pacific Northwest, *Tilth Prod. Q.* 13, 10–12.

[117] Lacey L. A., D. I. Shapiro – Ilan, (2008a). Microbial Control of Insect Pests in Temperate Orchard Systems: Potential for Incorporation into IPM. *Ann. Rev. Entomol.*, 53, 121-144.

[118] Lacey L. A., D. Thomson, C. Vincent, S. P. Arhturs (2008b). Codling moth granulovirus: a comprehensive review. *Biocontrol Science and Technology*, 18, 639-663.

[119] Lactin, D. J., N. J. Holliday, D. L. Johnson, and R. Craigen. (1995). Improved rate model of temperature-dependent development by arthropods. *Environ. Entomol.* 24, 68-75.

[120] Lardeux, F., and J. Cheffort. 1997. Temperature thresholds and statistical modelling of larval *Wuchereria bancrofti* (Filariidea: Onchocercidae) developmental rates. *Parasitology*, 111, 123-134.

[121] Litsinger J.A., Bandong J. P., B. L. Canapi (2011). Effect of multiple infestations from insect pests and other stresses on irrigated rice in the Philippines: I. Damage functions. *Int. J. Pest Manag.*, 57, 93-116.

[122] Liu, T. X. P. A. Stansly. (1997). Effects of Pyriproxyfen on three Species of *Encarsia* (Hymenoptera: Aphelinidae), Endoparasitoids of *Bemisia argentifolii* (Homoptera: Aleyrodidae). *J. Econ. Entomol.*, 90, 404-411.

[123] Logan, J. A., D. J. Wollkind, S. C. Hoyt, and L. K. Tanigoshi (1976). An analytic model for description of temperature dependent rate phenomena in arthropods. *Environ. Entomol.*, 5, 1133-1140.

[124] Logan, J. A. (1988). Toward an expert system for development of pest simulation models. *Environ. Entomol.*, 17, 359-376.

[125] Losel, P.M., Penners, G., Potting, R.P.J., Ebbinghaus, D., Elbert, A., and Scherkenbeck, J. (2000). Laboratory and field experiments towards the development of an attract and kill strategy for the control of the codling moth, *Cydia pomonella. Ent. Exp. Appl.* 95, 39-46

[126] Losel, P.M., Potting, R.P.J., Ebbinghaus, D., and Scherkenbeck, J. (2002). Factors affecting the field performance of an attracticide against the codling moth *Cydia pomonella. Pest Manage. Sci.* 58:1029-1037.

[127] Lo, P. L., L. M. Cole (2007). Impact of pheromone mating disruption and pesticides on oriental fruit moth (*Grapholita molesta*) on peaches. *New Zealand Plant Prot.,* 60, 67-71.

[128] Manetsch, T.J. (1976).Time-varying distributed delays and their use in aggregative models of large systems.*IEEE Transactions on Systems, Man, and Cybernetics*, 6 , 547–553.

[129] McCann S.F., G.D. Annis, R. Shapiro, et al. (2001). The discovery of indoxacarb: Oxadiazines as a new class of pyrazoline-type insecticides. *Pest Manage. Sci.*, 57, 153–64.

[130] Mertz, F.P., Yao, R.C., (1990). *Saccharopolyspora spinosa* sp.nov. isolated from soil collected in a sugar mill rum still. *Int. J. Syst. Bacteriol.*, 40, 34-39.

[131] Milonas, P. G., and M. Savopoulou-Soultani. 2000. Development, survivorship, andreproduction of *Adoxophyes orana* (Lepidoptera: Tortricidae) at constant temperatures. *Ann. Entomol. Soc. Amer.*, 93, 96-102.

[132] Milonas, G., M. Savopoulou-Soultani, M. (2006). Seasonal abundance and population dynamics of *Adoxophyes orana* (Lepidoptera: Tortricidae) in northern Greece. *Int. J. Pest Manage.*, 52, 45-51.

[133] Minks, A. K. (1976). Experience with Pheromone Trapping and Disruption of Moths. *Pestic. Sci.*,7 , 642-646.

[134] Miyamoto, J., Hirano, M., Takimoto, Y., M. Hatakoshi (1993). Insect growth regulators forpest control with emphasis on juvenile hormone analogs: present status and future prospects. *Pest Contr. Env. Saf.*, 524, 144-168.

[135] Molinari F, Cravedi P, Rama F, Reggiori F, Dal Pane M, Boselli M (2000a). L'uso del"disorientamento" per il controllo di *Cydia molesta* inserito nelle strategie di difesa integratadelle pomacee. In: *Atti delle Giornate Fitopatologiche* 2000, Perugia, April 16-20, 2000, 1, 333-340.

[136] Molinari, F., Cravedi, P., Rama, F., Reggiori, F., Pane, M. D. and Galassi, T. (2000b). L'uso deiferomoni secondo il metodo del "Disorientamento" nella difesa del pesco da *Cydia molesta* e *Anarsia lineatella.* GF-2000 Atti Giornate Fitopatologiche, Perugia, 16-20aprile 2000, 1, 341-347.

[137] Mordue, A. J., A. Blackwell (1993). Azadirachtin: an update. *J. Insect Phys.*, 39, 903-924.

[138] Neumann, U. (1997). Successful employment of pheromones in apple: exemplary results from Europe. *IOBC wprs Bulletin* 20, 73-78.

[139] Onstad D.W., (1987). Calculation of economic-injury levels and economic thresholds for pest management. *J. Econ. Entomol.*, 80, 297–303.

[140] Pedigo, L.P., Hutchins S.H., Higley L.G., (1986). Economic injury levels in theory and practice. *Annu. Rev.Entomol.*, 31, 341–368.

[141] Polavarapu S, G. Lonergan, H. Peng, K. Neilsen (2001). Potential for mating disruption of *Sparganothis sulfureana* (Lepidoptera: Tortricidae) in cranberries. *J. Econ. Entomol.*, 94, 658-665.

[142] Piñero J., S. Dorn (2007). Synergism between aromatic compounds and green leaf volatiles derived from the host plant underlies female attraction in the oriental fruit moth. *Ent. Exp. Appl.*, 125, 185-194.

[143] Poinar, G. O., Jr. (1991). Nematode parasites, pp. 273-281. *In*: L. P. S. van der Geest and H. H. Evenhius, (eds.), Tortricid Pests, their Biology, Natural Enemies and Control. Elsevier Science Publishers, Amsterdam, The Netherlands.

[144] Price, D., F. Summers (1961). Cyclical changes of moths and larvae of the peach twig borer in California. *J. Econ. Entomol.* 54, 933–936.

[145] Retnakaran, A., I. Gelbic, M. Sundaram, et al. 2001. Mode of action of the ecdysone agonist tebufenozide (RH-5992), and an exclusion mechanism to explain resistance to it. *Pest Manage. Sci.*, 57, 951–7.

[146] Rice R.E., P. Kirsch (1990). Mating disruption of oriental fruit moth in the United States. In: Ridgway RL, Silverstein RM, Inscoe MN (eds) Behavior-modifying chemicals for insect management. Marcel Dekker, New York, pp 193–211.

[147] Riedl H., B. A. Croft (1974). A study of pheromone trap catches in relation to codling moth (Lepidoptera:Olethreutidae) damage. *Can. Ent.*, 106, 525–537.

[148] Riddiford, L.M. (1994). Cellular and molecular actions of juvenile hormone: general considerations and premetamorphic actions. *Adv. Ins. Physiol., 24,* 213-274.

[149] Ridgway R. L., R. M. Silverstein, M. N. Inscoe MN (1990). Behaviour-modifying chemicals for insect management: applications of pheromones and other attractants. Marcel Dekker, New York.

[150] Roelofs, W. L., A. Comeau, R. Selle (1969a). Sex pheromone of the oriental fruit moth. *Nature*, 224, 723.

[151] Roelofs, W.L., A. Comeau, A. (1969b). Sex pheromone specificity, taxonomic and evolutionary aspects in Lepidoptera. *Science*, 165, 398-400.

[152] Roelofs, W. L., A. Comeau, A. Hill and G. Milicevic (1971). Sex attractant of the codling moth: Characterization with electroantennogram technique. *Science*, 174, 297–299.

[153] Roelofs, W.L., J. Kochansky, R. Carde, H. Arn., S. Rauscher (1973).Sex attractant of the grape vine moth *Lobesia botrana*. Schweiz. *Entomon. Ges.*, 46, 71-73.

[154] Roelofs, W., J. Kochansky, E. Anthon, R. Rice, and R. Carde´(1975). Sex pheromone of the peach twig borer moth (*Anarsia lineatella*). *Environ. Entomol.*, 68, 358-360.

[155] Roelofs, W.L., R. L. Brown (1982). Pheromones and evolutionary relationships of Tortricidae. *Annu. Rev. Ecol. Syst.*, 13, 395-422.

[156] Roehrich, R., (1977). Essais de pheromones pour la capture des raffles de la pyrale de la vigne *(Sparganothis pilleriana* Schiff.). C.R. de la r6union sur les ph6romones sexuelles des insectes, Montfavet 25-27 oct. 1977, INRA Montfavet, 18-19.

[157] Rot, M., M. Blazi (2005). Zatiranje breskovega zavijaca (*Cydia molesta* L.) z metodozbeganja. Lectures and Papers-Presented at the 7th Slovenian Conference on Plant Protection, Zrece, Slovenia, 8-10-March 2005, 175-181.

[158] Salgado, V. L. (1998). Studies on the mode of action of Spinosad: Insect symptoms and Physiological correlates. *Pect. Biochem. Physiol.,* 60, 91-102.

[159] Samietz J., B. Graf, H. Höhn, L. Schaub, H. U. Höpli (2007). Phenology modelling of major insect pests in fruit orchards from biological basics to decision support: the forecasting tool SOPRA, *Bulletin OEPP/EPPO Bulletin*, 37, 255–260.

[160] Savopoulou-Soultani, M., A. Hatzivassiliadis, H. J. Vlug, A. K. Minks, and M. E. Tzanakakis. (1985). First records of the summer fruit tortricid, *Adoxophyes orana* F.v.R. in Greece. *Entomol. Hellen.*, 3, 65-66.

[161] Schuster, D. J., P. A. Stansly (2000). Response of Two Lacewing Species to Biorational and Broad-spectrum Insecticides. *Phytoparasitica*, 28, 297-304.

[162] Schuster, D. J. and P. A. Stanley (2007). Biorational Insecticides for integrated pest management in tomatos. *Biol. Control*, 41, 99-109.

[163] Schmutterer, H. (1990). Properties and potential of natural pesticides from the Neem tree, Azadirachta indica. *Ann. Rev. Entomol.* 35:271-297.

[164] Shorey H. H, Gerber R.G. (1996). Use of puffers for disruption of sex pheromone communication of codling moths (Lepidoptera: Tortricidae) in walnut orchards. *Environ .Entomol.* 25, 1398-1400.

[165] Silverstein, R. M. (1990). Practical use of pheromones and other behavior-modifying compounds: overview. In Behavior-modifying chemicals for insect management. R. Ridgway R. Silverstein and M. Inscoe Ed. New York and Basel.

[166] Staal, G.B. (1972). Biological activity and bioassay of juvenile hormone analogs. In: *Insect Juvenile Hormones: Chemistry and Action.* Ed. Menn, J.J., Beroza, M. Academic Press. New York.

[167] Stansly, P. A., T.-X. Liu, D. J. Schuster and D. E. Dean (1996). Role of Biorational Insecticides in Management of *Bemisia*, pp. 605-615, In: *Bemisia* 1995: *Taxonomy, Biology, Damage, Control and Management.* D. Gerling and R. T. Mayer [eds.] Intercept Ltd., Andover, Hants, United Kingdom.

[168] Stansly, P. A. and T. X. Liu. (1997). Selectivity of Insecticides to *Encarsia pergandiella* (Hymenoptera: Aphelinidae), an Endoparasitoid of *Bemisia argentifolii* (Hemiptera: Aleyrodidae). *Bull. Entomol. Res.*, 87, 525-531.

[169] Stelinski L. L. · J. R. Miller · R. Ledebuhr ·P. Siegert · L. J. (2007). Gut Season-long mating disruption of *Grapholita molesta* (Lepidoptera: Tortricidae) by one machine application of pheromone in wax drops (SPLAT-OFM). *J. Pest Sci.*, 80, 109–117.

[170] Stern V.M., R.F.Smith, R. van den Bosch, K. S. Hagen (1959). The integration of chemical and biological control of the spotted alfalfa aphid. *The integrated control concept*. Hilgardia, 29, 81–101.

[171] Stetter, J. and Folker L., (2000). Innovation in crop protection: trends in research (review). *Angew Chem Int.* Ed. 39, 1724-1744.

[172] Stinner, R. E., A. P. Guitierez, and G. D. Butler, Jr. (1974). An algorithm for temperature-dependent growth rate simulation. *Can. Entomol.*, 106, 519-524.

[173] Tabashnik, B.E., J.M. Schwartz, N. Finson, M.W. Johnson (1992). Inheritance of resistance to *Bacillus thuringiensis* in diamondback moth (Lepidoptera: Plutellidae). *J. Econ. Entomol.*, 85, 1046–1055.

[174] Tamaki et al., (1971). Tamaki, H. Noguchi, T. Yushima and C. Hirano, Two sex pheromones of the smaller tea tortrix: isolation, identification, and synthesis, *Appl. Ent. Zool.,* 6, 139–141.

[175] Thede, B.M., (1995). Summary of Environmental Fate Studies of Fenoxycarb. Ciba-Geigy. Corporation. Data Package Report No. 163859-N. DPR No. 51317-077.

[176] Trematerra P., A.Schiaretta, E. Tamasi (1999). Sul metodo attratticida impiegato nel controllo di *Cydia pomonella* L. *Informatore Fitopatologico*, 5, 41-44.

[177] Trematerra, P., Sciarretta, A., P. Gentile (2000). Trials on combined mating disruption in *Anarsia lineatella* (Zeller) and *Cydia molesta* (Busck) using CheckMateReg. SF dispenser. Atti Giornate Fitopatologiche, Perugia.

[178] Trimble, R.M. (1995). Mating disruption for controlling the codling moth, *Cydia pomonella* (L.) (Lepidoptera: Tortricidae), in organic apple production in southwestern Ontario. *Can. Entomol.* 127, 493-505.

[179] Tumlinson, J. H., C.E. Yonce, R. E. Doolittle, R.R. Heath, C. R. Genry, E. R. Mitchell (1974). Sex pheromones and reproductive isolation of the lesser peach tree borer and the peach tree borer. *Science*, 185, 614-616.

[180] US-EPA, (2008). United States Environmental Protection Agency (US-EPA): Integrated Pest Management (IPM) Principles. http://www.epa.gov/pesticides/factsheets/ipm.htm.

[181] Uvarov, B, P. (1931). Insects and Climate. *Trans. Entomol. Soc. Lond.*, 79, 1-247.

[182] Vantieghem, H., G. Sterk, U. Neumann (1991). Mating disruption technique against leafrollers (Tortricidae) in apple and pear orchards in Belgium. *Meded. Fac. Landbouww.* Rijksuniv, 56, 295-304.

[183] Vickers, R., W.G. Thwaite, D.G. Williams, A. H. Nicholas (1993). Control of codling moth in small plots by mating disruption: alone and with limited insecticide. *Entomol. Exp. Appl.* 86, 229-239.

[184] Waldner, W. (1997). Three years of large-scale control of codling moth by mating disruption in the South Tyrol, Italy. IOBC wprs Bulletin 20, 35-44.

[185] Waldstein, D. E., W. H. Reissig (2001). Apple damage, pest phenology, and factors influencing the efficacy of tebufenozide for control of obliquebanded leafroller (Lepidoptera: Tortricidae). *J. Econ. Entomol.*, 94, 673–679.

[186] Walker K.R., S.C. Welter (2001). Potential for outbreaks of leafrollers (Lepidoptera: Tortricidae) in California apple orchards using mating disruption for codling moth suppression. *J. Econ. Entomol.*, 94, 373-380.

[187] Wall C., A.R. Greenway, P.E. Burt (1976). Electroantennographic and field responses of the pea moth, *Cydia nigricana*, to sex attractants and related compounds. *Physiol. Ent.*, 1, 151–157.

[188] Wauchope, R. D., T. M. Buttler, A.G. Hornsby, P. W. M Augustijn-Beckers, J.P. Burt (1992). SCS/ARS/CES pesticides properties database for environmental decision making. *Rev. Environ. Contam. Toxicol.*, 123, 1-157.

[189] Weatherston, I., (1990). Principles of design of controlled-release formulations, pp.93-112 in Ridgway RL, Silverstein RM, Inscoe MN (eds.) *Behavior-modifying chemicals for insect management: applications of pheromones and other attractants.* Marcel Dekker, New York.

[190] Wilhelmus, M. W. (1991). *Steinernema* (*Neoaplenctana*) and *Heterorhabditis* Species. *Manual of Agricultural Nematology*. 1991. W.R. Nickle [ed.] Marcel Dekker, Inc. New York, NY.

[191] Witzgall P., M. Bengtsson, A. El-Sayed , A-C Bäckman, S. Rauscher, A-K Borg-Karlson, C.R. Unelius, J. Löfqvist (1999). Chemical communication in codling moth: towards environmentally safe control methods. *IOBC wprs Bulletin*, 22, 57-65.

[192] Witzgall, P., L. Stelinski, L. Gut, D. Thomson (2001). Codling Moth Management and Chemical Ecology, *Ann. Rev. Entomol.*, 53, 503-522.

[193] Witzgall, P. (2001). Pheromones – future techniques for insect control? *IOBC wprs Bulletin,* 24, 114-122.

[194] Witzgall, P., T. Lindblom, M. Bengtsson, M. Tóth (2004). *The Pherolist.* www-pherolist.slu.se.

[195] Wood., A. C. Compendium of Pecticides, Common Names http://www.alanwood.net/pecticides/index.

[196] Yamamoto, A., K. Ogawa (1989). Chemistry and commercial production of pheromones and other behaviour-modifying chemicals, in: *Insect pheromones in plant protection*, Jutsum AR, Gordon RFS (eds.), pp. 123-148. John Wiley, London.

[197] Zaied K.A., Z.M. El-Diasty, K.S. Kash , M.E. Ismail (2003). Sensitivity of larvae recombinant products of *Bacillus thuringiensis* as a Bioinsecticide agent. *P. J. Bio. Sci.* 6, 273-313.

In: Moths: Types, Ecological Significance and Control... ISBN: 978-1-61470-626-7
Editor: Luis Cauterruccio, pp. 115-143

Chapter 4

Moth Sex-Pheromone Production: Biosynthetic Pathways, Regulatory Physiology, Inhibitory Processes and Disruption

*Ada Rafaeli**
Department of Food Quality and Safety, Volcani Center, Agricultural Research Organization, Bet Dagan 50250, Israel

Abstract

Pheromones are chemicals emitted to send messages to individuals of the same species. Much of the research on chemical communication systems in insects has focused on moths in the order Lepidoptera, which is the second largest insect order with well over a hundred thousand described species. Most of the hundreds of species studied have been found to use a long-distance chemical communication system for attracting mates and the most widely explored are the sex-pheromones of female moths. The discovery of pheromones and their binding proteins have impacted Lepidopteran biology, and neural encoding, processing and integration of olfactory signals from mates are areas in which Lepidoptera continue to serve as important models. Moreover, the exploitation and the use of molecular techniques in the post genomics era have led to many advances in several aspects of moth pheromone research *viz*. the elucidation of biosynthetic pathways; the identification of key enzymes therein; the regulatory physiology of pheromone biosynthesis; the role of G-protein coupled receptors in the initiation and inhibition of these pathways and the role that pheromones play in the speciation process. Reproductive behavior in moths relies on the synchronization of various environmental and physiological events that influence the timing of sexual activities between the males and females. Receptivity in most female moths is broadcasted by the release of a unique blend of fatty acid-derived volatile sex-pheromones when they extrude their pheromone glands thereby assuming typical calling behavior. This behavior occurs only at specific times of the photoperiod, typically during the night (scotophase) and only then is sex-pheromone biosynthesis initiated. The insect's neuroendocrine system is a major

*Tel: 972-3-9683729; Fax: 972-3-9604428; E-mail: vtada@volcani.agri.gov.il

regulator of many physiological functions including mating-behavior. Environmental and internal signals such as age, photoperiod, temperature, mating history and host plant volatiles signal the neuroendocrine system to induce downstream events affecting sex-pheromone production. The release of a pheromone-biosynthesis activating neuropeptide (PBAN) into the hemolymph up-regulates the biosynthesis of fatty acid derived compounds. Once the female sex-pheromones are emitted, the males perceive and orient towards the source of the volatile. On reaching the females, males of several species display their hair-pencil complexes. The hair-pencil complexes also contain pheromonal compounds that are related structurally to the female sex-pheromones. These compounds play an essential role for successful mating. Mating affects subsequent female reproductive behavior and the production and release of sex-pheromone is suppressed. This phenomenon has been attributed to the transfer of seminal peptides to the female during copulation. In this review I will provide background on the biosynthesis of sex-pheromones and will focus on elements of regulation with particular focus on the characterization and mode of action of the Pheromone-Biosynthesis Activating Neuropeptide (PBAN) and its receptor. I will delve on inhibitory processes such as the outcome of female receptivity after mating and the action of seminal peptides such as Sex Peptide on its receptor. With our increasing understanding of the regulatory physiology of reproductive behavior we encounter several avenues that could be utilized for the disruption of female receptivity and prevention of subsequent successful mating events, avenues that I will explore as possibilities in the future development of new mating control or disruption strategies for field application against moth pests.

1. Introduction: Moth Reproductive Behavior

Insects represent a group with a phenomenal reproductive success. Understanding the behavioral adaptations of mating and post-mating responses would contribute considerably to our knowledge concerning the evolutionary significance of these processes. One of the most common mate finding and recognition strategy is based on volatile chemical cues. Much of the research on chemical communication systems in insects has focused on moths belonging to the order Lepidoptera, which represents the second largest insect order with well over a hundred thousand described species, many of which are serious agricultural crop pests. Most of the hundreds of species studied have been found to use a long-distance chemical communication system for attracting mates, a predictable feature given the cryptic nature and the nocturnal activity of this group of Lepidoptera.

Receptivity in many female moths is broadcasted by the release of a unique, species-specific blend of fatty acid-derived volatile sex-pheromones (section 2). The females assume typical calling behaviors, which involves the extrusion of their ovipositor tip thereby exposing the sex-pheromone glands found in the inter-segmental membrane between the ultimate and penultimate segments of the abdomen (Rafaeli and Gileadi, 1995). This behavior is synchronized to occur at specific times of the photoperiod (section 4), typically during specific hours of the night, the scotophase (Roelofs and Carde, 1974). Synchrony is particularly prominent in polyandrous species that mate more than once during their lifespan and where sex-pheromone production and release follow a diel-periodicity with peaks arising during specific time intervals (Raina et al., 1986).

Once the sex-pheromones are released the males, whose sensitivity is also synchronized according to the diel-periodicity, perceive and orient towards the source of the volatile. Males perceive this species-specific sex-pheromone blend through the activation of G-protein coupled olfactory receptors (OR) on the dendritic membranes of specialist olfactory receptor cells (chemosensilla) in the antennae, each tuned to one compound. In general, each insect olfactory sensory neuron expresses one member of the olfactory receptor (OR) gene family along with a highly conserved co-receptor, Or83b and the two ORs form a heteromeric complex to function as a chemosensor (see review: Sato et al., 2009). A specific blend is represented by a pattern of excitations across the different types of specialist cells, processed centrally in the macroglomerular complex of the insect's brain (for an excellent recent review on olfactory processing the reader is encouraged to consult Galizia and Rossler, 2010).

Soluble proteins, belonging to two major classes, odorant-binding proteins (OBPs) and chemosensory proteins (CSPs), are highly concentrated in the lymph of chemosensilla of the antennae and are believed to play a role by modulating signal perception. In some species, it has been reported that OBPs or CSPs, depending on the species, are also expressed in pheromone producing glands and may be involved in the delivery of these semiochemicals. A great number of OBPs and CSPs have been characterized in several insect species and their sequence motifs (Scaloni et al., 1999; Leal et al., 1999) have been used in genome-wide identification and annotation (Zhou et al 2006; Vieira et al., 2007; Vogt et al., 2002).

The great number of genes encoding OBPs and CSPs in the genome of several insects and their expression also in non-sensory organs certainly suggests alternative functions for these soluble proteins. Known examples include CSPs expressed in the female pheromone glands of the moths *Mamestra brassicae* (Jacquin-Joly et al. 2001), *Agrotis segetum* (Strandh et al., 2008) and *B. mori* (Dani et al., 2011). Interestingly, in female *Heliothis virescens* moths, mRNA for both the receptor protein (HR13) and the related PBP (PBP2) that both mediate the male antennal response to the major female pheromone component (Z)-11-hexadecenal (Z11-16:Al), were also found in the ovipositor tip (Widmayer et al., 2009). In an immuno-histochemical analysis it was possible to visualize the receptor protein in distinct sensilla types surrounding the ovipositor tip thus implying a chemosensory responsiveness in these sensilla types (Widmayer et al., 2009).

Perception of sex-pheromones by the males leads to typical upwind anemotaxis where the flight path follows a zigzag with counter-turning in accordance to the flight tracks of filaments of sex-pheromone and pockets of clean air (see review: Murlis et al., 1992). On reaching the females and before mating, males of several moth species may display their hair-pencil complexes that also contain fatty acid derived "pheromonal" compounds, and, in some cases these compounds are structurally similar to female sex-pheromones (see section 3). These male-produced pheromones are hypothesized to have many functions but have been shown to be essential for successful copulation (see review: Birch et al., 1990).

After mating, females of many insects undergo drastic behavioral changes that are attributed to the transfer of male-derived seminal peptides. Several seminal peptides have been identified to various degrees in diverse insect species. They alter the female reproductive behavior by affecting receptivity and thereby inhibiting multiple matings for a period of time. They also increase the rate of egg development, ovulation and oviposition of fertilized eggs and play a vital role in the formation of a mating plug, in sperm storage and in sperm competition. In addition, the antimicrobial peptides in these secretions maintain a favorable environment in the female's genital tract for sperm transfer. The role of male-derived seminal peptides in modulating female reproductive behavior has been extensively reviewed (Avila et al. 2011; Gillot, 2003). In moths they also suppress (transiently or permanently) the production and release of sex-pheromones (see review: Rafaeli, 2009; section 5).

2. Female Sex-Pheromone Production

Most moth sex pheromone components are biosynthesized via the synthesis of saturated fatty acids, followed by specific metabolism of these acids, including desaturation, chain-shortening, reduction, acetylation or oxidation (Tillman et al., 1999). These linear chain components with 10-18 carbons in length belong to a Type I class of sex-pheromones that are biosynthesized *de novo* from the fatty acid pathway. The fatty acid biosynthetic pathway depends on the activity of acetyl-CoA-carboxylase (ACCase) which carboxylates acetyl-CoA to produce malonyl-CoA. The biosynthesis is followed by the fatty acid synthase (FAS) enzymes which extend the fatty acid chain using units of malonyl-CoA to generate palmitic and stearic acids. Acyl-CoA desaturases and limited chain-shortening by β-oxidation enzymes are key enzymes in the biosynthesis of the specific sex-pheromone compounds (see exemplary biosynthetic pathways depicted in Figure 3. below; see review Jurenka, 2004). These enzyme families have evolved novel functional properties in moth species leading to complex combinations of unique fatty acyl-moieties of variable chain lengths, regio- and stereo-specificities that account for the great diversity of sex-pheromone structures (Knipple and Roelofs, 2003). Another group of compounds that is utilized as sex-pheromones in moths, belonging to the Type II class, are long straight chains of 17-23 polyunsaturated hydrocarbons with epoxy derivatives lacking a terminal functional group. These are biosynthesized from linoleic and linolenic acids producing dienes and trienes (Roelofs and Bjostad, 1984).

3. Male Moth Pheromones

Male pheromone production has been studied in several insect species. Male insects often possess scent-releasing organs in the form of hair-pencils, coremata or androconial scales (Birch et al., 1990). In the Lepidoptera, studies have identified hair-pencil secretions produced by several species (Phelan et al., 1986; Teal and Tumlinson, 1989; Heath et al., 1992; Thibout et al., 1994; Huang et al., 1996). The behavioral role of these secretions is not well understood, but most often these odors have been deemed important in courtship behavior. In general male pheromones can be considered to have many functions: they can act as aphrodisiacs to stimulate female receptivity during courtship (Birch, 1974); they have been reported to induce female calling (Szentesi et al., 1975), females become immobile allowing copulation; and they have been reported to function as male-to-male inhibitory compounds, thereby minimizing male-competition (Hirai et al., 1978; Teal et al., 1984; Wu et al., 1991; Teal and Tumlinson, 1989; Huang et al., 1996). Odors released by male hair-pencils are important in male acceptance by the female and may play a role in mate choice and species isolation (Hillier and Vickers, 2004). Close-range chemical cues have also been proposed as a trait used by females to assess male quality (Eisner and Meinwald, 1995). In species of Lepidoptera belonging to the Arciidae and Danaidae, scent gland composition is related to host plant consumption and sequestration of compounds during the larval and adult stages (Birch and Hefetz, 1987). However, in other moth species, as in the Heliothine moths, these male odors are derived from the fatty acid biosynthetic pathway, and as such are similar to the female sex-pheromone blends (Teal and Tumlinson, 1989; Huang et al., 1996). The presence of sensory receptor proteins, PBPs and CSPs in the vicinity of the female pheromone gland, as discussed above, support the possibility of their functional role as possible mediators in the activity of these male-produced compounds in the female. Indeed, studies of a CSP in *M. brassicae* revealed binding to pheromone components and long-chain fatty acids (Jacquin-Joly et al. 2001) but such speculations await confirmatory affinity binding studies to the specific male-produced compounds.

4. REGULATORY PHYSIOLOGY

4.1 PBAN/PK Neuropeptide Family

Reproductive behavior in male and female moths is dependent on the synchronization of various environmental and physiological events (Figure 1). The insect's neuroendocrine system is the major regulator of many physiological functions which include behavior and, relevant to the present review, mating-behavior. Environmental signals such as photoperiod, temperature and host plant volatiles are perceived by the central nervous system (CNS) of the female to induce release of a neuropeptide, Pheromone Biosynthesis Activating Neuropeptide (PBAN) into the hemolymph (Figure 1).

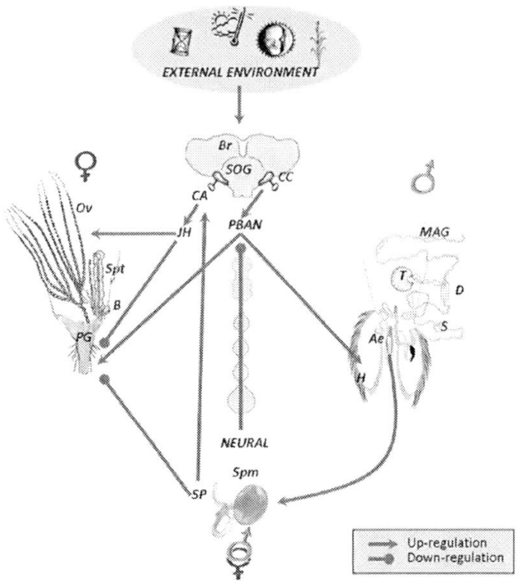

Figure 1. Schematic diagram showing factors influencing reproductive behavior in *Helicoverpa armigera*. (Male system adapted from Bober and Rafaeli, 2010). *Ae*=aedaegus; *B*=bursa copulatrix; *Br*=Brain; *CA*=corpora cardiaca; *D*=duplex; *H*=hair-pencils; *JH*=Juvenile hormone; *MAG*=male accessory glands; *Ov*=ovaries; *PBAN*= Pheromone biosynthesis-activating neuropeptide; *PG*=pheromone glands; *S*=simplex; *SOG*= suboesophageal gland; *SP*=Sex-peptide; *Spm*=spermatophore; *Spt*=spermatheca.

PBAN is a 33 amino-acid neuropeptide containing a conserved *C*-terminal, FXPRLa, which is the minimum sequence required to induce activity. To date, the sequences of 19 different moth species have been identified with a relatively high amino-acid sequence homology between them (Jurenka and Nusawardani, 2011). PBAN was first isolated (Raina and Klun, 1984) through the development of an *in vivo* bioassay where 1 day-old females are decapitated, kept for 24 hours so that endogenous PBAN is depleted and then injected with PBAN to test for induction of pheromonotropic activity, i.e. female sex-pheromone production. Utilizing an *in vitro* bioassay of isolated pheromone glands we were able to delineate the PBAN pheromonotropic response to the inter-segmental membrane between the ultimate and penultimate segments of the female ovipositor tip (Soroker and Rafaeli 1989, Rafaeli and Gileadi, 1995). Thus, we hypothesized that this tissue should contain a receptor for PBAN (see section 4.3).

Females injected with PBAN during the photophase, when endogenous PBAN levels are low also showed a transient stimulation of pheromone levels however, in *Helicoverpa armigera*, these intact females responded by suppression of the pheromone after 2 hours, probably due to an endogenous negative (possibly neural) feedback during the photophase (Rafaeli and Gileadi, 1995; 1996; Rafaeli et al., 1997). Indeed the neurohormone/neurotransmitter octopamine significantly inhibits PBAN stimulation (Rafaeli and Gileadi, 1995; 1996; Rafaeli et al., 1997). The release of PBAN from the neurohemal glands, the corpora cardiaca, into the hemolymph has been shown to be regulated by photoperiod in a number of moth species by its presence in the hemolymph only during the scotophase (Marco et al., 1996; Iglesias et al., 1999; 2002; Nagalakshmi et al., 2007). Moreover, ventral nerve cord (VNC) transaction during the scotophase results in a significant decrease in PBAN detection in the hemolymph (Iglesias et al., 1998) suggesting the involvement of neural inputs from the VNC controlling the release of PBAN (Rafaeli, 2009) (Figure 1).

4.2. PBAN Gene and Expression Profiles

PBAN is localized in the CNS with a predominant expression in 3 clusters of neurosecretory cells of the suboesophageal ganglion, although expression also occurs throughout the VNC including thoracic, abdominal and terminal abdominal ganglia (Kingan et al., 1992; Rafaeli et al., 1993; Ichikawa et al., 1995; Duportets et al., 1998; Ma et al., 1998). In addition, using immunoassay, the protein was found both in female and in male CNS throughout the photoperiod (Hirsch, J, 1991; Rafaeli et al., 1993). This is also true for non-Lepidopteran insects, for example, the presence of immuno-reactive PBAN-like peptides were detected in the CNS of larvae and adults of *Drosophila* (Choi et al., 2001). The gene encoding for PBAN also encodes for 4 other neuropeptides, the PBAN-gene-neuropeptides (PGNs) all bearing the FXPRLa *C*-terminal, *viz.* PGN 7, PGN 8, PGN 18 and PGN 24 according to the number of amino acids for each peptide (Davis et al., 1992; Ma et al., 1994). PGN 18 bears close similarity to a larval PBAN-like peptide, Pss-PT, isolated from *Pseudaletia separate* larval heads that has been shown to have melanizing effects on the cuticle (Matsumoto et al., 1992). PGN 24 bears resemblance to *B. mori* diapause hormone (Imai et al., 1991).

Thus, PBAN belongs to a family of neuropeptides and the peptide ligands in this family have been identified in a wide variety of species across all major orders of insects. They are all characterized by having an FXPRLa *C*-terminal that is critical for receptor activation (X = S, T, or G). Some of these peptides are pheromonotropic and others have various other functions in various species. The pyrokinins (PKs) were first identified in insects as stimulating gut muscle contraction (Nachman et al., 1986). In *Drosophila* the *capability* gene encodes 3 neuropeptides: Capa-1, 2 and 3, where Capa-3 is a member of the PBAN/pyrokinin (PK) family, referred to as Cap2b-3 (Kean et al., 2002). The PKs and Cap2b-3 peptides have a WFGPRLa *C*-terminal, as does the diapause hormone (DH) of moths. Additionally, two neuropeptides are encoded by the *hugin* gene of *Drosophila*: pyrokinin-2 (PK-2) and ecdysis triggering hormone (ETH) containing a PRLa *C*-terminal (Meng et al., 2002). The functions, other than pheromone biosynthesis, that have been attributed to the PBAN/PK peptide family include melanization in Lepidoptera larvae (Matsumoto et al., 1990); embryonic diapause in *B. mori* (Imai et al., 1991); acceleration of puparium formation in flies (Zdarek et al., 1997); and pupal diapause development in Heliothine moths (Sun et al., 2003). Pyrokinins have also been found in Crustacea (Torfs et al., 2001) and immuno-histochemical evidence indicates the presence of PKs in Arachnida (Neupert et al., 2001). A recent study using annotated gene resources outlines several genes encoding predicted peptides in this family (Jurenka and Nusawardani, 2011).

A PBAN-like peptide in the vertebrates is neuromedin U which bears the PRXa *C*-terminal and is a peptide that is involved in a variety of physiological functions including feeding behaviors. It is found in the CNS as well as peripherally in a variety of tissues including the digestive tract (Howard et al., 2000). Neuromedin U turned out to be pheromonotropic in our bioassays and will induce pheromone production in moths, indicating cross-reactivity of the FXPRXa peptides (Choi et al., 2003). However, when comparing the peptide structure of all these ligands across Insecta with those of vertebrates little similarity exists except in the *C*-terminal. Thus, PBAN-like peptides, found in other organisms, have different functions. This emphasizes the ubiquity and multifunctional nature of the PRXa peptides. Additionally, although there is cross-reactivity due to the common *C*-terminal, there are differences in efficacy (Stern et al., 2007). Clearly, the full length of the native ligand is required for maximal efficacy indicating the importance of a specific PBAN-receptor.

4.3. PBAN-Receptors and Receptor-Ligand Interactions

Having discovered that PBAN-like candidate ligands are present in *Drosophila* and that the vertebrate ligand, neuromedin U is pheromonotropic we mined the *Drosophila* genome for a neuromedin U-like receptor on the assumption that receptors and their ligands co-evolve. Three G-protein coupled receptors (GPCRs) in the *Drosophila* genome are homologous to the vertebrate neuromedin U-receptor (CG8784, CG8795 and CG9918). Using a PCR based cloning strategy we succeeded in identifying and characterizing the first PBAN-R (Choi et al., 2003). To date, receptors and ligands have been found in 12 *Drosophila* species, 8 moth species (Figure 2), 3 mosquito species, *Tribolium castaneum*, *Apis mellifera, Pediculus humanus, Nasonia vitripennis, Acyrthosiphon pisum, Dermacentor variabilis*, and *Ixodes scapularis*. The PBAN receptors in moths are more related to the PK-2 receptors in insects and more similar to fly PK-2 receptors than to the moth DH receptors (Bober and Rafaeli, 2009). This list of PBAN/PK receptors and ligands will invariably expand as more genome projects become available for annotation. In contrast to their ligands, the PBAN receptors exhibit considerable sequence similarity indicating a conserved primary and secondary structure. Phylogenetic analyses have been variously reported and have shown similar relationships (Figure 2; Bober and Rafaeli, 2009; Lee et al., 2010; Jurenka and Nusawardani, 2011).

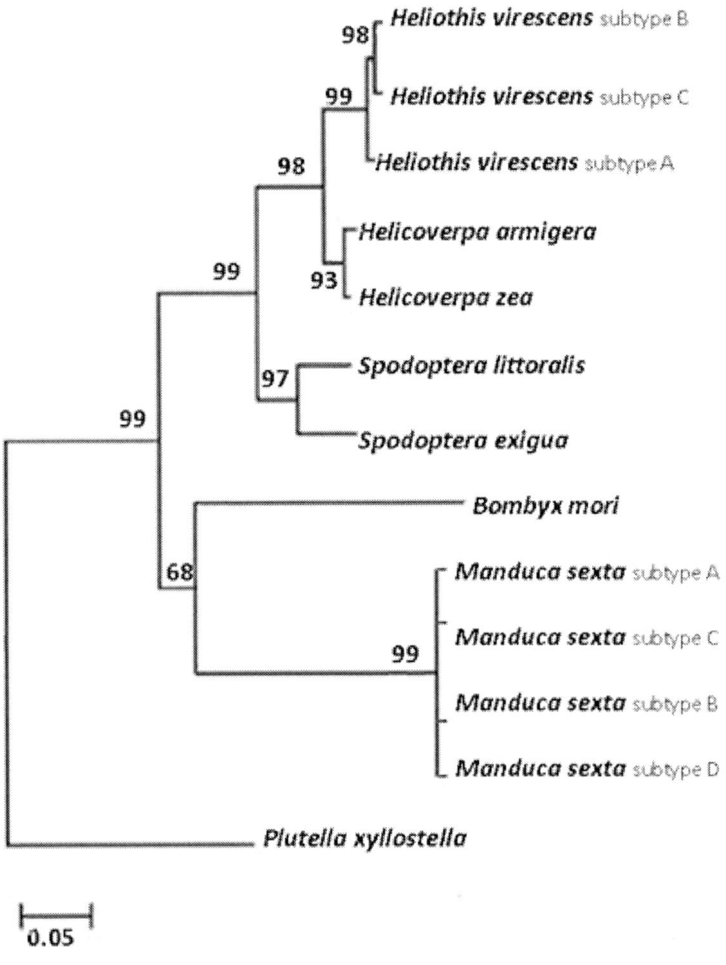

Figure 2. Phylogenetic tree based on deduced amino acids sequences of PBAN-Rs in 8 moth species. The tree was made with the neighbor-joining method using multiple alignments of amino acid sequences. The numbers above branches are bootstrap values (1000 repetitions). The bar indicates number of estimated amino acid changes per 100 amino acids. The GenBank Accession Nos. include: *Heliothis virescens* PBAN-Rs (subtype A, EU000525; subtype B EU000526, subtype C, EU000527); *Helicoverpa zea* PBAN-R (AY319852); *H. armigera* PBAN-R (AY792036); *Spodoptera littolalis* PBAN-R (HC321177); *S. exigua* PBAN-R (EU365878); *Bombyx mori* PBAN-R (AB181298); *Manduca sexta* PBAN-R (subtype A FJ240221; subtype B, FJ240222; subtype C, FJ240223; subtype D, FJ240224); *Plutella xylostella* PBAN-R (AY974334).

In *H. virescens* screening of a larval CNS cDNA library revealed three putative receptor subtypes and nucleotide sequence comparisons suggest that they are produced through alternative splicing at the 3' end (Kim et al., 2008). Reverse Transcriptase (RT)-PCR preferentially amplified PBAN-R subtype C from female pheromone glands, containing a longer 3' end *C* terminal tail. The receptor in pheromone glands of *B. mori* also contains a longer 3' end, similar to *H. virescens* subtype-B whereas in *H. zea* and *H. armigera* the shorter 3' end tail is found in the pheromone glands thereby bearing closer similarity to the *H. virescens* subtype-A (Figure 2). It has been suggested that the longer 3' terminal is involved in internalization of the receptor (Hull et al., 2005).

It is interesting that these different receptor subtypes also induce different downstream intracellular signal cascades. In *H. armigera*, both the cyclic-AMP and calcium pathways are activated on receptor-ligand interaction (Rafaeli and Soroker, 1989; Soroker and Rafaeli, 1995; Rafaeli and Gileadi 1996; Jurenka et al., 1996). In *H. armigera* sodium fluoride, a G-protein activator can induce intracellular cyclic-AMP and subsequent downstream events leading to pheromone production, independent of the ligand PBAN (Rafaeli and Gileadi, 1996). Additionally, the calcium-calmodulin inhibitor, W7, significantly inhibits these downstream events indicating the importance of calcium ligand-activated channels (Rafaeli and Gileadi, 1996). On the other hand, in *B. mori*, only calcium acts as the intracellular signal where W7 inhibits pheromone production, cyclic-AMP seems not to be involved (Hull et al., 2007a; 2007b; 2009; 2010). In addition, through pharmacological, biochemical and molecular methods it was demonstrated that PBAN signaling in *B. mori* proceeds through a Gq-mediated PLC activation (Hull et al., 2010) (Figure 3).

Although it may be apparent that there are two types of PBAN-Rs: One with calcium signaling cascade including cyclic-AMP and the other dependent only on calcium having a longer 3' end *C* terminal tail, allocating these characteristics may be too premature until more evidence becomes available as to the involvement of cyclic-AMP in the signal transduction of other PBAN-Rs. For example, PBAN induces calcium elevations by the *P. xyllostella* PBAN-R, structurally much closer to *H. zea* than to the *B. mori* PBAN-Rs (Figure 2) however cyclic-AMP levels were not analyzed in the *P. xyllostella* study (Lee et al., 2010).

With our knowledge concerning the calcium dependent signal cascade and protein binding of the PBAN-R, we were able to incorporate functional expression studies using a calcium indicator in Sf9 insect cell lines. Thus we could demonstrate that PBAN is the "preferred" ligand for the PBAN-R, and the *Drosophila* PK-2 receptor rather than the *Drosophila* PK-1 receptor (PK1-R), in which Cap2b-3 (PK-1) is the "preferred" ligand (Choi et al., 2003). In addition, the expressed *H. zea* PBAN-R can be activated by neuromedin U but the *D. melanogaster* PK1-R cannot (Choi et al., 2007) indicating discrimination between the two receptors.

The superfamily of G-protein coupled receptors (GPCRs) is one of the largest families of proteins found in all animals and is involved in signal transduction in most cell types, characterized as having seven transmembrane domains with an extracellular and intracellular domain. A conserved disulfide bridge is found between the 1st and 2nd extracellular loops. To learn more about the PBAN ligand-receptor interactions we expressed chimera receptors in Sf9 cells (Choi et al., 2007). This was done by exchanging extracellular domains (3 loops) between the *Hez*PBAN-R and the *Drm*PK1-R. Chimera II, having the second extracellular loop exchanged, lost all activity although its expression was unaffected thus showing that loss of activity was not due to poor expression. The other two chimera receptors, with first and third loops interchanged, exhibited reduced activity. By exchanging Loop 3, selectivity was conferred toward the specific peptide ligands either PBAN-like or PK/DH-like (Choi et al., 2007; Choi and Jurenka, 2010).

The recent crystal structures of rhodopsin and several other GPCRs have been reported, which greatly aides in the functional characterization of these membrane bound proteins. We used these structures to model a putative structure for the PBAN-R using molecular dynamics (Stern et al., 2007). The seven transmembrane helices of the *H. zea* PBAN-R were identified, built, packed and oriented correctly after multiple sequence alignment with several other GPCRs using the X-ray structure of rhodopsin as a template. Molecular dynamics simulations were run on three different β-turn types of the *C*-terminal hexapeptide of PBAN and the results were clustered into 12 structurally distinct groups. The lowest energy conformation from each group was used for computer simulated docking with the model of the *H. zea* PBAN-R. Highest scoring complexes were examined and putative binding sites were identified and compared to the chimera receptors with swapped extracellular domains between the *H. zea* PBAN-R and the *D. melanogaster* PK1-R (CG9918). Here too, with chimera II the lowest scores were obtained, confirming the biochemical data.

4.4. PBAN Mode of Action

What enzyme in the biosynthetic pathway is affected by the signal cascade brought about through PBAN binding to its receptor? Demonstration of the enzymatic key regulatory step in the biosynthesis of the sex-pheromones primarily relies on the tracking of labeled precursors and intermediates into pheromone molecules in the absence and presence of PBAN. Thus, if production of a labeled pheromone component from incorporation of labeled precursor occurs in the absence of PBAN to the same extent as in its presence the labeled precursor must be acting downstream of the regulatory enzyme and therefore regulation must occur upstream. The specific step in the biosynthetic pathway of Lepidopteran sex-pheromones that is influenced by PBAN has only been investigated in a limited number of species and these studies resulted in contrasting conclusions. It appears that several different enzymes can be affected depending on the species of moth but studies of this nature have so far indicated that PBAN does not influence desaturase activity (Martinez et al., 1990; Fabrias et al., 1994; 1995; Osawa et al., 1993; Fang et al., 1995).

We used both stable isotopes and specific enzyme inhibitors to determine the rate limiting step of PBAN regulation of pheromone biosynthesis in *H. armigera* and *Plodia interpunctella* (Tsfadia et al., 2008). These studies showed that only incorporation of labeled acetate is affected by PBAN and that this incorporation can be inhibited by the Acetyl Coenzyme A Carboxylase (ACCase) inhibitor, tralkoxydim. Levels of incorporation of labeled malonyl CoA or palmitic acid (downstream of acetate, Figure 3) were unaffected by the presence or absence of PBAN (Tsfadia et al., 2008). Thus, in *H. armigera,* the rate limiting step for PBAN control of pheromone biosynthesis is by regulating the ACCase which is the rate limiting enzyme of fatty acid biosynthesis, prior to the action of FAS (Figure 3). PBAN is also thought to influence ACCase activity in *Argyrotaenia velutinana* (Tang et al., 1989) and *H. zea* (Jurenka et al., 1991). In contrast, in *B. mori*, PBAN has been shown to stimulate the reductase that converts an acyl-CoA to an alcohol precursor (Osawa et al., 1993). The calcium-calmodulin (calcineurin) complex will apparently de-phosphorylate an acyl-CoA reductase which catalyzes the formation of bombykol, the main pheromone component of *B. mori* (Matsumoto et al., 1995) (Figure 3).

It is interesting to note that with the different receptor subtypes and the different signal transduction mechanisms there are also differences in the rate limiting step that is affected by PBAN stimulation. However, although the sequence of the PBAN-Rs of *S. littoralis* is phylogenetically related more to the *H. zea* and *H. armigera* PBAN-Rs (Figure 2), PBAN has been shown to up-regulate the fatty acid reduction in this species (Martinez et al., 1990; Fabrias et al., 1994), as determined for *B. mori*. However, the use of labeled precursors was not employed to determine the rate limiting step for PBAN stimulation in these studies (Martinez et al., 1990; Fabrias et al., 1994). In *Manduca sexta* PBAN is thought to regulate the mobilization of fatty acids from triacylglycerols (Fang et al., 1995; Tumlinson et al., 1997) and indeed the sequence of its PBAN-R subtypes are in a sub-branch of their own, branching from the *B. mori* PBAN-R (Figure 3). In *Sesamia nonagrioides* the acetyl transferase activity is proposed to be controlled by PBAN but the PBAN-R has not yet been identified (Mas et al., 2000). Clearly, as more PBAN-Rs become characterized at the cellular and molecular levels in the future it will be possible to unambiguously correlate function with sequence structure.

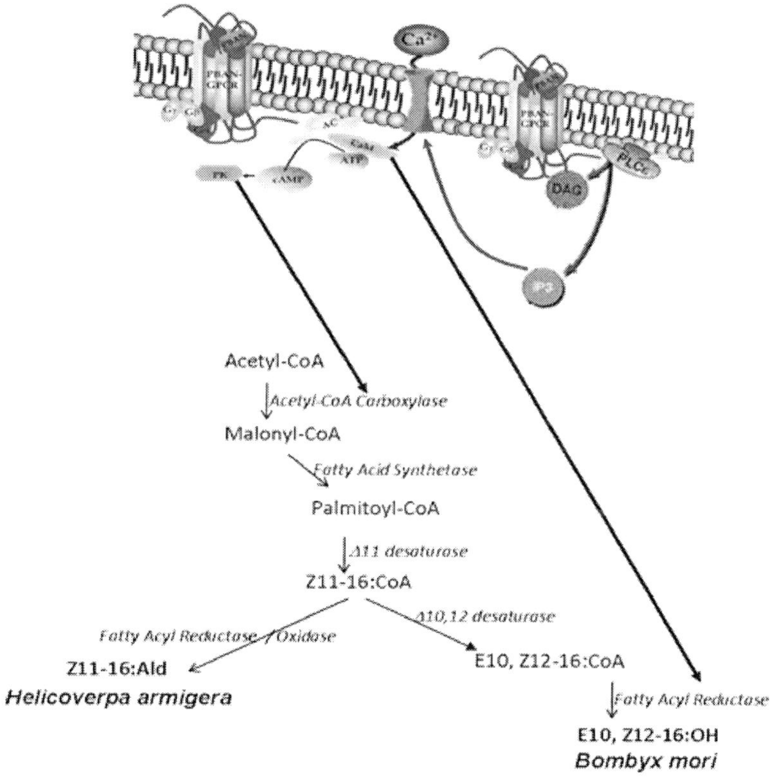

Figure 3. Contrasting signal transduction mechanisms and rate-limiting steps for PBAN stimulation involved in the up-regulation of the pheromone biosynthetic pathways of the *H. armigera* (Rafaeli, 2009) and *B. mori* (Hull et al., 2010) main sex-pheromone components. *AC*=adenylate cyclase; *CaM*=calcium-calmodulin; *cAMP*=cyclic-AMP; *DAG*=diacylglycerol; IP_3=inositol 1,4,5 triphosphate *PK*=protein kinase; *PLC*=phospholipase C.

4.5. PBAN-Receptor Differential Gene Expression Profiles

The ubiquity and multifunctional nature of the PBAN/PK family of peptides suggests that the PBAN receptor proteins could also be present in a variety of insect tissues with possible alternative functions from that of sex-pheromone biosynthesis.

In order to visualize the presence of the PBAN-R at the protein level we developed a binding assay based on the synthesis of a photoaffinity-biotinylated PBAN analog and preparation of membranes of various tissues that might contain the PBAN-R (Rafaeli and Gileadi, 1997; Rafaeli et al., 2003).

The membranes are incubated in buffer containing the PBAN analog in the presence and absence of unlabeled PBAN to determine competitive displacement. The incubation is subsequently irradiated using UV to react the photoaffinity group on the labeled PBAN ligand and the membranes are run on SDS PAGE, transferred on nitrocellulose which is treated with avidin-horse-radish peroxidase for biotin detection. In this way, specific binding to the PBAN analog is achieved.

Using the above methodology, we isolated a specific protein bound to its ligand at approximately 50kD from membrane preparations of female pheromone glands. The same protein was also isolated from membrane preparations of Sf9 cells expressing the PBAN-R (Rafaeli et al., 2007). Using this binding assay to identify the PBAN-R at the protein level and real-time quantitative-PCR (qPCR) to identify the mRNA for the PBAN-R, we showed the presence of the PBAN-R (protein and gene transcript) in various tissues (Rafaeli et al., 2007).

In addition to the adult female pheromone glands, the protein was also detected in neural tissues including the brain-suboesophageal-corpora-cardiaca and corpora-allata complexes (Br-complex), thoracic ganglia, and the VNC including the ventral abdominal ganglia. The gene transcript was also identified in these neural tissues and, in addition, it was also detected in the male counterpart tissue to the pheromone gland, the aedaegus-hair-pencil complex (male-complex). This male-complex is structurally homologous to the female pheromone gland and is found in the abdominal tip of the male. The presence of the gene transcript in this tissue was surprising since the protein could not be detected using the ligand binding assay (Rafaeli et al., 2007). Whilst the presence of PBAN in males, although a mystery since its discovery two decades ago, can be understood in terms of the ubiquitous and pleiotropic characteristic of this peptide family, the presence as well as the transcriptional regulation (see below) of its receptor in the male-complex tempted us to re-examine a possible function for PBAN and its receptor in this tissue.

As discussed above, and, since fatty acid-derived pheromonal compounds were identified in male-complexes of several moth species (Birch 1974; Szentesi et al., 1975; Hirai et al., 1978; Teal et al., 1984; Phelan et al., 1986; Teal and Tumlinson, 1989; Wu et al., 1991; Heath et al., 1992; Thibout et al., 1994; Huang et al., 1996; Hillier and Vickers, 2004), we sort to determine whether PBAN plays a role in the regulation of the biosynthesis of these compounds. We utilized both physiological bioassays and RNA interference (RNAi) technology to identify several male pheromone components that are responsive to PBAN stimulation and to photoperiod, and are also significantly affected by silencing of the PBAN-R (Bober and Rafaeli, 2010).

We hypothesize that these components are key in the chemical communication between females and males during copulation. Our study clearly demonstrated the power of RNAi technology for re-examining unanswered questions from two decades ago but it remains to be demonstrated that those PBAN up-regulated male produced compounds are indeed responsible for a successful mating encounter by either arresting females for copulation, increasing their receptivity, or deterring co-specific males from competing. A temporal differential expression study revealed that the PBAN-R gene transcript levels are age dependent in both the male-complexes and female pheromone glands of two moth species, *H. armigera* and *B. mori* (Bober et al., 2010).

In female *H. armigera*, precise temporal quantifications reveal that levels reach their peak at a critical period of 5 hours post-eclosion. Adults of the latter species do not emerge fully competent for reproductive behavior and females are only fully vitellogenic after 24 hours. Hence, this temporal pattern of the PBAN-R gene expression closely parallels the rise in Juvenile hormone (JH) levels and reproductive maturation that occurs post-eclosion in this species (Fan et al., 1999a). On the other hand, in *B. mori* females, that emerge fully vitellogenic (Ohnishi, 1987), peak transcript levels of the PBAN-R are attained by 24 hours pre-eclosion thereby also in accordance with the reproductive maturation of this species. The same temporal pattern can also be seen in the males of both of these species. Thus, it is clear that PBAN-R temporal regulation occurs at the transcriptional level of both these species and in both sexes.

The observation that there is a temporal correlation of both JH and the PBAN-R, initiated an examination of the role of JH in the transcriptional regulation of the PBAN-R. JH has been implicated in the control of pheromone production (Fan et al., 1999a; Rafaeli et al., 2003) but its exact role has not been clearly elucidated. Contrary to our hypothesis that JH may be responsible for the transcriptional regulation of the PBAN-R, removal of the head and consequently the source of JH, before peak PBAN-R transcripts are reached did not prevent normal gene expression levels (Bober et al., 2010). These results indicated that up-regulation, at the critical time, is not dependent on JH (or any other factor) originating from the head. We thus suggest that the PBAN-R transcript may be regulated early during adult maturation by a developmental program that primes the tissues for the later PBAN-R expression. Although JH is not involved in the up-regulation of the PBAN-R, it does significantly down-regulate PBAN-R transcript levels when injected *in vivo* at the critical period of 5 hours post-emergence (Bober et al., 2010). This inhibition is consistent with previous data that showed inhibition of sex-pheromone production as well as inhibition in PBAN binding assays after treatment with a topical application of JH or Fenoxycarb (FX), a JH analogue (Rafaeli and Bober, 2005) (Figure 1). Thus, in *H. armigera*, it is the absence of JH which is critical for normal up-regulation of the PBAN-R gene expression levels that occur post-eclosion and steady states of JH titers and PBAN-R expression levels are both reached after 5-6 hours post-eclosion.

Under natural conditions, increased levels of JH, after the initial post-eclosion rise, only occur after mating, directly by the transfer of JH in the seminal fluid (Ramaswamy et al., 1997) and/or due to an allatotropic influence of male accessory gland products (Fan et al., 1999b; 2000) (Figure 1). Thus JH's inhibitory role on the PBAN-R gene expression levels may be relevant in terms of its natural effect through mating-induced suppression of sex-pheromone production.

5. Inhibitory Processes

5.1. Post-Mating Behavior and Role of Seminal Fluid Peptides from Male Accessory Glands

During copulation insect males transfer seminal peptides (Acps), produced in Male Accessory Glands (MAGs) that are implicated in the female post-mating responses. It is not clear whether these processes are regulated by one or many factors and how they are associated. Such substances have been reported in several insect species (see reviews by Gillot, 2003; Avilla et al., 2011). In *Drosophila*, sex-peptide (*D. melanogaster* Acp70A (*Drm*SP) is considered to be very important in eliciting post-mating non-receptivity and increased fecundity (Chen et al., 1988; Saudan et al., 2002; Herndon and Wolfner, 1995). *Drm*SP is a 36-amino-acid peptide produced in the male *Drosophila* accessory gland (Chen et al., 1988) and passes from the female reproductive tract into the hemolymph (Peng et al., 2005; Pilpel et al., 2008), ultimately acting directly on targets in the nervous system of the female (Kubli, 2003; Häsemeyer 2009; Yang et al., 2009). Additional to its effects on non-receptivity, *Drm*SP stimulates juvenile hormone (JH)-III-bisepoxide release from the corpora allata in *D. melanogaster* (Moshitzky et al., 1996; Soller et al., 1997; 1999), thereby possibly modulating oocyte maturation and egg-laying.

Mating-induced termination of sex pheromone production has been studied in a number of moth species: *Helicoverpa zea* (Raina, 1989); *Heliothis virescens* (Ramaswamy et al., 1996); *Lymantria dispar* (Giebultowicz et al., 1991); *Argyrotaenia velutinana* (Jurenka et al., 1994); *Epiphyas postvittana* (Foster, 1993); *Bombyx mori* (Ando et al., 1996) and *Plodia interpunctella* (Rafaeli and Gileadi, 1999). In these studies a permanent or transient decline of sex pheromone production after mating was observed and correlated to whether the species exhibited single or multiple mating. In addition to the termination of sex pheromone production, calling behavior ceases and oviposition behavior is initiated (Riddiford and Ashenhurst, 1973). In *H. zea,* a pheromone suppressive peptide (*Hez*PSP), causing depletion of sex pheromone, was found in extracts of male accessory glands (Raina 1989; Kingan et al., 1993). *Hez*PSP was isolated and characterized from the accessory gland and duplex of male *H. zea* (Kingan et al., 1995) but bears no sequence homology to *Drm*SP. A synthetic form was subsequently shown to be pheromonostatic in *H. armigera* (Eliyahu et al., 2003).

5.2. SP-Like Peptide and Its Receptor in Moths

Similar to its allatotropic effect in *D. melanogaster*, *Drm*SP stimulates JH-II production by isolated corpora allata of *H. armigera* virgin females in a dose dependant manner (Fan et al., 1999b) (Figure 1). As mentioned above, JH titers significantly increase in females of some moth species after mating and the involvement of JH in mating-induced pheromone suppression has been reported (Ramaswamy et al., 1997; Delisle et al., 2000), but its precise role in this respect is rather unclear.

We have shown in *H. armigera* that the level of PBAN, present in the hemolymph during the scotophase in virgin females, is drastically reduced after mating (Nagalakshmi et al., 2007). Extracts of *H. armigera* MAGs containing Acps, when injected into virgin females, terminate pheromone production. We also demonstrated that synthetic *Drm*SP can terminate PBAN-stimulated pheromone production by female *H. armigera* (Fan et al., 1999b). Structure activity experiments in *H. armigera,* using truncated fragments of *Drm*SP showed that *Drm*SP's *N*-terminal peptides are allatotropic and *C*-terminal fragments are pheromonostatic in virgin females (Fan et al., 2000) (Figure 1).

We demonstrated the presence of factors in moth MAGs that are immuno-reactive with an antibody to *Drm*SP (Nagalakshmi et al., 2004). At least one fraction of the *H. armigera* MAG extract is both immuno-reactive to *Drm*SP antibody and is pheromonostatic (Nagalakshmi et al., 2004). In addition, we have found that different sets of *Drm*SP-like immuno-reactive peptides (called here *Hea*SP) are up-regulated during the scotophase in both male accessory glands and in the mated female CNS (Nagalakshmi et al., 2007).

These findings indicate that target receptors for these seminal-peptides might reside both in the pheromone gland as well as the CNS of the female moth. Indeed, a receptor for *Drm*SP was identified (Yapici et al., 2008) in the *D. melanogaster* female's reproductive tract and nervous system. Female flies that lack this sex peptide receptor (SP-R, also known as CG16752; called *Drm*SP-R here), failed to respond to *Drm*SP and continued to show virgin behavior even after mating. Subsequent comparisons with genome projects resulted in the annotation of the SP-R in several *Drosophila* species and in *Aedes* and *Anopheles* mosquitoes, *Tribolium* beetles, in the moth *B. mori,* the pea aphid *Acyrthosiphon pisum*, the human louse *Pediculus humanus*, the tick *Ixodes scapularis* and in the Gastropod *Aplysia californica* (Kim et al., 2010). On the basis of sequence homologies deposited in the GenBank, we also identified a putative SP-R in *H. armigera* with 99% homology to the *B. mori* SP-R (Hanin et al., 2011).

In an attempt to draw some light on the physiological significance of SP-like peptides in this moth, we also conducted a differential expression study of this receptor where we compared gene expression levels in relation to different photoperiods, sex and mating status of the moth (Hanin et al., 2011). Both photoperiod and mating influenced the SP-R gene expression levels.

Moreover, sexual dimorphic changes were observed in neural tissues due to the different physiological states. After mating SP-R transcript levels in female neural tissues and pheromone glands are up-regulated. Physiological studies *in vivo* confirmed this up-regulation of gene expression levels in pheromone glands isolated from mated females (Hanin et al., 2011). These observations comply with our physiological observations concerning the cross-reactivity of *Drm*SP and SP-like moth factor/s from *H. armigera* MAGs and are consistent with the claim that SP has a regulatory role in reproductive behavior in the moths.

Although flies and moths are evolutionarily distant groups, the cross-reactivity of their seminal-peptides suggests that these post-mating responses may be strongly conserved during evolution amongst the various insect species. The structural conservation of SP-R genes from several insects, together with the observations that *Drm*SP can induce post-mating responses in the moth *H. armigera* and that up-regulation of putative *H. armigera* SP-R gene expression in the pheromone gland and the female brains occurs after mating, indicates that this receptor family is likely to mediate post-mating changes in female reproductive behavior in many different insect orders.

Future comparative studies involving the SP-R family will no doubt provide evolutionary insight into its functions. It is unclear whether Acps affect post-mating behavior directly, indirectly (as a result of their effects on the female CNS), or due to a combination of both direct and indirect influences. No studies have been undertaken that focus on defining the physiological, biochemical and molecular effects of Acps in moths and their precise roles in the manifestation of these behavioral changes. Moreover, the interactions of seminal peptides with the regulatory peptides controlling receptivity and the biosynthesis of pheromones have yet to be studied.

6. Potential Bio-Rational Strategies Targeting Reproductive Behavior

Disrupting the chemical communication system has been an approach taken since the discovery of sex-pheromones. This strategy has been largely successful as part of an integrated pest management (IPM) protocol for monitoring up-coming surges of pest populations and, in some cases, for mating disruption (for review see Welter et al., 2005). These methods target odor perception, the receptor neurons.

For such purposes the development of para-pheromones have a high potential as alternative material in IPM strategies, particularly when the natural pheromones are expensive to produce, present longevity problems, or are quickly degraded under field conditions (see review Renou and Guerrero, 2000). The following section will consider alternative methodologies that await further investigation but carry promising potential as possible new strategies for field implementation.

6.1. Targeting the Biosynthetic Pathway

The development of specific inhibitors of the enzymes involved in the biosynthetic pathway is a potential tool that could be exploited in the future as a new strategy for insect control. Pheromone analogs have been shown to inhibit the two key steps, β-oxidation and desaturation. Cyclopropene fatty acids inhibited delta-9 and delta-11 desaturases in the pheromone glands of *S. littoralis* and *Thaumetopoea pityocampa* (Fabrias et al., 1995; 1996). A number of fluorinated, acetylenic and cyclopropane fatty acids (Camps et al., 1992; Rosell et al., 1992) that are structurally related to palmitic acid have also been shown to inhibit pheromone biosynthesis in the latter moth species. In addition, the application of 2-halofatty acids inhibited isolated pheromone glands of a number of moth species (Hernanz et al., 1997).

With regard to the key enzymes involved in the up-regulation of the pheromone biosynthetic pathway of *H. armigera*, we demonstrated the inhibitory influence of the herbicides from two different families; the 2-aryloxyphenoxypropionate (FOP) and the cyclohexandione-oxime (DIM) (Eliyahu et al., 2003; Hanin et al., 2008). These commercial herbicides specifically target the plastidic acetyl coenzyme A carboxylase (ACCase) of Gramineae family of plants. Their inhibitory effects on sex-pheromone biosynthesis were duplicated in two other moth species *Plodia interpunctella* (Tsfadia et al., 2008) and the codling moth *Cydia pomonella* (Kleinman, 2008). These findings suggest the potential for a novel alternative method of insect pest management, based on inhibition of fatty acid biosynthesis in females, which will affect sex-pheromone production and subsequent mating success, thereby reducing population growth.

6.1. Regulatory Peptide Mimics and Antagonists

Peptides have many disadvantages for their use in bio-rational design for insecticides. Their instability due to peptidases in the insect's hemolymph and tissues and their poor bioavailability due to an inability to penetrate the cuticle are the major obstacles. However, mimetic analogs of peptides with enhanced bioavailability and amphiphilic agonists and antagonists are being developed as selective agents that are capable of disrupting neuropeptide regulated processes (see review Nachman, 2009). These provide promising tools for new pest insect management strategies in the future. Such analogs of the PK/PBAN neuropeptide class have shown efficient penetration and potent activity when administered through both topical and oral routes and were shown to maintain an aqueous solubility required to reach their target receptor once they encounter the hemolymph (Nachman et al., 1996). Nachman and co-workers (2009) also report on the design of an amphiphilic PK/PBAN analog that selectively inhibits pheromonotropic activity of PBAN and demonstrates an ability to penetrate the cuticle. These reports demonstrate the promising use of peptide mimetics as antagonists of critical physiological processes, such as reproductive behavior that may be effective in pest control strategies in the future.

A recent evolutionary trace analysis on the PBAN-receptor sequences was conducted to predict the putative ligand recognition and binding sites (Jurenka and Nusawardani, 2011). In this study, class-specific amino acid residues which could be involved in ligand recognition and thereby indicate putative binding domains were identified. Thus, by combining molecular modeling and biochemical studies the design of small molecules that will disrupt ligand binding in these domains becomes feasible. Progress in developing target specific antagonists for PBAN awaits the refinement and further bioassays of potential antagonists.

6.2. Silencing of Key Regulatory Genes

RNAi has already proved its usefulness in functional genomic research on insects, but it also has considerable potential for the control of pest insects (for an extensive discussion on the potential and concerns of RNAi and highlights on implementation of RNAi in insect control the reader is referred to some excellent recent reviews: Bellès, X. 2010; Huvenne and Smaghee, 2010; Mito et al., 2011). Here, my discussion will focus on the use of RNAi for the silencing of the pheromone biosynthetic pathway.

RNAi, utilizing the injection of dsRNA against PBAN-Rs, has been successfully implemented in a number of moth species to suppress the expression of the PBAN-R in female pheromone glands of *B. mori* (Ohnishi et al., 2006) and *P. xyllostella* (Lee et al., 2010). In these studies females showed 25-50% reduction in pheromone titers and, in the case of *P. xyllostella* a 20-40% reduction in mating rates (Lee et al., 2010) showing that a reduction in PBAN-R expression subsequently affects male attractiveness. Additionally, silencing of the PBAN-R was successfully utilized to reveal PBAN's crucial function in *H. armigera* male pheromone production (Bober and Rafaeli, 2010). Moreover, silencing of several genes involved in the pheromone biosynthetic pathway was studied in *B. mori* where the fatty acyl reductase, the 10, 12 desaturase, the acyl-CoA binding protein and the fatty acid transport protein were all effectively silenced and resulted in a significant knockdown of pheromone production (Ohnishi et al., 2006; 2009). In all of the above studies, however, the dsRNA was delivered by injections either to adults or to pupae. In all cases there were no effects on development or eclosion but to make this a feasible strategy for targeted insect control the success in feeding of the dsRNA during the larval stage awaits future evaluation: Would the silencing affect the adult stage after delivery through feeding? Future research focused on the efficacy of dsRNA delivery will undoubtedly enlighten on novel strategies for pest control specifically targeting the reproductive potential of pest species.

6.3. Utilizing Seminal Peptides to Augment the Sterile Insect Technique

The sterile insect technique (SIT) is a well-established strategy for insect population suppression that is based on the introduction of large numbers of sterile males. Wild, fertile females that will mate with sterile males will produce no viable offspring, resulting in a reduction of progeny for the next generation. This strategy is effective in cases where females mate only once. For polyandrous species, where females will mate several times, the effectiveness of the strategy is diluted by the probability of females mating with wild fertile males in a subsequent mating. Thus, the mating behaviors of the target species and, specifically, the re-mating behavior of females is one of many important factors in determining the efficacy of this method.

As discussed in section 5, the important role that seminal fluid proteins play in insect reproduction, and specifically in suppressing female attractiveness, indicates an important area that can be exploited for future targeted insect control strategies. In this regard, programs that use sterilized transgenic insects carrying a non-mobilized fluorescent-protein-based marker are being implemented for preliminary studies on transgenic insects and their impact on the environment using sterile insects for minimum and controlled exposure to the environment (see review: Wimmer, 2003). Thus, the development of new technologies that could eliminate or reduce female re-mating behavior in those species where it occurs would provide opportunities to increase the efficiency and efficacy of the SIT.

Conclusions

From the above discussions it is evident that the reproductive behavior of male and female moths is influenced by various biotic and physiological factors. Moreover, the reproductive behavior of mated females is affected by a plethora of male-derived peptides, some perhaps possessing redundant and overlapping synergistic functions. The future clarification of their structures and functions and their interactions with their receptors and with the regulatory peptides that are responsible for maintaining receptivity in virgin female moths will enlighten our scientific knowledge concerning the evolution and functional significance of these signaling peptides. Beyond the significance of this research to insect physiology and biology as a whole, the outcome of future studies will provide insights into the mechanisms of signaling peptides in the animal kingdom. Notably, such research has the potential to provide vital information that will open new horizons for effective insecticide design that could be exploited for diverse practical applications in the agro-chemical industries. Specifically, the knowledge advanced will provide insight that could be directed at the reproductive potential of insects to prevent population outbursts of important agricultural pests, whilst avoiding any harmful effects to humans, livestock or the environment as a whole. Through the integration of several interdisciplinary strategies future advances will also pave the way to the discovery of new technologies.

ACKNOWLEDGMENTS

I gratefully acknowledge the US-Israel Binational Agricultural Research and Development (BARD) (Grants Nos. IS 2978-98R; IS 3634-04C and IS 4163-08C) for supporting the studies on the PBAN-R and the Israel Science Foundation (Grants Nos. 539/02 and 876/06) for supporting the studies on seminal peptides. I thank Mr. Avi Azrielli for dedicated technical assistance and all my graduate students and postdoctoral fellows, past and present, for their contributions. This is contribution No. 612/11 from the Agricultural Research Organization, Volcani Center, Bet Dagan, Israel.

REFERENCES

Ando, T. Kasuga, K. Yajima, Y. Kataoka, H. and Suzuki, A. (1996). Termination of sex pheromone production in mated females of the silkworm moth. *Archives of Insect Biochemistry and Physiology 31*, 207-218.

Avila, F. W. Sirot, L. A. LaFlamme, B. A. Rubinstein, C. D. and Wolfner, M. F. (2011). Insect Seminal Fluid Proteins: Identification and Function. *Annual Review of Entomology 56*, 21-40.

Bellès, X. (2010). Beyond Drosophila: RNAi in vivo and functional genomics in insects. *Annual Review of Entomology 55*, 111-128.

Birch, M. C. and Hefetz, A. (1987). Extrusible organs in male moths and their role in courtship behavior. *Bulletin of the Entomological Society of America 33*, 222-229.

Birch, M. C. Poppy, G. M. and Baker, T. C. (1990). Scents and eversible scent structures of male moths. *Annual Review of Entomology 35*, 25–58.

Birch, M. C. (1974). Aphrodisiac pheromones in insects. In M. C. Birch (Ed.), *Pheromones*. (pp. 115-134). Amsterdam, North Holland

Bober, R. and Rafaeli, A. (2009). Pheromone Biosynthesis Activating Neuropeptide (PBAN) and its G-protein coupled receptor. In M. Krishnan and K. Chandrasekar (Eds.) *Short views on Insect Molecular Biology* (Chapter 7 pp. 131-145). International Book Mission, Academic Publisher, South India.

Bober, R. and Rafaeli, A. (2010). Gene-silencing reveals the functional significance of pheromone biosynthesis activating neuropeptide receptor (PBAN-R) in a male moth. *Proceedings of the National Academy of Sciences, USA 107*, 16858-16862.

Bober, R. Azrielli, A. and Rafaeli, A. (2010). Developmental regulation of the Pheromone Biosynthesis Activating Neuropeptide-Receptor (PBAN-R): Re-evaluating the role of Juvenile Hormone. *Insect Molecular Biology 19*, 77-86.

Camps, F. Hospital, S. Rosell, G. Delgado, A. and Guerrero, A. (1992). Synthesis of biosynthetic inhibitors of the sex pheromone of *Spodoptera littoralis*. Part II: Acetylenic and cyclopropane fatty acids. *Chemistry and Physics of Lipids 61*,157–167.

Chen, P. S. Stumm-Zollinger, E. Aigaki, T. Balmer, J. Bienz, M. and Bohlen, P. (1988). A male accessory gland peptide that regulates reproductive behavior of female *D. melanogaster*. *Cell 54*, 291-298.

Choi, M.-Y. and Jurenka, R. A. (2010) Site-directed mutagenesis and PBAN activation of the *Helicoverpa zea* PBAN-receptor *FEBS Letters 584*, 1212–1216.

Choi, M. Y. Fuerst, E. J. Rafaeli, A. and Jurenka, R. A. (2003). Identification of a pheromone biosynthesis-activating neuropeptide G protein-coupled receptor in pheromone glands of *Helicoverpa zea*. *Proceedings of the National Academy of Sciences, USA 100*, 9721-9726.

Choi, M. Y. Fuerst, E. Rafaeli, A. and Jurenka, R. A. (2007). Role of extracellular domains in PBAN/Pyrokinin GPCRs from insects using chimera receptors. *Insect Biochemistry and Molecular Biology 37*, 296–306.

Choi, M-Y. Rafaeli, A. and Jurenka, R. A. (2001). Pyrokinin/PBAN-like peptides in the central nervous system of *Drosophila melanogaster*. *Cell and Tissue Research* 306, 459-465.

Dani, F. R. Michelucci, E. Francese, S. Mastrobuoni, G. Cappellozza, S. La Marca, G. Niccolini, A. Felicioli, A. (2011). Odorant-binding proteins and chemosensory proteins in pheromone detection and release in the silkmoth *Bombyx mori*. *Chemical Senses doi:10.1093/chemse/bjq137*.

Davis, M. B. Vakharia, V. N. Henry, J., Kempe T. G. and Raina, A. K. (1992). Molecular cloning of the pheromone biosynthesis-activating neuropeptide in *Helicoverpa zea*. *Proceedings of the National Academy of Sciences, USA 89*, 142–146.

Delisle, J. Picimbon, J. F. and Simard, J. (2000). Regulation of pheromone inhibition in mated females of *Choristoneura fumiferana* and *C. rosaceana*. *J. Insect Physiol. 46*, 913-921.

Duportets, L. Gadenne, C. Dufour, M. C. Couillaud, F. (1998). The pheromone biosynthesis activating neuropeptide (PBAN) of the black cutworm moth, *Agrotis ipsilon*: immunohistochemistry, molecular characterization and bioassay of its peptide sequence. *Insect Biochemistry and Molecular Biology 28*, 591–599.

Eisner, T. and Meinwald, J. (1995). The chemistry of sexual selection. *Proceedings of the National Academy of Sciences, USA 92*, 50-55.

Eliyahu, D. Nagalakmish, V. Kubli, E. C. Choffat, Y. Applebaum, S. W. and Rafaeli, A. (2003). Inhibition of pheromone biosynthesis in *Helicoverpa armigera* by pheromonostatic peptides. *Journal of Insect Physiology 49,* 569-574.

Fabrias, G. Barrot, M. and Camps, F. (1995). Control of the sex pheromone biosynthetic pathway in *Thaumetopoea pityocampa* by the pheromone biosynthesis activating neuropeptide. *Insect Biochemistry and Molecular Biology 25*, 655–660.

Fabrias, G. Gosalbo, L. Quintana, J. and Camps, F. (1996). Direct inhibition of (*Z*)-9-desaturation of (*E*)-11-tetradecenoic acid by methylenehexadecenoic acids in the biosynthesis of *Spodoptera littoralis* sex pheromone. *Journal of Lipid Research 37*, 1503–1509.

Fabrias, G. Marco, M. P. and Camps, F. (1994). Effect of the pheromone biosynthesis activating neuropeptide on sex pheromone biosynthesis in *Spodoptera littoralis* isolated glands. *Archives of Insect Biochemistry and Physiology 27*, 77–87.

Fan, Y. Rafaeli, A. Gileadi, C. and Applebaum, S. W. (1999a). Juvenile Hormone induction of pheromone gland PBAN-responsiveness in *Helicoverpa armigera* females. *Insect Biochemistry and Molecular Biology 29*, 635-641.

Fan, Y. Rafaeli, A. Gileadi, C. Kubli, E. and Applebaum, S. W. (1999b). *Drosophila melanogaster* sex peptide stimulates JH-synthesis and depresses sex pheromone production in *Helicoverpa armigera. Journal of Insect Physiology 45*, 127-133.

Fan, Y. Rafaeli, A. Moshitzky, P. Kubli, E. Choffat, Y. Applebaum, S. W. (2000) Common functional elements of *Drosophila melanogaster* seminal peptides involved in reproduction of *Drosophila melanogaster* and *Helicoverpa armigera* females. *Insect Biochemistry and Molecular Biology 30*, 805-812.

Fang, N. Teal, P. E. A. and Tumlinson, J. H. (1995). PBAN regulation of pheromone biosynthesis in female tobacco hornworm moths, *Manduca sexta* (L.). *Archives of Insect Biochemistry and Physiology 29*, 35–44.

Foster, S. P. (1993). Neural inactivation of sex pheromone production in mated light brown apple moths, *Epiphyas postvittana* (Walker). *Journal of Insect Physiology 39*, 267-273.

Galizia, C. G. and Rossler, W. (2010). Parallel Olfactory Systems in Insects: Anatomy and Function. *Annual Review of Entomology 55*, 399-420.

Giebultowicz, J. M. Raina, A. K. Uebel, E. C. and Ridgway, R. L. (1991). Two-step regulation of sex-pheromone decline in mated gypsy moth females. *Archives of Insect Biochemistry and Physiology 16*, 95-105.

Gillott, C. (2003). Male accessory gland secretions: modulators of female reproductive physiology and behavior. *Annual Review of Entomology 48*, 163-184

Hanin, O. Rubin, B. Applebaum, S.W. and Rafaeli, A. (2008). Structure-activity relationships of pheromonostasis induced by ACCase-inhibitor herbicides in the moth *Helicoverpa armigera. Pesticide Biochemistry and Physiology 91*, 153-15.

Hanin, O. Azrielli, A. Zakin, V. Applebaum, S. and Rafaeli, A. (2011). Identification and differential expression of a sex-peptide receptor in *Helicoverpa armigera. Insect Biochemistry and Molecular Biology* doi: 10.1016/j.ibmb.2011.03.004

Häsenmeyer, M. Yapici, N. Heberlein, U. and Dickson, B. J. (2009). Sensory neurons in the *Drosophila* genital tract regulate female reproductive behavior. *Neuron 61*, 511-518.

Heath, R. R. Landolt, P. J. Dueben, B. D. Murphy, R. E. and Schneider, R. E. (1992). Identification of male cabbage looper sex pheromone attractive to females. *Journal of Chemical Ecology 18*, 441–453.

Hernanz, D. Fabrias, G. Camps, F. (1997). Inhibition of sex pheromone production in female lepidopteran moths by 2-halofatty acids. *Journal of Lipid Research 38*, 1988–1994.

Herndon, L. A. and Wolfner M. F. (1995). A *Drosophila* seminal fluid protein, Acp26Aa, stimulates egg-laying in females for one day following mating. *Proceedings of the National Academy of Sciences USA 92*, 10114-10118.

Hillier, N. K. and Vickers, N. J. (2004). The role of Heliothine hairpencil compounds in female *Heliothis virescens* (Lepidoptera: Noctuidae) behavior and mate acceptance. *Chemical Senses 29*, 499–511.

Hirai, K. Shorey, H. H. and Gaston, L. K. (1978). Competition among courting male moths: male-to-male inhibitory pheromone. *Science 202*, 844-845.

Hirsch, J. (1991). Distribution and mode of action of pheromone biosynthesis activating neuropeptide (PBAN) in the moth *Helicoverpa (Heliothis) armigera*. M.Sc. Thesis. The Hebrew University of Jerusalem, Faculty of Agriculture.

Howard, A. D. Wang, R. Pong, S. -S. Mellin, T. N. Strack, A. Guan, X. -M. Zeng, Z. Williams, Jr, D. L. Feighner, S. D. Nunes, C. N. and et al. (2000). Identification of receptors for neuromedin U and its role in feeding. *Nature 406*, 70-74.

Huang, Y. P. Xu, S. F. Tang, X. H. Zhao, Z. W. and Du, J. W. (1996). Male orientation inhibitor of cotton bollworm: Identification of compounds produced by male hairpencil glands. *Sinica 3,* 172-182.

Hull, J. J. Kajigaya, R. Imai, K. and Matsumoto, S. (2007a) Sex pheromone production in the silkworm, *Bombyx mori*, is mediated by store-operated Ca2+ channels. *Bioscience, Biotechnology and Biochemistry 71,* 1993–2001.

Hull, J. J. Kajigaya, R. Imai, K. and Matsumoto, S. (2007b) The *Bombyx mori* sex pheromone biosynthetic pathway is not mediated by cAMP. *Journal of Insect Physiology 53*, 782–793.

Hull, J. J. Lee, J. M. and Matsumoto, S. (2010) Gqa-linked phospholipase Cb1 and phospholipase Cg are essential components of the pheromone biosynthesis activating neuropeptide (PBAN) signal transduction cascade. *Insect Molecular Biology 19*, 553–566.

Hull, J. J. Lee, J. M. Kajigaya, R. and Matsumoto, S. (2009) *Bombyx mori* homologs of STIM1 and Orai1 are essential components of the signal transduction cascade that regulates sex pheromone production. *Journal of Biological Chemistry 284*, 31200–31213.

Hull, J. J. Ohnishi, A. and Matsumoto, S. (2005). Regulatory mechanisms underlying pheromone biosynthesis activating neuropeptide (PBAN)-induced internalization of the *Bombyx mori* PBAN receptor. *Biochemical and Biophysical Research Communications 334*, 69–78.

Huvenne, H. and Smaghee, G. (2010) Mechanisms of dsRNA uptake in insects and potential of RNAi for pest control: A review. *Journal of Insect Physiology 56*, 227–235.

Ichikawa, T. Hasegawa, K. Shimizu, I. Katsuno, K. Kataoka, H. Suzuki, A. (1995). Structure of neurosecretory cells with immunoreactive diapause hormone and pheromone biosynthesis activating neuropeptide in the silkworm, *Bombyx mori. Zoological Science 12*, 703-712.

Iglesias, F. Jacquin-Joly, E. Marco, M. P. Camps, F. and Fabrias, G. (1999). Temporal distribution of PBAN-like immunoreactivity in the hemolymph of *Mamestra brassicae* females in relation to sex pheromone production and calling behavior. *Archives of Insect Biochemistry and Physiology 40*, 80–87.

Iglesias, F. Marco, P. Francois, M. C. Camps, F. Fabrias, G. and Jacquin-Joly, E. (2002). A new member of the PBAN family in *Spodoptera littoralis*: molecular cloning and immunovisualisation in scotophase hemolymph. *Insect Biochemistry and Molecular Biology 32*, 901–908.

Iglesias, F. Marco, M. P. Jacquin-Joly, E. Camps, F. and Fabrias, G. (1998). Regulation of sex pheromone biosynthesis in two noctuid species, *S. littoralis* and *M. brassicae*, may involve both PBAN and the ventral nerve cord. *Archives of Insect Biochemistry and Physiology 37*, 295–304.

Imai, K. Konno, T. Nakazawa, T. Komiya, T. Isobe, M. Koga, K. Goto, T. Yaginuma, T. Sakakibara, K. Hasegawa, K. and Yamashita, O. (1991). Isolation and structure of diapause hormone of the silkworm, *Bombyx mori. Proceedings of the Japanese Academy 67B*, 98–101.

Jacquin-Joly, E. Vogt, R. G. Francois, M. C. Nagnan-Le Meillour, P. (2001). Functional and expression pattern analysis of chemosensory proteins expressed in antennae and pheromonal gland of *Mamestra brassicae. Chemical Senses 26*, 833–844.

Jurenka, R.A. (1996). Signal transduction in the stimulation of sex pheromone biosynthesis in moths. *Archives of Insect Biochemistry and Physiology. 33*, 245-258.

Jurenka, R. (2004) Insect Pheromone Biosynthesis. *Topics in Current Chemistry 239*, 97-132.

Jurenka, R. A. Fabrias, G. DeVoe, L. and Roelofs, W. L. (1994). Action of PBAN and related peptides on pheromone biosynthesis in isolated pheromone glands of the redbanded leafroller moth, *Argyrotaenia velutinana. Comparative Biochemistry and Physiology C 108*, 153–160.

Jurenka, R. A. Jacquin, E. and Roelofs, W. L. (1991). Control of the pheromone biosynthetic pathway in *Helicoverpa zea* by the pheromone biosynthesis activating neuropeptide. *Archives of Insect Biochemistry and Physiology 17*, 81-91.

Jurenka, R. and Nusawardani, T. (2011). The pyrokinin/ pheromone biosynthesis-activating neuropeptide (PBAN) family of peptides and their receptors in Insecta: evolutionary trace indicates potential receptor ligand-binding domains. *Insect Molecular Biology* doi: 10.1111/j.1365-2583.2010.01065.

Kean, L. Cazenave, W. Costes, L. Broderick, K. E. Graham, S. Pollock, V. P. and et al. (2002). Two nitridergic peptides are encoded by the gene capability in *Drosophila melanogaster. American Journal of Physiology - Regulatory, Integrative and Comparative Physiology 282,* R1297–R1307.

Kim, Y.-J. Bartalska, K. Audsley, N. Yamanaka, N. Yapici, N. Lee, J.-Y. Kim, Y.C. Markovic, M. Isaac, E. Tanaka, Y. and Dickson, B. J. (2010). MIPs are ancestral ligands for the sex peptide receptor. *Proceedings of the National Academy of Sciences USA 107*, 6520-6525.

Kim, J.- L. Nachman, R. J. Karlygash, A. Sarjeet, G. and Adams, M. E. (2008). Pheromone Biosynthesis Activating Neuropeptide (PBAN) Receptor of *Heliothis virescens*: identification, functional expression, and structure-activity relationships of ligand analogs. *Peptides 29*, 268-275.

Kingan, T. G., Blackburn, M. B., Raina, A. K. 1992. The distribution of pheromone biosynthesis-activating neuropeptide (PBAN) immunoreactivity in the central nervous system of the corn earworm moth, *Helicoverpa zea. Cell and Tissue Research 270*, 229–240.

Kingan, T. G. Bodnar, W. M. Raina, A. K. Shabanowitz, J. and Hunt, D. F. (1995). The loss of female sex pheromone after mating in the corn earworm moth *Helicoverpa zea*: Identification of a male pheromonostatic peptide. *Proceedings of the National Academy of Sciences USA 92*, 5082-5086.

Kingan, T. G. Thomas-Laemont, P. A. Raina, A. K. (1993). Male accessory gland factors elicit change from 'virgin' to 'mated' behaviour in the female corn earworm moth *Helicoverpa zea. Journal of Experimental Biology 183*, 61-76.

Kleinman, Z. (2008). Reproductive behaviour and pheromone production of the codling moth *Cydia pomonella* (Lepidoptera:Tortricidae): stimulatory and inhibitory influences. MSc Thesis, The Hebrew University, Faculty of Agriculture.

Knipple, D. C. and Roelofs, W. L. (2003). Molecular biological investigations of pheromone desaturases. In: G. Blomquist, and R. Vogt, (Eds.) *Insect Pheromone Biochemistry and Molecular Biology* pp. 81-106. Elsevier, Amsterdam.

Kubli, E. (2003). Sex-peptides: seminal peptides of the *Drosophila* male. *Cellular and Molecular Life Sciences 60*, 1689-1704.

Leal, W. S. Nikonova, L. Peng, G. (1999). Disulfide structure of the pheromone binding protein from the silkworm moth, *Bombyx mori. FEBS Letters* 464, 85–90.

Lee, D.-W. Shrestha, S. Kim, A.Y. Park, S. J. Yang, C. Y. Kim, Y. Koh, Y. H. (2011). RNA interference of pheromone biosynthesis-activating neuropeptide receptor suppresses mating behavior by inhibiting sex pheromone production in *Plutella xylostella* (L.) *Insect Biochemistry and Molecular Biology 41*, 236-243.

Ma, P. W. K. Knipple, D. C. and Roelofs, W. L. (1994). Structural organization of the *Helicoverpa zea* gene encoding the precursor protein for pheromone biosynthesis activating neuropeptide and other neuropeptides. *Proceedings of the National Academy of Sciences, USA 91*, 6506-6510.

Ma, P. W. K. Knipple, D. C. and Roelofs, W. L. (1998). Expression of a gene that encodes multiple neuropeptides in the central nervous system of corn earworm, *Helicoverpa zea. Insect Biochemistry and Molecular Biology 28,* 373-385.

Marco, M. P. Fabrias, G. Lazaro, G. and Camps, F. (1996). Evidence for both humoral and neural regulation of sex pheromone biosynthesis in *Spodoptera littoralis*. *Archives of Insect Biochemistry and Physiology 31*, 157-167.

Martinez, T. Gabrias, G. and Camps, F. (1990). Sex pheromone biosynthetic pathway in *Spodoptera littoralis* and its activity by a neurohormone. *Journal of Biological Chemistry 256*, 1381–1387.

Mas, E. Lloria, J. Quero, C. Camps, F. Fabrias, G. (2000). Control of the biosynthetic pathway of *Sesamia nonagrioides* sex pheromone by the pheromone biosynthesis activating neuropeptide. *Insect Biochemistry and Molecular Bio*logy *30*, 455-459.

Matsumoto, S. Fonagy, A. Kurihara, K. Uchiumi, T. Nagamine, M. Chijimatsu, M. Mitsui, T. (1992). Isolation and primary structure of a novel pheromonotropic neuropeptide structurally related to leucopyrokinin from the armyworm larvae, *Pseudaletia separate. Biochemical and Biophysical Research Communications 182*, 534-539.

Matsumoto, S. Kitamura, A. Nagasawa, H. Kataoka, H. Orikasa, C. Mitsui, T. Suzuki, A. (1990). Functional diversity of a neurohormone produced by the suboesophageal ganglion: molecular identity of melanization and reddish colouration hormone and pheromone biosynthesis activating neuropeptide. *Journal of Insect Physiology 36*, 427-432.

Matsumoto, S. Ozawa, R. Nagamine, T. Kim, G. H. Uchiumi, K. Shono, T. Mitsui, T. (1995). Intracellular transduction in the regulation of pheromone biosynthesis of the silkworm, *Bombyx mori*: suggested involvement of Calmodulin and phosphoprotein phosphatase. Bioscience, Biotechnology, *and Biochemistry 59*, 560–562.

Meng, X. Wahlstrom, G. Immonene, T. Kolmer, M. Tirronen, M. Predel, R. Kalkkinen, N. Heino, T. Sariola, H. Roos, C. (2002) The *Drosophila* hugin gene codes for myostimulatory and ecdysis-modifying neuropeptides. *Mechanisms of Development 117*, 5-13.

Mito, T. Nakamura, T. Bando, T. Ohuchi, H. Noji, S. (2011). The advent of RNA interference in Entomology. *Entomological Science 14*, 1-8.

Moshitzky, P. Fleischmann, I. Chaimov, N. Saudan, P. Klauser, S. Kubli, E. Applebaum, S.W. (1996) Sex-peptide activates juvenile hormone biosynthesis in the *Drosophila melanogaster* corpus allatum. *Archives of Insect Biochemistry and Physiology 32*, 363-374.

Murlis, J. Elkinton, J. S. Carde, R. T. (1992). Odor plumes and how insects use them. *Annual Review of Entomology 37*, 505-532.

Nachman, R. J. (2009). Agonists/Antagonists of the insect kinin and pyrokinin/PBAN neuropeptide classes as tools for rational pest control. I. Ishaaya and A. R. Horowitz, (Eds.) *Biorational Control of Arthropod Pests: Application and Resistance Management* pp. 21-48. The Netherlands Springer.

Nachman, R. J. Holman, G. M. Cook, B. J. (1986). Active fragments and analogs of the insect neuropeptide leucopyrokinin: structure–function studies. *Biochemistry Biophysics Research Communications 137*, 936–942.

Nachman, R. J. Teal, P. E. A. Ben-Aziz, O. Davidovitch, M. Zubrzak, P. Altstein, M. (2009). An amphiphilic, PK/PBAN analog is a selective pheromonotropic antagonist that penetrates the cuticle of a heliothine insect. *Peptides 30*, 616-621

Nachman, R. J. Teal, P. E. A. Radel, P. Holman, G. M. Abernathy, R. L. (1996) Potent pheromonotropic/myotropic activity of a carboranyl pseudotetrapeptide analog of the insect pyrokinin/PBAN neuropeptide family administered via injection and topical application. *Peptides 17*, 747–752.

Nagalakshmi, V. K. Applebaum, S. W. Kubli, E. Choffat, Y. Rafaeli, A. (2004). The presence of *Drosophila melanogaster* sex peptide-like immunoreactivity in the accessory glands of male *Helicoverpa armigera. Journal of Insect Physiology 50*, 241–248.

Nagalakshmi, V. K. Applebaum, S. W. Azrielli, A. Rafaeli, A. (2007). Female sex pheromone suppression and the fate of sex-peptide-like peptides in mated moths of *Helicoverpa armigera. Archives of Insect Biochemistry and Physiol*ogy *64*, 142-155.

Neupert, S. Russell, W. K. Predel, R. Russell, D. H. Strey, O. F. Teel, P. D. Nachman, R. J. (2001). The neuropeptidomics of *Ixodes scapularis* synganglion. *Journal of Proteomics 72*, 1040-1045.

Ohnishi, A. Hashimoto, K. Imai, K. Matsumoto, S. (2009). Functional Characterization of the *Bombyx mori* Fatty Acid Transport Protein (BmFATP) within the Silkmoth Pheromone Gland. *The Journal of Biological Chemistry 284*, 5128-5136.

Ohnishi, A. Hull, J. J. Matsumoto, S. (2006). Targeted disruption of genes in the *Bombyx mori* sex pheromone biosynthetic pathway. *Proceedings of the National Academy of Sciences, USA 103*, 4398-4403.

Ohnishi, E. (1987). Growth and maturation of ovaries in isolated abdomens of *Bombyx mori*: response to ecdysteroids and other steroids. *Zoological Science 4*, 315–321.

Ozawa, R. A. Ando, T. Nagasawa, H. Kataoka, H. Suzuki, A. (1993). Reduction of the acyl group: the critical step in bombykol biosynthesis that is regulated *in vitro* by the neuropeptide hormone in the pheromone gland of *Bombyx mor*i. Bioscience, *Biotechnology and Biochemistry 57*, 2144-2147.

Peng, J. Chen, S. Busser, S. Liu, H. F. Honeggar, T. Kubli, E. (2005). Gradual release of sperm bound sex-peptide controls female postmating behavior in *Drosophila. Current Biology 15*, 207-213.

Phelan, P.L. Silk, P.J. Northcott, C.J. Tan, S.H. Baker, T.C. (1986). Chemical identification and behavioral characterization of male wing pheromone of *Ephesia elutella* (Pyralidae). *Journal of Chemical Ecology 12,* 135–146.

Pilpel, N. Nezer, I. Applebaum, S.W. Heifetz, Y. (2008). Mating-increases trypsin in female Drosophila hemolymph. *Insect Biochemistry and Molecular Biology 38*, 320-330.

Rafaeli A. (2002). Neuroendocrine control of pheromone biosynthesis in moths. *International Review of Cytology 213*, 49-91.

Rafaeli, A. (2009). Pheromone Biosynthesis Activating Neuropeptide (PBAN): Regulatory Role and Mode of Action. *General and Comparative Endocrinology 162*, 69–78.

Rafaeli, A., Bober, R. (2005). The effect of the juvenile hormone analog, fenoxycarb on the PBAN-receptor and pheromone production in adults of the moth *Helicoverpa armigera*: an "aging" hormone in adult females? *Journal of Insect Physiology 51*, 401-410.

Rafaeli, A. Bober, R. Becker, L. Choi, M.-Y. Fuerst, E.-J. Jurenka, R. A. (2007). Spatial distribution and differential expression of the receptor for pheromone-biosynthesis-activating neuropeptide (PBAN-R) at the protein and gene levels in tissues of adult *Helicoverpa armigera (Lepidoptera: Noctuidae). Insect Molecular Biology 16*, 287-293.

Rafaeli, A. and Gileadi, C. (1995). Modulation of the PBAN-stimulated pheromonotropic activity in *Helicoverpa armigera. Insect Biochemistry and Molecular Biology 25*, 827-834.

Rafaeli, A. and Gileadi, C. (1996). Down regulation of pheromone biosynthesis: Cellular mechanisms of pheromonostatic responses. *Insect Biochemistry and Molecular Biology 26*, 797-807.

Rafaeli, A. and Gileadi, C. (1997). Neuroendocrine control of pheromone production in moths. *Invertebrate Neuroscience 3*, 223-229.

Rafaeli, A. and Gileadi, C. (1999). Synthesis and biological activity of a photoaffinity-biotinylated pheromone-biosynthesis activating neuropeptide (PBAN) analog. *Peptides 20*, 787-794.

Rafaeli, A. Gileadi, C. Fan, Y. Cao M. (1997). Physiological mechanisms of pheromonostatic responses: effects of adrenergic agonists and antagonists on moth pheromone biosynthesis. *Journal of Insect Physiology 43*, 261-269.

Rafaeli, A. and Soroker, V. (1989). Cyclic-AMP mediation of the hormonal stimulation of [^{14}C]-acetate incorporation by *Heliothis armigera* pheromone glands *in vitro. Molecular and Cellular Endocrinology 65*, 43-48.

Rafaeli, A. Soroker, V. Hirsch, J. Kamensky, B. Raina, A. K. (1993). Influence of photoperiod and age on the competence of pheromone glands and on the distribution of immunoreactive PBAN in *Helicoverpa spp. Archives of Insect Biochemistry and Physiology 22*, 169-180.

Rafaeli, A. Zakarova, T. Lapsker, Z. Jurenka, R. A. (2003). The identification of an age- and female-specific putative PBAN membrane-receptor protein in pheromone glands of *Helicoverpa armigera*: Possible up regulation by Juvenile Hormone. *Insect Biochemistry and Molecular Biology 33*, 371-380.

Raina, A. K. (1989). Male induced termination of sex pheromone production and receptivity in mated females of *Heliothis zea. Journal of Insect Physiology 35*, 821–826.

Raina, A. K. and Klun, J. A. (1984) Brain factor control of sex pheromone production in the female corn earworm moth. *Science 225*, 531-532.

Raina, A. K. Klun, J. A. Stadelbacher, E. A. (1986). Diel periodicity and effect of age and mating on female sex pheromone titer in *Heliothis zea* (Lepidoptera: Noctuidae). *Annals of the Entomological Society of America 79*, 128–131.

Ramaswamy, S. B. Qiu, Y. Park, Y. I. (1996). Neuronal control of post-coital pheromone production in the moth *Heliothis virescens*. *Journal of Experimental Zoology 274*, 255-263

Ramaswamy, S. B. Shu, S. Park, Y.I. Zeng, F. (1997). Dynamics of Juvenile Hormone-mediated gonadotropism in the Lepidoptera. *Archives of Insect Biochemistry and Physiology 35*, 539–558.

Renou, M. and Guerrero, A. (2000). Insect parapheromones in olfaction research and semiochemical-based pest control strategies. *Annual Review of Entomology 48*, 605-630.

Riddiford, L. M. and Ashenhurst, J. B. (1973). The switchover from virgin to mated behavior in female Cecropia moths: The role of the bursa copulatrix. *Biological Bulletin 144*, 162-171.

Roelofs, W. L. and Bjostad, L. B. (1984). Biosynthesis of Lepidopteran pheromones. *Bioorganic Chemistry 12*, 279-298.

Roelofs, W. L. and Cardé, R. (1974). Sex pheromones in the reproductive isolation of Lepidopterous species. In: *Pheromones* M. C. Birch (Ed.) pp. 96-114. Elsevier Publication Co.

Rosell, G. Hospital, S. Camps, F. Guerrero, A. (1992). Inhibition of a chain shortening step in the biosynthesis of the sex pheromone of the Egyptian armyworm *Spodoptera littoralis*. *Insect Biochemistry and Molecular Biology 22*, 679–685.

Sato, K. Touhara, K. Insect olfaction: receptors, signal transduction, and behavior (2009). *Results and Problems in Cell Differentiation 47*, 121-138.

Scaloni, A. Monti, M. Angeli, S. Pelosi, P. (1999). Structural analyses and disulfide-bridge pairing of two odorant-binding proteins from *Bombyx mori*. B*iochemical and Biophysical Research Communications 266*, 386–391.

Soller, M. Bownes, M. Kubli, E. (1997). Mating and sex peptide stimulate the accumulation of yolk in oocytes of *D. melanogaster*. *European Journal of Biochemistry 243*, 732-738.

Soller, M. Bownes, M. Kubli, E. (1999). Control of oocyte maturation in sexually mature *Drosophila* females. *Developmental Biology 208*, 337-351

Soroker, V. and Rafaeli, A. (1989). *In vitro* hormonal stimulation of [^{14}C] acetate incorporation by *Heliothis armigera* pheromone glands. *Insect Biochem*istry *19*, 1-5.

Soroker, V. and Rafaeli, A. (1995). Multi-signal transduction of the pheromonotropic response by pheromone gland incubations of *Helicoverpa armigera*. *Insect Biochemistry and Molecular Biology 25*, 1-9.

Stern, P. S. Yu, L. Choi, M.-Y. Jurenka, R.A. Becker, L. Rafaeli, A. (2007). Molecular modeling of the binding of pheromone biosynthesis-activating neuropeptide (PBAN) to its receptor. *Journal of Insect Physiology 53*, 803-818.

Strandh, M. Johansson, T. Ahrén, D. Löfstedt C. (2008). Transcriptional analysis of the pheromone gland of the turnip moth, *Agrotis segetum* (Noctuidae), reveals candidate genes involved in pheromone production. *Insect Molecular Biology 17*, 73–85.

Sun, J. S. Zhang, T. Y. Zhang, Q. R. Xu, W. H. (2003). Effect of the brain and suboesophageal ganglion on pupal development in *Helicoverpa armigera* through regulation of FXPRLamide neuropeptides. *Regulatory Peptides 116*, 163-171.

Szentesi, A. Toth, M. Dodrovolsky, A. (1975). Evidence and preliminary investigation on a male aphrodisiac and a female sex pheromone in *Mamestra brassicae* (L.) *Acta Phytopathologica Academiae Scientiarum Hungaricae 10*, 425-429.

Tang, J. D. Charlton, R. E. Jurenka, R. A. Wolf, W. A. Phelan, P. L. Streng, L. Roelofs, W. L. (1989). Regulation of pheromone biosynthesis by a brain hormone in two moth species. *Proceedings of the National Academy of Sciences, USA 86*, 1806-1810.

Teal, P. E. A. Tumlinson, J. H. (1989). Isolation, identification and biosynthesis of compounds produced by male hairpencil glands of *Heliothis virescens* (F.) (Lepidoptera: Lepidoptera). *Journal of Chemical Ecology 15*, 413-427.

Teal, P. E. A. Tumlinson, J. H. Mclaughlin, F. R. Heath, R. R. Rush, R. A. (1984). (Z)-11-hexadecanol: A behavioral modifying chemical present in the pheromone gland of female *Heliothis zea* (Lepidoptera: Noctuidae). *Canadian Entomologist 116*, 777-779.

Thibout, E. Ferary, S. Auger, J. (1994). Nature and role of sexual pheromones emitted by males of *Ascrolepiopsis assectella* (Lep.). *Journal of Chemical Ecology 20*, 1571-1581.

Tillman, J. A. Seybold, S. J. Jurenka, R. A. Blomquist, G. J. (1999). Insect pheromones- an overview of biosynthesis and endocrine regulation. *Insect Biochemistry and Molecular Biology 29*, 481–514.

Torfs, P. Nieto, J. Cerstiaens, A. Boon, D. Baggerman, G. Poulos, C. Waelkens, E. Derua, R. Calderon, J. De Loof, A. (2001). Pyrokinin neuropeptides in a crustacean: isolation and identification in the white shrimp *Penaeus vannamei*. *European Journal of Biochemistry 268*, 149-154.

Tsfadia, O. Azrielli, A. Falach, L. Zada, A. Roelofs, W. L. Rafaeli, A. (2008). Pheromone biosynthetic pathways: PBAN-regulated rate-limiting steps and differential expression of desaturase genes in moth species. *Insect Biochemistry and Molecular Biology 38*, 552-567

Tumlinson, J. H. Fang, N. Teal, P. E. A. (1997). The effect of PBAN on conversion of fatty acyls to pheromone aldehydes in female. In: *Insect pheromone research: new directions*. R. T. Cardé and A. K. Minks, (Eds.) pp. 54-55. Chapman and Hall, New York.

Vieira, F.G. Sánchez-Gracia, A. Rozas, J. (2007). Comparative genomic analysis of the odorant-binding protein family in 12 *Drosophila* genomes: purifying selection and birth-and death evolution. *Genome Biology* 8, R235.

Vogt, R.G. Rogers, M.E. Franco, M.D. Sun, M. (2002). A comparative study of odorant binding protein genes: differential expression of the PBP1-GOBP2 gene cluster in *Manduca sexta* (Lepidoptera) and the organization of OBP genes in *Drosophila melanogaster* (Diptera). *Journal of Experimental Biology 205*, 719–744.

Welter, S. C. Pickel, C. Millar, J. Cave, F. Van Steenwyck, R. A. Dunley, J. (2005). Pheromone mating disruption offers selective management options for key pests. *California Agriculture 59*, 16-22.

Widmayer, P. Heifetz Y. Breer H. (2009). Expression of a pheromone receptor in ovipositor sensilla of the female moth (*Heliothis virescens*). *Insect Molecular Biology 18*, 541–547.

Wimmer, E. A. (2003). Applications of insect transgenesis. *Nature Reviews 4*, 225-232.

Wu, W. Q. Tang, X. H. Xu, S. F. Du, J. W. (1991). The diel rhythm of calling behavior and sex pheromone production of *Helicoverpa armigera* (Hubner) (Lepidoptera: Noctuidae). *Contributions from Shanghai Institute of Entomology 10*, 57-62.

Yang, C.H. Rumpf, S. Xiang, Y. Gordon, M. D. Song, W. Jan, L.Y. Jan, Y. N. (2009). Control of the postmating behavioral switch in Drosophila females by internal sensory neurons. *Neuron. 61*, 519-526.

Yapici, N. Kim, Y. J. Ribeiro, C. Dickson, B. J. (2008). A receptor that mediates the post-mating switch in *Drosophila* reproductive behaviour. *Nature 451*, 33-37.

Zdarek, J. Nachman, R. J. Hayes, T. K. (1997). Insect neuropeptides of the pyrokinin/PBAN family accelerate pupariation in the fleshfly (*Sarcophaga bullata*) larvae. *Annals of the New York Academy of Sciences 814*, 67-72.

Zhou, J- J. Kan, Y. Antoniw, J. Pickett, J. A. Field, L. M. (2006). Genome and EST analyses and expression of a gene family with putative functions in insect chemoreception. *Chemical Senses 31*, 453–465.

In: Moths: Types, Ecological Significance and Control... ISBN: 978-1-61470-626-7
Editor: Luis Cauterruccio, pp. 145-167

Chapter 5

HOST PLANT SELECTS FOR EGG SIZE IN THE MOTH *LOBESIA BOTRANA*: INTEGRATING REPRODUCTIVE AND ECOLOGICAL TRADE-OFFS IS NOT A SIMPLE MATTER

Luis M. Torres-Vila[1,*], Eva Cruces-Caldera[1] and M. Carmen Rodríguez-Molina[2]

[1] Luis M. Torres-Vila and Eva Cruces-Caldera, Servicio de Sanidad Vegetal, DG de Explotaciones Agrarias y CA, Consejería de Agricultura y Desarrollo Rural, Avda. de Portugal s/n, E-06800 Mérida, Badajoz, Spain.

[2] M. Carmen Rodríguez-Molina, Departamento de Fitopatología, Centro de Investigación Agraria Finca La Orden-Valdesequera, E-06187 Guadajira, Badajoz, Spain.

ABSTRACT

Life history theory attempts to define the rules controlling female reproductive effort, the trade-off between fecundity and egg size and the associated trade-off between egg size and offspring performance. In the tortricid moth *Lobesia botrana* Den. and Schiff., egg size is a highly labile trait depending on several proximate – environmental – factors, which correlates positively with offspring performance in adverse habitats but not in favorable ones. Host plant and its environment could then modulate the trade-off between egg size and larval performance and thereby between egg size and female fitness. The ultimate – adaptive – effect of host plant on egg size and related reproductive variables was investigated by comparing thirteen field populations of *L. botrana* derived from a cultivated (*Vitis vinifera* L.) and a wild (*Daphne gnidium* L.) host plant. We selected these hosts because of their ecological, historical and economic connotations. Daphne is considered the ancestral host plant of the moth and hence has been colonized since ancient times, while vine has been recently colonized on a historical scale, in spite of moth is currently the major vine pest worldwide. Results showed that larger females produced larger eggs, were more fecund, lived longer and had a greater reproductive effort irrespective of host plant. Heritability estimates supported the occurrence of

*E-mail: luis.torres@adr.juntaex.es, Tel.: +34 924 00 25 30, Fax.: +34 924 00 22 80

heritable variation for egg size and fecundity, but not for longevity. Unlike with female fecundity and longevity, selective pressures imposed by host plant affected significantly the size of eggs, daphne females producing smaller eggs than vine females. Host-mediated egg size differences were consistent despite the huge variation among populations within host plants, which was interpreted because of local adaptation processes under a weak gene flow. Results overall do not support however the expected trade off between egg size and number since fecundity did not differ significantly between host plants. Host-mediated selective forces driving egg size of *L. botrana* and their related reproductive and ecological trade-offs are finally discussed from an evolutionary perspective regarding host plant quality, insect-plant relationships, moth reproductive physiology and natural enemy pressure.

INTRODUCTION

Energetic resources are costly to acquire and life history theory attempts to define the rules controlling the reproductive effort, defined as the fraction of resources assigned to reproduction within the total energetic balance (Schaffer 1974a,b, Stearns 1977). A related reproductive decision is to adjust the pattern of investment in each individual offspring (Smith and Fretwell 1974, Pianka 1976, Parker and Begon 1986, McGinley et al. 1987, Berrigan 1991) which is currently viewed as evolutionary concomitant with - rather than subsidiary of - reproductive effort (Winkler and Wallin 1987, Caley et al. 2001, Fischer et al. 2006, Karl et al. 2007). Most life-cycle models assume that female reproductive effort per individual offspring (egg size) involves a critical selective choice coupled with the trade-off between egg size and fecundity (Smith and Fretwell 1974, Begon and Parker 1986, Winkler and Wallin 1987). Whether a female's reproductive effort is prefixed in a given environment, female have two reproductive options to maximize fitness: produce fewer large or more small eggs.

A compromise between fecundity and egg size -relative to adult size- has been showed to be species-specific across butterfly species (García-Barros 1994, 2000) despite that a huge variation in egg size within species is rather the rule, both environmental and genetic in origin (Fox and Czesak 2000, Parry et al. 2001, Torres-Vila and Rodríguez-Molina 2002, Fischer et al. 2006, Zovi et al. 2008). Intraspecific geographic variation in egg size is particularly widespread (Harvey 1983a, b, García-Barros 1992, Fischer and Fiedler 2001, Parry et al. 2001, Zovi et al. 2008) supporting diverging reproductive strategies shaped by natural selection (Dobzhansky 1970). Variation may particularly arise because egg size is both a maternal and an offspring trait, so that egg size may be subjected to opposite selection forces in parent and progeny generations (Fox and Czesak 2000, but see Marshall and Uller 2007). Recent studies have questioned, however, that a trade-off between egg size and fecundity is universal. In some Lepidoptera species, mostly income breeders, there is evidence that the trade-off between egg size and fecundity is sensitive not only to genetic background, but also to physiological and environmental factors including adult feeding, female age, female size or egg composition. Hence, females might have physiological mechanisms to optimize egg size without compromising fecundity (Fischer et al. 2006, Karl et al. 2007).

Egg size is important too because an inverse trade-off between offspring number and offspring performance appears to be widespread in nature. Such an important evolutionary trade-off was early postulated from a Darwinian perspective (Williams 1966) and theories concerning optimal reproductive strategies assume that egg size correlates positively with offspring performance (Smith and Fretwell 1974, Parker and Begon 1986, McGinley et al. 1987, Fox and Czesak 2000). A general trade-off between egg size and larval performance in the Lepidoptera is however controversial since it has been experimentally proved in some species but not in others (Fischer and Fiedler 2001, Torres-Vila and Rodríguez-Molina 2002 and references therein).

In our species model, the tortricid moth *Lobesia botrana* Den. and Schiff., egg size is a highly labile trait. A number of proximate factors (environmental and physiological) have the potential to shape egg size, including among them female age at mating, water availability, pupal weight loss, larval feeding, female body weight and female age. In *L. botrana* egg size correlates positively with offspring performance; large neonate larvae (emerging from large eggs) have higher starvation endurance than small ones, and larval settlement and survival of large neonates are significantly improved in adverse habitats unlike favorable ones (Torres-Vila and Rodríguez-Molina 2002). In the former instance, potential fitness gain for females predicts that whether large larvae have selective advantage over small ones, then selection for large egg size should prevail over high fecundity.

These considerations on reproductive resource allocation strongly suggest that the host plant of L. botrana could have the potential for selecting egg size (Braby 1994) and hence modulate the trade-off between egg size and fecundity. The moth is polyphagous, but despite the wide host range recorded that includes both cultivated and wild plants (Bovey 1966, Coscollá 1997, Torres-Vila 2000), just vine (Vitis vinifera L.) and daphne (Daphne gnidium L.) are largely considered the two most important host plants, at least in Mediterranean areas. The remainder host plants recorded in the literature are either rare, naturally unselected (they have been tested only in the laboratory) or just collaterally exploited because their phenology not allow to complete the entire life cycle of L. botrana (Torres-Vila 2000).

Regarding the cultivated host plant, *L. botrana* is largely considered the most important vine pest in the Palearctic region (Roehrich and Boller 1991, Torres-Vila 2000). Because of human action, moth also has colonized some areas of central Africa and western Asia, but it has been in recent years when the sudden colonization of the Americas (Chile, Argentina and the USA) has led to making of *L. botrana* a global vine pest. When looking at the current worldwide pest status, it is hard to believe that a little over two centuries ago the species was completely unknown in vineyards. The starting point of *L. botrana* as vine pest remains unclear as shown by the following brief historical overview. Jacquin (1788) first cited *L. botrana* on grapevine in late 18th century, few years after the original description of the species by Denis and Schiffermüller [1775]. Throughout the 19th century, some authors reported the occasional presence of *L. botrana* on pergola and arbor grapevines in Central Europe (mostly in periurban areas) but the species continued to be virtually absent in vineyards (Froelich 1828, Köllar 1837, Audouin 1842, Frey 1880, Mayet 1890). The large-scale invasion and initial severe outbreaks in European vineyards did not begin until about a century after Jacquin's (1788) seminal paper, in late 19th and early 20th centuries (Berlese 1894, 1924, Marchal 1912, Clarió-Soulán and Nonell Comas 1912, Silvestri 1912, Feytaud 1920, 1924, Voukassovitch 1924, González de Andrés 1935, Balachowsky and Mesnil 1935-1936, Bovey 1966).

Regarding the wild host plant, the flax-leaved daphne or spurge flax is a thymeleaceous, evergreen and sclerophylous shrub up to 2 m high, with small lanceolate leaves and inflorescences in terminal panicles. The plant, especially leaves, contains toxic secondary compounds with multiple activity. Daphne typically populates Mediterranean shrublands and open woodlands, the typical Mediterranean forest known as *dehesa*, being often also present in the margins of cultivated fields and vineyards. Daphne was already known as host plant of *L. botrana* decades before the beginning of vineyard colonization (Cuní y Martorell 1874, Millière 1875) and was early proposed as the ancestral host plant of *L. botrana* (Marchal 1912, Grassé 1928, but see Bovey 1966). Daphne is currently considered the most important wild host of *L. botrana* because of the important ecological, evolutionary and historical interconnections between them (Nuzzaci and Triggiani 1982, Torres-Vila 2000, Maher and Thiéry 2006).

In this chapter, we explore whether host plant-imposed selective pressures in the wild could shape trade-offs between *L. botrana* life-history traits, and more specifically between egg size and related reproductive traits. Host plant may entail both proximate (environmental) and ultimate (adaptive) effects on egg size and this chapter deals with the second aspect, for which daphne (a host likely colonized since ancient times) and vine (a recently colonized host on a historical scale) were compared. As Forister et al. (2009) stressed, herbivorous insects that have recently incorporated novel hosts provide unique opportunities for understanding factors that promote or constrain the evolution of host plant range. Research in this scenario may improve our understanding of selective pressures driving resource allocation and reproductive patterns in *L. botrana*, but also might help to elucidate the selective forces involved in host plant shifting, vineyard colonization in the past, and the worldwide adaptive success of this moth to grapevine.

MATERIALS AND METHODS

Insect Life History

The European grapevine moth, *L. botrana* achieves three generations on vine in southern temperate areas, while a complete fourth additional generation is increasingly frequent due to global warming (Torres-Vila et al. 2010). Female lays single eggs and after hatching, larvae develop on inflorescences, unripe berries and ripening-ripe berries during the first, second and third generation, respectively. When *L. botrana* populations are associated with daphne, the three larval generations develop on shoots, inflorescences and fruiting inflorescences, respectively. Individuals from the last generation overwinter as diapausing pupae from autumn to early spring. Adults do not exhibit migratory habits and show reduced active dispersal (Roehrich and Carles 1981, but see Torres-Vila et al. 2006).

Field-Derived Populations of *L. Botrana*

We initially compared two populations of L. botrana (populations V1 and D1, see Table 1) derived from vine and daphne (first trial). The preliminary results obtained lead us to work with an extensive array of geographical populations derived from daphne and vine that were collected over several years (second trial). Wild-derived populations of L. botrana were selected after an intensive field inspection throughout Extremadura region (southwestern

Spain), looking for a compromise between a sufficient sample size and a suitable location of host plant areas. Populations living on daphne are often naturally scarce and the same occurs in intensively sprayed commercial vineyards, which limits the number of founding parents and may bias the genetic pool of target populations. Sampling protocol expressly avoided proximity between vineyards and daphne stands (at least 1 km apart) in order to prevent, or at least minimize, between-population gene flow. To guarantee that target populations were representative of conditions suffered in cultivated-vine and wild-daphne environments, we avoided isolated marginal vineyards and minor daphne stands. Larvae were collected in vine plots (>1 ha) inside large vine-growing areas or in daphne stands (>50 plants) within large shrublands near or under holm (Quercus ilex L.) or cork oak (Q. suber L.) open woodlands.

We finally tested thirteen *L. botrana* populations, covering an area of about 80 x 100 square kilometers, seven derived from vine and six from daphne (Table 1). Populations were collected in larval stage (mostly 4th-5th instar larvae) usually during spring and early summer on vine inflorescences or daphne shoots. Larval development was completed in the laboratory on a general-purpose artificial diet at 25±1°C and 60±10% r.h., under a L16:D8 photoperiod. Pupae were isolated in glass tubes (70 x 9 mm in diameter) stoppered with cardboard plugs to ensure virginity of moths.

Adults obtained from each wild population were used as parents to produce under the same rearing conditions the F_1 individuals for tests. We ended up larval development on artificial diet to facilitate rearing and to produce similar-sized adults, as different plant diets, vine cultivars and vine phenology strongly affect female body size and egg size in *L. botrana* (Torres-Vila et al. 1999, Torres-Vila and Rodríguez-Molina 2002, Thiéry and Moreau 2005, Moreau et al. 2007).

Table 1. Summary of the field-derived *L. botrana* populations used in this study

Population	Host plant	Population (Site)	Collection date	Collected larvae [a]	Tested females [b]
V1	Vine	Guadajira	June-2003	21	26
V2	Vine	Guareña I	June-2004	59	40
V3	Vine	Guareña II	June-2004	28	32
V4	Vine	Mérida	June-2004	44	48
V5	Vine	Guareña III	July-2004	32	54
V6	Vine	Calamonte I	July-2004	53	23
V7	Vine	Calamonte II	July-2004	52	19
D1	Daphne	Cornalvo [c]	June-2003	43	16
D2	Daphne	Manchita	June-2004	86	50
D3	Daphne	Madroñera	June-2004	30	25
D4	Daphne	Jaraicejo I [d]	June-2004	245	31
D5	Daphne	Jaraicejo II	July-2004	50	27
D6	Daphne	Arroyo de San Serván [e]	Sept-2004	64	57

[a] Number of larvae collected in the field used as parents when adults

[b] Number of F_1 females tested in the laboratory

[c] Protected area of the Cornalvo Natural Park

[d] Field-derived larvae were tested directly when reached adulthood because of the large sample size available

[e] Diapausing population, tested in spring 2005.

Laboratory Tests

All tests with F_1 adults were performed in a controlled environment room at 25±1°C, 60±10% r.h. and a L(15+1):D8 long day photoperiod. The first 15 photophase hours were at a 1000-lux luminosity and the last one (dusk) at 25-lux. Dusk simulation was necessary as this is the daily period during which main activity of *L. botrana* takes place, including flight, feeding, calling, mating and egg laying. Newly emerged F_1 females (<1-day-old) were collected daily, weighed on a Ohaus precision electrobalance to the nearest 0.1 mg, caged individually in 30-ml clear plastic containers as mating and oviposition chambers and provided with water *ad libitum* through a soaked cotton wick. Note that *L. botrana* is a capital breeder typically included among those requiring water but no sugar in adult stage to maximize reproductive output (Torres-Vila 1996). A fresh leaf of daphne or a leaf cutting of vine (according to the origin of the population tested) was added to each container for oviposition and replaced daily. One 2-day-old virgin male was confined with each female 1 h before dusk and containers were thereafter observed. If mating occurred, male was removed the next morning; if not, unmated female was eliminated from the trial. Females were then examined for egg laying and all eggs laid daily on leaves or containers were marked and counted until female death. Egg size variation was studied over the whole oviposition period just with V1 and D1 populations (fist trial) as preliminary results lead us to concentrate study effort in the size of first eggs, much more numerous and expected to have a higher adaptive value in the wild (Begon and Parker 1986, Torres-Vila and Rodríguez-Molina 2002 and references therein). We hence studied size variation of first eggs laid using a large array of populations originated from vine and daphne (second trial). In this case, just 10-12 eggs per female randomly chosen among the compliment deposited on the first day after mating were marked. After female death, eggs were incubated at 25°C for 10 days to be sure that hatching was fulfilled and egg size, fecundity (total eggs), percent fertility, longevity and reproductive effort were recorded (or estimated) for each female as follows. Egg size was assessed as the elliptic surface of their orthogonal projection, $S = \pi \cdot a \cdot b$ (mm^2), where a and b are the major and minor ellipse semiaxis respectively. Surface was chosen as an index of egg size because an accurate egg weight or volume assessment was very difficult given the special features of *L. botrana* flat-type eggs: small size, frailty, reduced thickness and high adherence to the substrate. Mean egg size was estimated for each female from five isolated, undeformed, hatched eggs (chorions) randomly chosen among those marked on the first oviposition day (second trial) and successive days (first trial) or from all eggs if fewer eggs were available, particularly in the last oviposition days. Eggs were measured using a Nikon DS-U1 digital camera connected to a Leica S6D stereomicroscope and resulting images analyzed with Eclipse Net 1.20 free software. Lastly, reproductive effort was estimated for each female (V1 and D1 populations) by multiplying daily mean egg size with daily fecundity and summing all daily values (see Fischer et al. 2006). Mated but unfertilized females (100% unhatched eggs) were excluded from statistical analysis.

Heritability Estimates

To ascertain in *L. botrana* the genetic background of the studied traits and to confirm that the variation observed is heritable - at least in part- in this species, we estimated heritability

(h^2) for egg size, fecundity and longevity. Experiment was designed according to the offspring-parent regression method (offspring-mother regression as two studied traits were female linked) using a laboratory strain. We initially established 36 families from random-mated parents but in three of them progeny was not obtained or was lost. Thereby just 33 families were retained for statistical computations. Egg size data were obtained from five eggs per female as usual, five eggs per mother in the parental generation and five eggs per daughter (equal-sized families of five daughters) in the progeny generation. Adult handling, larval rearing and data collection were conducted as usual.

Heritability of the studied traits (h^2) was estimated as twice the slope of the offspring-mother linear regression (b_{OM}), and standard error of heritability as twice the standard error of the regression SE (h^2) = 2 SE (b_{OM}), since all three regressions were on just one parent (Falconer and Mackay 1996, Roff 1997).

Data Analysis

Analysis of Covariance (ANCOVA) was used to explore the effect of host plant on egg size, fecundity, longevity and reproductive effort, using female size as covariate (first trial). Although mean female size did not significantly differ between F_1 tested females derived from daphne and vine, which completed their larval development on artificial diet in the laboratory (ANOVA, $F_{1,11} = 0.98$, $P = 0.34$ ns), we use anyway female size as covariate when appropriate to avoid any body size-related bias in the variables analyzed. A two-way model I Analysis of Variance (ANOVA) was used to assess the effects of host plant, egg-laying day and their interaction on egg size and fecundity, considering host and day factors with fixed effect (first trial). A nested ANCOVA was used to explore the effect of host plant on egg size (eggs laid on the first day), fecundity and longevity, using female size as covariate. In this experimental design we included thirteen populations of *L. botrana* with random effect (seven collected on vine and six on daphne) and thereby population factor was nested to host plant as main factor (second trial). Distribution normality and ANCOVA prerequisites were confirmed prior to computation: independence between host plant (main factor) and female size, correlation between the variables and female size (covariate) and lack of interaction between host plant and female size (slopes of regression lines between every variable and female size were not significantly different among populations). F-statistics with AN(C)OVAs was computed as described in Sokal and Rohlf (1995).

RESULTS

Detailed Comparison between Vine (V1) and Daphne (D1) Populations (First Trial)

Female size had a strong significant effect on the four studied variables: egg size, fecundity, longevity and reproductive effort (Table 2). Larger females produced larger and more eggs, lived longer and had a greater reproductive effort. Host plant significantly affected egg size and fecundity but in an opposite way. While vine females produced significantly larger eggs than daphne females, fecundity was significantly lower in the former than in the latter. Host plant did not affect female longevity nor reproductive effort (Table 2).

Table 2. Analysis of Covariance (ANCOVA) of the effect of host plant on egg size, fecundity, longevity and reproductive effort in *L. botrana* using female size as covariate

Variable	Source	df	MS	F	*P*
egg size	host	1	0.00354	8.20	<0.01
	female size	1	0.00337	7.80	<0.01
	error	39	0.00043		
fecundity	host	1	9577.15	5.54	<0.05
	female size	1	101352.00	58.64	<0.001
	error	39	1728.30		
longevity	host	1	0.07	0.03	0.87 ns
	female size	1	115.29	46.03	<0.001
	error	39	2.50		
reproductive	host	1	930.35	3.33	0.08 ns
effort	female size	1	15064.05	53.96	<0.001
	error	39	279.15		

Analysis includes two *L. botrana* populations derived from vine (V1) and daphne (D1) (see Table 1 for population details and Figure 1 where data analyzed are plotted).

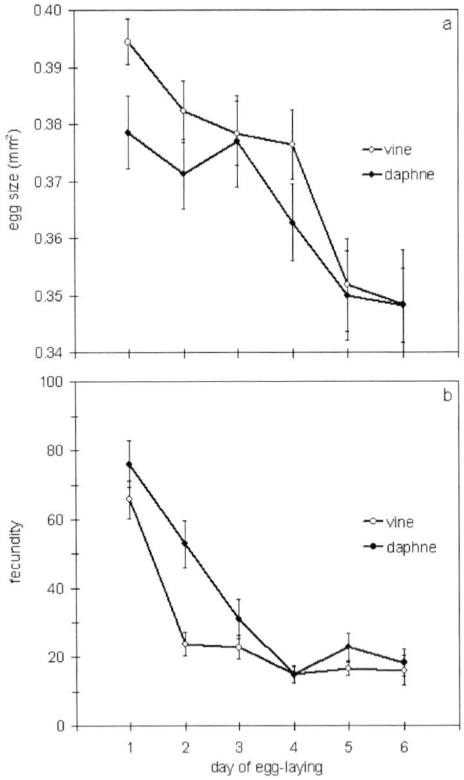

Figure 1. Variation in egg size (a) and fecundity (b) over oviposition period (means ± SE) depending on host plant in two *L. botrana* populations derived from vine (V1: white circles) and daphne (D1: black circles). See Table 1 for details on V1 and D1 populations and Table 3 for a complete ANOVA. SE: standard error.

Table 3. Two-way model I Analysis of Variance (ANOVA) of the effects of host plant, egg-laying day and their interaction on egg size and fecundity in *L. botrana*

Variable	Source	df	MS	F	*P*
egg size	host	1	0.00424	4.24	<0.05
	day	5	0.00906	9.06	<0.001
	host x day	5	0.00018	0.18	0.97 ns
	error	196	0.00100		
fecundity	host	1	4155.67	11.21	<0.001
	day	5	16024.69	43.23	<0.001
	host x day	5	924.95	2.50	<0.05
	error	196	370.69		

Analysis includes two *L. botrana* populations derived from vine (V1) and daphne (D1) (see Table 1 for population details and Figure 2 where data analyzed are plotted). Host plant and egg-laying day were analyzed with fixed effect. A two-way ANOVA including egg-laying day as fixed factor confirmed the significant effect of host plant on egg size and fecundity (Table 3), with vine females producing larger but fewer eggs than daphne females. Egg-laying day also significantly affected egg size and fecundity, both reproductive variables decreasing with female age (Figure 1). Fecundity was significantly affected by the host x day interaction that evidenced a steeper decline pattern in vine females with female age, at least initially (Figure 1).

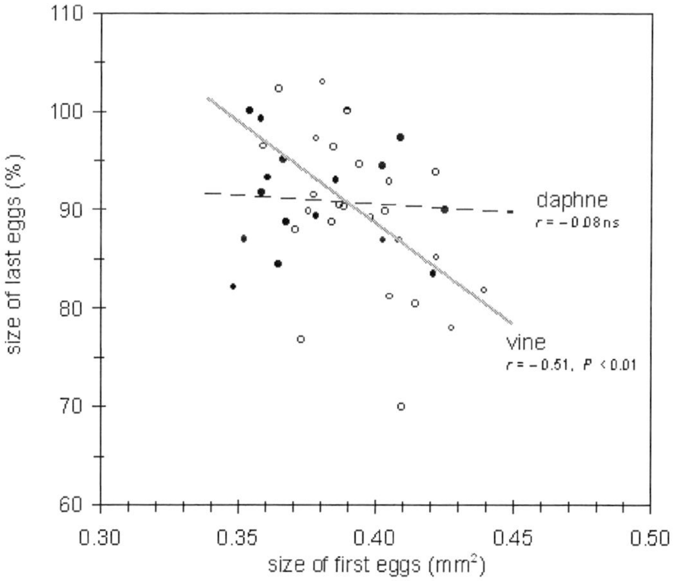

Figure 2. The relationship between mean size of first eggs (mm^2) and the last eggs (as a percentage of first egg size) depending on host plant in two populations of *L. botrana* derived from vine (V1: white circles, solid line) and daphne (D1: black circles, dashed line). Five eggs measured per female per egg class. See Table 1 for details on V1 and D1 populations. Line slope was negative in both cases, but regression was only significant with vine females: Vine (V1): $y = -206.43x + 171.21$, $r = -0.51$, $F_{1,24} = 8.19$, $P < 0.01$, $n = 26$; Daphne (D1): $y = -17.67x + 97.70$, $r = -0.08$, $F_{1,14} = 0.09$, $P = 0.78$ ns, $n = 16$.

Table 4. Nested Analysis of Covariance (ANCOVA) of the effect of host plant on egg size, fecundity and longevity in *L. botrana* using female size as covariate

Variable	Source	df	MS	Error MS	F	*P*
Egg size	(a) host plant	1	0.0120	(b)	4.89	<0.05
	(b) population{host plant}	11	0.0025	(d)	5.01	<0.001
	(c) female size	1	0.0366	(d)	74.61	<0.001
	(d) error	434	0.0005			
Fecundity	(a) host plant	1	8508.81	(b)	0.92	0.36 ns
	(b) population{host plant}	11	9217.46	(d)	5.36	<0.001
	(c) female size	1	1161898.70	(d)	676.05	<0.001
	(d) error	434	1718.67			
Longevity	(a) host plant	1	3.09	(b)	0.22	0.65 ns
	(b) population{host plant}	11	14.04	(d)	5.74	<0.001
	(c) female size	1	837.14	(d)	341.95	<0.001
	(d) error	434	2.45			

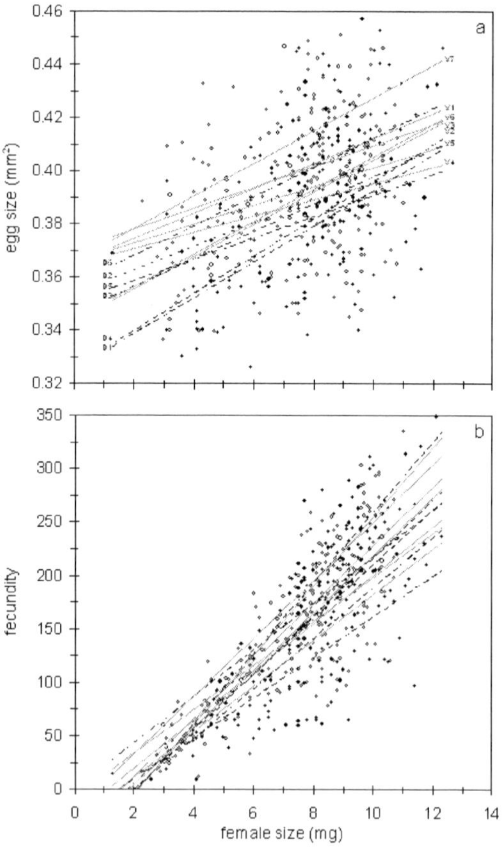

Figure 3. Effect of host plant on first egg size (a) and fecundity (b) in *L. botrana* using female size as covariate. Linear regressions of seven populations derived from vine (white circles, solid lines) and six from daphne (black circles, dashed lines) are plotted. Lines are labeled with population codes in the upper chart as there were significant differences in egg size between host plants. See Table 1 for details on tested populations and Table 4 for a complete ANCOVA.

Analysis include seven *L. botrana* populations derived from vine and six from daphne (see Table 1). Population was analyzed with random effect and nested to host plant. F-statistics were computed as described in Sokal and Rohlf (1995)

Although at first sight egg size decline pattern was also dissimilar between host plants, lack of significant host x day interaction did not support such view (Table 3). Nevertheless, when egg size variation was studied for each individual female in relative terms by comparing the size of the first (mm^2) and last eggs (as a percentage of first egg size), egg size decline pattern with female age was noticeably host plant-dependent (Figure 2). In vine females the greater the size of first eggs, the smaller the size of the last eggs. By contrast, in daphne females there was no correlation between the size of first and last eggs laid as the slope of regression line did not differ significantly from zero (Figure 2) .

Overall Comparison between Vine and Daphne Populations (Second Trial)

Results from the more robust design involving thirteen populations (second trial) corroborated that female size was strongly correlated with egg size, fecundity and longevity, so that larger females produced larger eggs, were more fecund and lived longer (Table 4). The most important result arising from the ANCOVA was that host plant significantly affected the size of eggs produced by *L. botrana*. This was consistent with results from the first trial and confirmed that *L. botrana* populations derived from vine produced larger eggs than those derived from vine (Table 4, Figure 3). Such differences occurred despite there was also huge variation in egg size among populations within host plants (nested factor in Table 4). A high variation among populations within hosts also occurred with female fecundity and longevity. Interestingly, host plant-mediated egg size variation just emerged when female size was used as covariate through ANCOVA, since an equivalent ANOVA was unable to detect a significant effect of host plant on egg size ($F_{1,11} = 2.21$, $P = 0.17$ ns). Further ANCOVAs using fecundity or longevity as a second covariate in addition to female size did not increase analytic power, presumably due to the high colinearity among all three variables. Unlike egg size, fecundity and longevity were not significantly affected by host plant, surely a result of the large variation among populations within host plants (Table 4).

Heritability estimates were significant just for egg size and fecundity, indicating that these important reproductive traits are heritable in *L. botrana* (Table 5). Heritability estimates were however unexpectedly too high to be traits closely connected with female fitness (Roff 1997), which was interpreted as a result of uncontrolled maternal effects inherent to the offspring-mother regression approach (Falconer and Mackay 1996).

Table 5. Heritability estimates for egg size, fecundity and longevity in a laboratory strain of *L. botrana* using the offspring-parent regression method

Variable	Heritability ($h^2 \pm$ SE)
Egg size	0.47 ± 0.21
Fecundity	0.57 ± 0.13
Longevity	0.59 ± 0.36 ns

$h^2 = 2\ b_{OM}$ and SE (h^2) = 2 SE (b_{OM}), being b_{OM} the slope of the offspring-mother linear regression. SE: standard error. Data from 33 equal-sized families of five daughters; five eggs measured per female.

CONCLUSION

Does Host Plant Select for Egg Size?

Results confirmed that larger females produced larger eggs, were more fecund and lived longer (Torres-Vila et al. 1999, Torres-Vila and Rodríguez-Molina 2002, Torres-Vila et al 2002), and showed that larger females had greater reproductive effort. The most important finding was that vine females consistently produced larger eggs than daphne females irrespective of female size. Egg size differences between vine and daphne populations were strong enough to remain significant despite the huge variation in egg size among populations within host plants. Such variation in egg size (as well as in fecundity and longevity) may derive from local adaptation processes triggered by selective pressures other than host plant. This situation can occur whether gene flow is unable to counteract the effects of selection on local populations, even if these populations could actually be part of a true metapopulation (Hanski and Gilpin 1997). Egg size variation among populations exploiting different host plants has been also found in other lepidopterans (Braby 1994).

Results suggest that *L. botrana* females living on vine and daphne have chosen different life strategies in allocating nutritional resources to eggs in order to maximize fitness. We emphasize that experiments were expressly designed to explore variation driven by selective pressures imposed by host plant in the wild, that is, we were concerned with ultimate rather than proximate factors. The occurrence of heritable variation for egg size and fecundity in *L. botrana* (as showed by heritability estimates) highlighted that both reproductive traits may be the target of natural selection, even if the selective effect of host plant was only apparent in the case of egg size. Hence, we tentatively interpret the observed differences in egg size because of genetic variation between moth populations living on different host plants even if a plastic response can not be discarded. A number of lepidopterans exhibit phenotypic plasticity in egg size, and some of them are even able to quickly manipulate the size of eggs produced depending on host plant to which females are facing (Gibbs et al 2005, Mizumoto and Nakasuji 2007).

Egg Size and Fecundity Trade-Off

Life-cycle models often assume a trade-off between egg size and fecundity, mostly when reproductive effort is prefixed in larval stage as typically occurs in a capital breeder as *L. botrana*. Although there is extensive experimental evidence supporting such a reproductive trade-off (Fox and Czesak 2000) its universal relevance has been questioned and it is currently under debate (see references in the Introduction). First trial showed that reproductive effort did not differ significantly (although near the significance limit) between vine and daphne populations, and supported a trade-off between egg size and fecundity. Vine females produced larger eggs than daphne females and fecundity of vine females was lower to offset that extra resource allocation into larger eggs. Results differed noticeably when a large array of vine and daphne geographical populations was analyzed jointly in the second trial. Vine females again produced larger eggs but in this case fecundity did not differ significantly between host plants. The huge variation in fecundity within host plants made that between-host differences were diluted to become no significant. This outcome seemingly challenged the expected trade-off between fecundity and egg-size in *L. botrana*, since vine females were able to produce larger eggs than daphne females without compromising fecundity. Longevity-related somatic cost did not affect that trade-off because host plant did not significantly affect female longevity. This is not a trivial issue as longevity requires a high energetic inversion (longevity is perhaps the highest somatic cost) that could interact with reproductive effort within the total energetic balance.

How Does Vine Females Produce Larger Eggs without Jeopardizing Fecundity?

Reproductive effort and longevity were similar in vine and daphne populations, so that, other things being equal, reproductive output of vine and daphne females should be balanced in energetic terms. This premise did not hold, however, as vine females were overall able to produce larger eggs without jeopardizing fecundity. As showed, there was substantial variation in egg size and fecundity, not only among populations within hosts, but also among females within populations, which ultimately resulted in a lack of trade-off between egg size and fecundity. This circumstance suggests that there should be additional underlying mechanisms accounting for the trade-off between egg size and fecundity (Fischer et al. 2006, Karl et al. 2007).

Daily egg-laying pattern could partially account for what seems a reproductive energetic unbalance. As expected, *L. botrana* fecundity and egg size significantly declined over the oviposition period (Torres-Vila et al. 1999, Torres-Vila and Rodríguez-Molina 2002), but results also showed that decline pattern tended to differ between host plants. Vine and daphne females both produced eggs progressively smaller with female age, egg size decline patterns being roughly similar, except that daphne females produced smaller first eggs and entered later into the trajectory of egg size decline. The most interesting finding arose when egg size variation along time was studied for each individual female, as there were significant differences between host plants. In vine females the greater the size of the first eggs produced, the smaller the size of the last eggs laid, as previously reported (Torres-Vila and Rodríguez-Molina 2002). Distinctly, daphne females were more prone to keep egg size regardless of female age.

Despite the potential interest in adaptive terms, these noticeable host-mediated differences in temporal resource allocation were however not enough to account for the unbalance in reproductive effort between both host plants. Thereby, to explain the lack of a trade-off between egg size and fecundity in *L. botrana*, it is mandatory to evoke underlying mechanisms other than egg size decline, as could be variation in egg composition or water content.

Further biochemical analyses of *L. botrana* eggs of different ages, sizes and origins (vine or daphne) would be necessary to confirm this point. The situation observed in the moth *L. botrana* (a capital breeder) could be strikingly similar to that reported in the rotting fruit-feeding butterfly *Bicyclus anynana* Butler (an income breeder), species in which females producing large eggs do not have to sacrifice fecundity due to adjustments in egg composition and water content (Fischer et al. 2006, Karl et al. 2007).

Is Egg Size Decline with Female Age Adaptive?

Begon and Parker (1986) postulated an adaptive strategy to explain egg size decline, that first eggs are larger because they are most likely to be laid and then they may contribute relatively more to offspring production.

A bet-hedging strategy has been also proposed to explain egg size decline with maternal age as adaptive (Fox and Czesak 2000) since natural selection could promote egg size variation among offspring produced by a single female to ensure that at least some progeny sizes are well adapted to fluctuating and unpredictable environments (Philippi and Seger 1989, Forbes 1991, Marshall and Uller 2007). Alternatively, egg size decline with female age has also been often viewed as a non-adaptive trait and attributed to either resource depletion or poor vitellogenesis control when first eggs are unnecessarily too large (Wiklund and Karlsson 1984, Fox and Czesak 2000, Torres-Vila and Rodríguez-Molina 2002 and references therein).

The situation with *L. botrana* is rather complex to interpret as there is evidence supporting both the adaptive and non-adaptive viewpoints. The adaptive hypothesis fits well with the reproductive strategy of *L. botrana* since first eggs laid are the most numerous (about 50% of the eggs are laid on the first oviposition day) and these eggs are decisive to maximize female fitness (Fischer et al. 2006 and references therein). Torres-Vila and Rodríguez-Molina (2002) reported however that vine-derived diapaused females in which reproductive effort (body weight and reproductive reserves) was constrained after overwintering, managed to maintain a high-realized fecundity at the expense of reducing egg size, instead of reducing fecundity to produce larger eggs. Such female response seems to fit with the non-adaptive hypothesis as eggs could be unnecessarily too large and therefore egg size may be reduced to maintain fecundity, which interestingly also supports in turn a trade-off between egg size and fecundity.

Egg Size and Offspring Performance Trade-Off

Host plant is considered a major potential factor selecting for egg size in insects. A negative correlation between host plant quality and the strength of selection favoring larger offspring size appears to be widespread (Braby 1994, Fox and Czesak 2000, Agosta 2008). The rational is that a poor host quality should promote an increase in egg size because large neonate larvae (emerging from large eggs) should take adaptive advantage when faced to an unfavorable host plant. This trade-off has been specifically showed with *L. botrana*, as large neonate larvae perform better than small ones to endure starvation, which in turn enhances larval settling and survival at least in adverse habitats (Torres-Vila and Rodríguez-Molina 2002). Results of this study show that vine populations of *L. botrana* consistently produce larger eggs than daphne populations so that, according to the optimal resource allocation theory, one could conclude that daphne is a better host than vine. Recent research with *L. botrana* also suggests that daphne is better than vine in nutritional terms (Thiéry and Moreau 2005) and that females prefer to lay eggs on daphne fruits rather than on grapes (Maher and Thiéry 2006).

That vine is a poorer host than daphne also seems to make sense from a historical perspective. Vine is a relatively new host plant for *L. botrana* to which moth has had to pay an adaptive cost (maybe a larger egg size) while daphne is considered to be the ancestral host plant, allegedly exploited for millennia (see Introduction). Forister et al. (2009) reported a similar case of recent host adaptation in the butterfly *Lycaeides melissa* Edwards, species in which the native host was a better larval resource and was preferred by ovipositing females in choice experiments. However, when flowers were available in addition to plant foliage, the native and novel hosts received similar numbers of eggs. These results illustrate that an additional selection pressure can reverse an apparently well-established trade-off optimizing larval performance.

Greater abundance and predictability of vine has also been proposed as a factor driving host plant shift of *L. botrana* from daphne to vine (Thiéry and Moreau 2005, Maher and Thiéry 2006). Host plant selection model by West and Cunningham (2002) lends credence to that successful exploitation of a poorer host plant by an insect can be explained merely in terms of host abundance and predictability, mostly when insect has low dispersal potential and is short-lived; and this picture reflects well what occurs with vine and vine moth in our study case.

Is Daphne Really Better Than Vine as Host Plant?

The above scenario, contrary to what might seem, remains puzzling. A number of important aspects still suggest that daphne is not a plant as optimal as it looks for *L. botrana*. Host plant quality is a complex concept difficult to realize since it is multifactorially determined by a wide set of variables, not only nutritional (including both energetic compounds and secondary metabolites) but also morphological, microclimatic and ecological, among which phenological coincidence, seasonal availability, abundance, interspecific competition and host-plant specific predation risk are specially important (Mira and Bernays 2002). As Janzen (1985) claimed “a host plant is more than its chemistry”, which means that host plant is not just something to fed on but something to live on. Among the main observational, circumstantial and experimental evidences supporting that daphne is not better than vine as host plant for *L. botrana* we found:

First, daphne leaves are tougher and harder than vine inflorescences, which may difficult leaf consumption, silk spinning and larval settling in first generation. Since egg size increases in the Lepidoptera in response to host physical defenses as toughness and hardness (Braby 1994, García-Barros 2000, Zalucki et al. 2002) daphne females should produce larger eggs, just the opposite of what was observed. Second, daphne has secondary compounds with insecticide and antifeedant activity (Stout et al. 1970, Maher and Thiéry 2006 and references therein) that may constraint larvae living on this host. A hidden energetic cost may even occur whether moth populations living on daphne require an additional metabolic pathway to digest, detoxify and cope with these secondary compounds (Scriber and Slansky 1981, Harborne 1982, Caprio and Tabashnik 1992). Third, unlike cultivated vine, daphne has a sympodic growth pattern determining that young leaves, flowers and fruits are always exposed on the top of the plant canopy. Since *L. botrana* use exclusively these daphne plant parts as substrate for egg laying, a higher egg exposure may promote predation and parasitism rates as well as egg desiccation in summer generations. Fourth, experimental results from multi-site artificial infestations in the field using *L. botrana* neonate larvae sustain that offspring settling and survival (two central aspects of female and offspring fitness) are often lower on daphne than on vine (Torres-Vila et al. unpublished results) irrespective that there is also variation enough in larval survival depending on host phenology (Torres-Vila et al. 1992, Gabel and Roehrich 1995). Fifth, from a strict nutritional quality perspective, it is still unclear whether daphne is a better host than vine for *L. botrana*. Recent research using plant powder-based agarized diets in the laboratory points out that daphne flowers are nutritionally richer than vine unripe grapes as revealed in terms of larval growth rate, pupal mass and female fecundity (Thiéry and Moreau 2005). Even assuming that agarized diets are nutritionally equivalent to fresh plants (especially regarding denaturation of daphne secondary compounds) more than a single phenological comparison between host plants would be necessary to support that daphne is nutritionally better than vine. It is well known that both vine and daphne nutritional quality changes substantially with plant phenology over season. For instance, vine quality for *L. botrana* (measured in terms of female fecundity) may be more than tripled from flowering to ripening (Torres-Vila et al. 1999). Sixth, closely linked to host nutritional quality, host phenology and resource predictability are major factors to be considered. Daphne has a complex reproductive pattern, in which flowering season extends through six months but fruiting is scarce due to continuous abortion of developing fruits. Nutritional quality of daphne leaves and flowers becomes increasingly relevant since these organs are in practice the main available resource for *L. botrana* over the season. In early season (first generation), leaves are the only resource available because daphne is not flowering yet. In middle-late season (summer larval generations), leaves are still the main resource as flowering is very unpredictable, often incomplete and even absent in many suboptimal daphne habitats. Even if daphne plants finally become flowering (June-October) and fruit set is successful, warm and dry summer conditions often promote a remarkably high fruit abortion rate, with reported values of 75%, 87% and even up to 98% fruit loss during August-September (Herrera 1987, Guitián and Guitián 1990), so that mature fruits are in practice largely unavailable for larvae developing in summer. Subsequently, when in October-November fruit ripening becomes prevalent (Herrera 1987, 1998, Guitián and Guitián 1990) it is too late for feeding on daphne fruits because individuals of the last generation have already entered diapause. Seventh, *L. botrana* populations living on daphne are subjected to a much higher predation (mainly spiders Salticidae, Gnafosidae, Thomisidae and Clubionidae) and parasitism pressures on either larvae and pupae (wasps Ichneumonidae and Chalcididae) or eggs (micro-wasps Trichogrammatidae) than populations living on vine (Torres-Vila et al. unpublished results). Even though this study was not designed to investigate this issue, our own results supported this notion since mean (± SE) larval parasitism recorded on the thirteen *L. botrana* populations was significantly higher on daphne (30.1 ± 7.0 %) than on vine (8.5 ± 3.3 %) (Mann-Whitney test, $U = 5.0$, df = 1, $P < 0.05$).

It follows that after considering all these nutritional and ecological aspects, the idea that daphne is overall better than vine as host plant for *L. botrana* is questionable.

Why Do Populations Living on Daphne Lay Eggs so Small?: A Hypothesis

This is really an intriguing question arising from the above considerations whose answer is not straightforward at all. Regrettably our study does not allow to face the problem with direct experimental evidence, but we hypothesize that there should be a selective pressure that tends to maintain an egg size smaller than expected in populations living on daphne that could be triggered by egg parasitoid wasps. Host plant environment and particularly natural enemies may be equally or even more important than host plant itself in adaptive terms (Bernays and Graham 1988, Jaenike and Papaj 1992, Mira and Bernays 2002). In such a tritrophic system (daphne-moth-wasp) the higher performance of large larvae emerging from large eggs could be under opposing selection whether large egg size carry a cost of increased parasitism. Whether female fitness gain due to escape from egg parasitoids prevails over benefits of producing larger larvae (more suitable to cope with a low-quality host plant) then small eggs could have a net adaptive advantage over large eggs. For instance, Zovi et al. (2008) found that egg parasitoids mediate selection for smaller eggs in the pine processionary moth, *Thaumetopoea pityocampa* Den. and Schiff., which may promote in turn adaptation of local populations to host plant.

Parasitism of *L. botrana* eggs by a number of *Trichogramma* species is widespread in Europe and these micro-wasps have even been proposed for biological control (Coscollá 1997, Torres-Vila 2000). Parasitism rate of *Trichogramma* spp. is known to be higher on large eggs of the target host insect (Barnay et al. 1999 and references therein) and specifically in *L. botrana* (Moreau et al. 2009). Small eggs may be visually and chemically more cryptic than large eggs as kairomonal cues exploited by parasitoid females are egg size-dependent (Bruines et al. 1994). Field experience indicates that *L. botrana* populations living on daphne are subjected to a much higher pressure by natural enemies than vine populations, surely because daphne and vine populate respectively wild (insecticide-free) and cultivated (often sprayed) habitats differing in biodiversity. In a scenario with a high egg parasitoid pressure (as daphne stands may be) and provided that parasitoid females locate more easily large eggs, selection predicts that females living on daphne will increase fitness by laying smaller eggs. In a quite distinct scenario in which parasitoid pressure is relaxed (as vineyards may be) host plant quality itself may become increasingly important, and then the best adaptive response would be to produce larger eggs to improve larval performance. The two scenarios depicted are very appealing but to date remain speculative and merit additional research. The distinctive oviposition pattern observed in individual females derived from daphne and vine lends some credence to the parasitoid-based hypothesis. Unlike vine females, daphne females not only produced smaller eggs but also were prone to maintain a steady egg size over time, which could involve significant fitness gains face to egg parasitoids. Intriguingly, greater homogeneity in egg size suggests in turn that daphne stands would be more predictable than vineyards according with the above mentioned bet-hedging strategy. Whether the trade-off between egg size and fecundity is taken up in this context, again the side question would be: Why fecundity does not increase in daphne populations if females produce smaller eggs? As previously discussed, perhaps daphne females are able to produce smaller but more yolk-rich and costly eggs. In addition, ancillary selective pressures could be involved as for instance time limitation for ovipositing females. These females may become constrained if the number of mature eggs available in the ovarioles exceeds the number of current oviposition occasions. As seen, *L. botrana* is a crepuscular, short-lived and highly proovigenic moth, so that the narrow dusk window for egg laying could itself constraint fecundity rising. A trade-off between time limitation and egg size or fecundity has been reported for instance in *Pararge aegeria* L. In this butterfly, females suffering during oviposition a thermally-induced time limitation prevent fecundity rising (Gotthard et al. 2007), but if time limitation is due to male harassment, then females made a trade-off between egg size and number, laying a larger number of small eggs (Gibbs et al. 2005).

In conclusion, this chapter shows that host plant selects for egg size in *L. botrana* and suggests that variation in egg size among populations may be adaptive, but also emphasizes that underlying selective pressures triggering changes in egg size are the outcome of a still obscure set of reproductive and ecological trade-offs whose complex interaction in the wild remains to be elucidated.

ACKNOWLEDGEMENTS

We thank the staff of the Sección de Protección Integrada (Servicio de Sanidad Vegetal): A. Sánchez González, J.J. Ferrero-García, D. Martín-Vertedor, F. Ponce Escudero, F. Barrena Galán, J. Ángel Rebollo Nieto, C. Aza Barrero and E. Delgado Valiente, for collecting field populations and providing technical assistance in both laboratory and field.

REFERENCES

Agosta, S.J. (2008). Selection on offspring size varies within and among families in relation to host nutritional quality. *Evolutionary Ecology, 22,* 71-83.

Audouin, V. (1842). *Histoire des insectes nuisibles à la vigne et particulièrement de la Pyrale.* Paris: Fortin-Masson et Cie.

Balachowsky, A.S. and Mesnil, L. (1935-1936). *Les insectes nuisibles aux plantes cultivées* (2 vol.). Paris: Busson-L. Méry.

Begon, M. and Parker, G.A. (1986). Should egg size and clutch size decrease with age?. *Oikos, 47,* 293-302.

Berlese, A. (1894). *Le Tignuole della vite (Cochylis ambiguella Hubn. ed Eudemis botrana Schiffm.) e modo di combatterle [...] con l'aggiunta di alcuni documenti sull' efficacia della rubina da prove recenti* [Extr. Bollettino di Entomologia Agraria e Patologia Vegetale, 17]. Padova: Stab. Prosperini.

Berlese, A. (1924). *Entomologia agraria. Manuale sugli insetti nocivi alle piante coltivate, campestri, orticole ed ai loro prodotti e modo di combatterli.* Firenze: Tip. M. Ricci [R. Stazione di Entomologia Agraria in Firenze].

Barnay, O., Pizzol, J., Gertz, C., Kienlen, J. C., Hommay, G. and Lapchin, L. (1999). Host density-dependence of discovery and exploitation rates of egg patches of *Lobesia botrana* (Lepidoptera: Tortricidae) and *Ephestia kuehniella* (Lepidoptera: Pyralidae) by the parasitoid *Trichogramma cacoeciae* (Hymenoptera: Trichogrammatidae). *Journal of Economic Entomology, 92,* 1311–1320.

Bernays, E.A. and Graham, M. (1988). On the evolution of host specificity in phytophagous arthropods. *Ecology, 69,* 886-892.

Berrigan, D. (1991). The allometry of egg size and number in insects. *Oikos, 60,* 313-321.

Bovey, P. (1966). Super-famille des Tortricoidea. In A.S. Balachowsky (Ed.), *Entomologie appliquée à l'agriculture,* 2 (1) (pp. 456-893). Paris: Masson et Cie.

Braby, M.F. (1994). The significance of egg size variation in butterflies in relation to host plant quality. *Oikos, 71,* 119-129.

Bruines, E.B.A., Wajnberg, E. and Pak, G.A. (1994). Genetic variability in the reactive distance in *Trichogramma brassicae* after automatic tracking of the walking path. *Entomologia Experimentalis et Applicata, 72,* 297-303.

Caley, M.J., Schwarzkopf, L. and Chine, R. (2001). Does total reproductive effort evolve independently of offspring size?. *Evolution, 55,* 1245-1248.

Caprio, M.A. and Tabashnik, B.E. (1992). Evolution of resistance to plant defensive chemicals in insects. In B.D. Roitberg and M.B. Isman (Eds.), *Insect chemical ecology* (pp. 179-215). New York: Chapman and Hall.

Clarió-Soulán, I.V. and Nonell Comas, J. (1912). *Cochylis, eudemis, piral y altisa de la vid. Medios más eficaces para combatir cada una de estas enfermedades.* Pamplona: Congreso Nacional de Viticultura.

Coscollá, R. (1997). *La polilla del racimo de la vid (Lobesia botrana Den. y Schiff.).* Paterna, Valencia: Generalitat Valenciana, Conselleria de Agricultura, Pesca y Alimentación.

Cuní y Martorell, M. (1874). *Catálogo metódico y razonado de los lepidópteros que se encuentran en los alrededores de Barcelona, de los pueblos cercanos y de otros lugares de Cataluña [...].* Barcelona: T. Gorchs.

Dobzhansky, T. (1970). *Genetics of the evolutionary process.* New York: Columbia University Press.

Falconer, D.S. and Mackay, T.F. (1996). *Introduction to quantitative genetics.* Harlow, London: Prentice Hall.

Feytaud, J. (1920). Sur l'extension de l'eudémis en France. *Le Progrès Agricole et Viticole, 75(38),* 281-284.

Feytaud, J. (1924). Notes biologiques sur les vers du raisin (cochylis et eudémis). *Revue de Zoologie Agricole et Appliquée, Numéro spécial,* 1-36.

Fischer, K. and Fiedler, K. (2001). Egg weight variation in the butterfly *Lycaena hippothoe*: more small or fewer large eggs?. *Population Ecology, 43,* 105-109.

Fischer, K., Bot, A.N.M., Brakefield, P.M. and Zwaan, B.J. (2006). Do mothers producing large offspring have to sacrifice fecundity? *Journal of Evolutionary Biology, 19,* 380-391.

Forbes, L.S. (1991). Optimal size and number of offspring in a variable environment. *Journal of Theoretical Biology, 150,* 299-304.

Forister, M.L., Nice, C.C., Fordyce, J.A. and Gompert, Z. (2009). Host range evolution is not driven by the optimization of larval performance: the case of *Lycaeides melissa* (Lepidoptera: Lycaenidae) and the colonization of alfalfa. *Oecologia, 160,* 551-561.

Fox, C.W. and Czesak, M.E. (2000). Evolutionary ecology of progeny size in arthropods. *Annual Review of Entomology, 45,* 341-369.

Frey, H. (1880). *Die Lepidopteren der Schweiz.* Leipzig: Verlag von W. Engelmann.

Froelich, F. von (1828). *Enumeratio Tortricum L. [in] regno Würtembergico [Enumeratio Tortricum Würtembergiae] indigenarum sistens specierum differentias, synonyma selecta, eaurum domicilia et tempus cum descriptionibus.* Tubingae [Tübingen]: Typis Schoenhardtianis.

Gabel, B. and Roehrich, R. (1995). Sensitivity of grapevine phenological stages to larvae of European grapevine moth, *Lobesia botrana* Den. et Schiff. (Lep., Tortricidae). *Journal of Applied Entomology, 119,* 127-130.

García-Barros, E. (1992). Evidence for geographic variation of egg size and fecundity in a satyrine butterfly, *Hipparchia semele* (L.) (Lepidoptera-Nymphalidae-Satyrinae). *Graellsia, 48,* 45-52.

García-Barros, E. (1994). Egg size variation in European Satyrine butterflies (Nymphalidae, Satyrinae). *Biological Journal of the Linnean Society, 51,* 309-324.

García-Barros, E. (2000). Body size, egg size, and their interspecific relationships with ecological and life history traits in butterflies (Lepidoptera: Papilionoidea, Hesperioidea). *Biological Journal of the Linnean Society, 70,* 251-284.

Gibbs, M., Lace, L.A., Jones, M.J. and Moore, A.J. (2005). Egg size-number trade-off and a decline in oviposition site choice quality: Female *Pararge aegeria* butterflies pay a cost of having males present at oviposition. *Journal of Insect Science, 5,* 39 (9 pp.).

González de Andrés, C. (1935). *La polilla de las uvas (Polychrosis botrana, Schiff.).* La Coruña: Estación de Fitopatología Agrícola de Galicia.

Gotthard, K., Berger, D. and Walters, R.J. (2007). What keeps insects small? Time limitation during oviposition diminishes the fecundity benefit of female size in a butterfly. *American Naturalist, 169,* 768-779.

Grassé, P.P. (1928). Eudémis et plantes sauvages. *Le Progrès Agricole et Viticole, 90(49),* 541-544.

Guitián, J. and Guitián, P. (1990). Fenología de la floración y fructificación en plantas de un espinal del Bierzo (León, noroeste de España). *Anales del Jardín Botánico de Madrid, 48,* 53-61.

Hanski, I.A. and Gilpin, M.E. (Eds.) (1997). *Metapopulation biology: ecology, genetics, and evolution.* London: Academic Press.

Harborne, J.B. (1982). *Introduction to ecological biochemistry.* London: Academic Press.

Harvey, G.T. (1983a). A geographical cline in egg weights in *Choristoneura fumiferana* (Lepidoptera: Tortricidae) and its significance in population dynamics. *Canadian Entomologist, 115,* 1103-1108.

Harvey, G.T. (1983b). Environmental and genetic effects on mean egg weight in spruce budworm (Lepidoptera: Tortricidae). *Canadian Entomologist, 115,* 1109-1117.

Herrera, C.M. (1998). Long-term dynamics of mediterranean frugivorous birds and fleshy fruits: a 12-yr study. *Ecological Monographs, 68,* 511-538.

Herrera, J. (1987). Biología reproductiva de algunas especies del matorral de Doñana. *Anales Jardín Botánico de Madrid, 44,* 483-497.

Jacquin, N.J. (1788). *Collectanea ad botanicam, chemiam, et historiam naturalem.* Vindobonae [Viena]: Officina Wappleriana.

Jaenike, J. and Papaj, D. (1992). Behavioral plasticity and patterns use by insects. In B.D. Roitberg and M.B. Isman (Eds.), *Insect chemical ecology* (pp. 245-264). New York: Chapman and Hall.

Janzen, D.H. (1985). A host plant is more than its chemistry. *Illinois Natural History Survey Bulletin, 33,* 141-174.

Karl, I., Lorenz, M.W. and Fischer, K. (2007). Energetics of reproduction: consequences of divergent selection on egg size, food limitation, and female age for egg composition and reproductive effort in a butterfly. *Biological Journal of the Linnean Society, 91,* 403-418.

Köllar, V. (1837). *Naturgeschichte der schädlichen Insecten in Beziehung auf Landwirthschaft und Forstcultur.* Wien: Druck F. Ulrich.

Maher, N. and Thiéry, D. (2006). *Daphne gnidium*, a possible native host plant of the European grapevine moth *Lobesia botrana*, stimulates its oviposition. Is a host shift relevant?. *Chemoecology, 16,* 135-144.

Marchal, P. (1912). *Rapport sur les travaux accomplis par la Mission d'Études de la cochylis et de l'eudémis*. Paris: Librairie Polytechnique, Paris et Liège.

Marshall, D.J. and Uller, T. (2007). When is a maternal effect adaptive?. *Oikos, 116,* 1957-1963.

Martín-Vertedor, D., Ferrero-García, J.J. and Torres-Vila, L.M. (2010). Global warming affects phenology and voltinism of *Lobesia botrana* in Spain. *Agricultural and Forest Entomology, 12,* 169-176.

Mayet, V. (1890). *Les insectes de la vigne*. Montpellier: C. Coulet, G. Masson.

McGinley, M.A., Temme, D.H. and Geber, M.A. (1987). Parental investment in offspring in variable environments: theoretical and empirical considerations. *American Naturalist, 130,* 370-398.

Millière, P. (1875). Catalogue raisonné des Lépidoptères [du Département] des Alpes-Maritimes [III]. *Mémoires de la Société des Sciences Naturelles et Historiques des Lettres et des Beaux-arts de Cannes et de l'Arrondissement de Grasse, 5,* 51-216.

Mira, A. and Bernays, E.A. (2002). Trade-offs in host use by *Manduca sexta*: plant characters vs natural enemies. *Oikos, 97,* 387-397.

Mizumoto, M. and Nakasuji, F. (2007). Egg size manipulation in the migrant skipper, *Parnara guttata guttata* (Lepidoptera: Hesperiidae), in response to different host plants. *Population Ecology, 49,* 135-140.

Moreau, J., Thiéry, D., Troussard, J.P. and Benrey, B. (2007). Grape variety affects female but also male reproductive success in wild European grapevine moths. *Ecological Entomology, 32,* 747-753.

Moreau, J., Richard, A., Benrey, B. and Thiéry, D. (2009). Host plant cultivar of the grapevine moth *Lobesia botrana* affects the life history traits of an egg parasitoid. *Biological Control, 50,* 117-122.

Nuzzaci, G. and Triggiani, O. (1982). Note sulla biocenosi in Puglia della *Lobesia (Polychrosis) botrana* (Schiff.) (Lepidoptera: Tortricidae) infeudata a *Daphne gnidium* L.. *Entomologica, 17,* 47-52.

Parker, G.A. and Begon, M. (1986). Optimal egg size and clucth size, effects of environment and maternal phenotype. *American Naturalist, 128,* 573-592.

Parry, D., Goyer, R.A. and Lenhard, G.J. (2001). Macrogeographic clines in fecundity, allocation, and offspring size of the caterpillar *Malacosoma disstria. Ecological Entomology, 26,* 281-291.

Philippi, T. and Seger, J. (1989). Hedging ones evolutionary bets, revisited. *Trends in Ecology and Evolution, 4,* 41-44.

Pianka, E.R. (1976). Natural selection of optimal reproductive tactics. *American Naturalist, 109,* 775-784.

Roehrich, R. and Boller, E. (1991). Tortricids in vineyards. In L.P.S. Van der Geest and H.H. Evenhuis (Eds.), *Tortricid pests their biology, natural enemies and control* (pp. 507-514). Amsterdam: Elsevier.

Roehrich, R. and Carles, J.P. (1981). Observations sur les déplacements de l'eudémis, *Lobesia botrana. Bollettino di Zoologia Agraria e di Bachicoltura, Ser. II, 16,* 10-11.

Roff, D.A. (1997). *Evolutionary quantitative genetics*. New York: Chapman and Hall.

Schaffer, W.M. (1974a). Optimal reproductive effort in fluctuating environments. *American Naturalist, 108,* 783-790.

Schaffer, W.M. (1974b). The evolution of optimal reproductive strategies: the effects of age structure. *Ecology, 55,* 291-303.

Scriber, J.M. and Slansky, F. (1981). The nutritional ecology of immature insects. *Annual Review of Entomology, 26,* 183-211.

Silvestri, F. (1912). Contribuzioni alla conoscenza degli insetti dannosi e dei loro simbionti. III. La tignoletta dell'uva (*Polychrosis botrana* Schiff.) con un cenno sulla tignola dell'uva (*Conchylis ambiguella* Hb.). *Bollettino del Laboratorio di Zoología Generale e Agraria della R. Scuola Superiore d'Agricoltura in Portici, 6,* 246-307.

Smith, C.C. and Fretwell, S.D. (1974). The optimal balance between size and number of offspring. *American Naturalist, 108,* 499-506.

Sokal, R.R. and Rohlf, F.J. (1995). *Biometry*. New York: Freeman and Co.

Stearns, S.C. (1977). The evolution of life-history traits: a critique of the theory and a review of the data. *Annual Review of Ecology and Systematics, 8,* 145-171.

Stout, G.H., Balkenhol, W.G., Poling, M. and Hickernell, G.L. (1970). The isolation and estructure of Daphnetoxin, the poisonous principle of *Daphne* species. *Journal of the American Chemical Society, 92,* 1070-1071.

Thiéry, D. and Moreau, J. (2005). Relative performance of european grapevine moth (*Lobesia botrana*) on grapes and other hosts. *Oecologia, 143,* 548-557.

Torres-Vila, L.M. (2000). *Lobesia botrana* Den. and Schiff. (Lepidoptera: Tortricidae) datasheet. In CAB-International (Ed.), *Crop protection compendium* (2nd edn) (CD-ROM). Wallingford-Oxon, UK: Commonwealth Agricultural Bureau-International.

Torres-Vila, L.M., Schmitz, V. and Stockel, J. (1992). Efecto de la evolución fenológica y de la variedad de vid en la instalación y supervivencia en primera generación de las orugas de la polilla del racimo (*Lobesia botrana* Den. y Schiff.; Lepidoptera: Tortricidae). *Boletín de Sanidad Vegetal Plagas, 18,* 755-764.

Torres-Vila, L.M., Stockel, J. and Rodríguez-Molina, M.C. (1996). Efecto de la indisponibilidad de agua sobre el potencial biótico de la polilla del racimo *Lobesia botrana* Den. y Schiff. (Lepidoptera: Tortricidae). *Boletín de Sanidad Vegetal Plagas, 22,* 443-449.

Torres-Vila, L.M., Rodríguez-Molina, M.C., Roehrich, R. and Stockel, J. (1999). Vine phenological stage during larval feeding affects male and female reproductive output of *Lobesia botrana* (Lepidoptera: Tortricidae). *Bulletin of Entomological Research, 89,* 549-556.

Torres-Vila, L.M., Rodríguez-Molina, M.C. and Stockel, J. (2002). Delayed mating reduces reproductive output of female European grapevine moth, *Lobesia botrana* (Lepidoptera: Tortricidae). *Bulletin of Entomological Research, 92,* 241-249.

Torres-Vila, L.M. and Rodríguez-Molina, M.C. (2002). Egg size variation and its relationship with larval performance in the Lepidoptera: the case of the European grapevine moth *Lobesia botrana*. *Oikos, 99,* 272-283.

Torres-Vila, L.M., McMinn, M., Rodríguez-Molina, A. and Rodríguez-Molina, M.C. (2006). Primera cita de *Lobesia botrana* Den. et Schiff. (Lepidoptera: Tortricidae) en la isla de Cabrera, Islas Baleares. *Bolletí de la Societat d'Història Natural de Les Balears, 49,* 45-49.

Voukassovitch, P. (1924). *Contribution à l'étude de l'eudémis (Polychrosis botrana Schiff.) de la pyrale de la Vigne (Oenophthira pilleriana Schiff.) et de leurs parasites*, (Thèse Doctoral). Toulouse: Faculté des Sciences de l'Université de Toulouse.

West, S.A. and Cunningham, J.P. (2002). A general model for host plant selection in phytophagous insects. *Journal of Theoretical Biology, 214,* 499-513.

Wiklund, C. and Karlsson, B. (1984). Egg size variation in satyrid butterflies: adaptive vs. historical "Bauplan", and mechanistic explanations. *Oikos, 43*, 391-400.

Williams, G.C. (1966). *Adaptation and natural selection.* Princeton: Princeton University Press.

Winkler, D.W. and Wallin, K. (1987). Offspring size and number: a life history model linking effort per offspring and total effort. *American Naturalist, 129,* 708-720.

Zalucki, M.P., Clarke, A.R. and Malcolm, S.B. (2002). Ecology and behavior of first instar larval Lepidoptera. *Annual Review of Entomology, 47,* 361-393.

Zovi, D., Stastny, M., Battisti, A. and Larsson, S. (2008). Ecological costs on local adaptation of an insect herbivore imposed by host plants and enemies. *Ecology, 89,* 1388-1398.

In: Moths: Types, Ecological Significance and Control... ISBN: 978-1-61470-626-7
Editor: Luis Cauterruccio, pp. 169-186

Chapter 6

SUBLETHAL EFFECTS OF PESTICIDES: THEIR IMPAIRMENT OF BIOLOGY AND PHYSIOLOGY OF EXPOSED MOTHS AND THEIR UNEXPOSED PROGENY

***Zbigniew Adamski*[1] *and Helen Ghiradella*[2]**
[1] Electron and Confocal Microscope Laboratory/Department of Animal Physiology and Developmental Biology, Faculty of Biology, Institute of Experimental Biology, Adam Mickiewicz University, Poznan, Poland; corresponding author.
[2] Department of Biological Sciences, State University of New York, Albany, NY, U.S.A.

ABSTRACT

Pesticides affect not only target species but also non-target organisms that live in exposed areas. Particular species, populations and individuals vary in terms of their resistance; some may survive pesticide exposure while others are more vulnerable. Moth larvae in particular are important pests, but due to various factors affecting exposure, some that are exposed may reach the mature (imago) stage and produce offspring.

Most research on pesticides has focused on their lethal effects, especially ovicidal and larvicidal activity on developmental stages. Data concerning sublethal effects are limited, due to the huge number and vast range of such effects and the fact that until fairly recently the complexity of these systems was not realized. These effects can now be classified in ascending categories, starting from basic biological levels (genetic, biochemical), through more complex phenomena (cellular, histological, and physiological), and on to the levels of single organism and whole population (malformations, behavioral anomalies, reproductive alterations, reduced resistance to endogenous and exogenous stress, etc.).

It is now becoming clear that toxic effects (developmental alterations, decreased survival of larvae or pupae) may not be limited to the exposed generation but may reach into subsequent generations as well, and in their natural environments may subject the animals to intense selective pressure and risk of extinction, leading in turn to serious disturbances within foodwebs.

The authors review current knowledge concerning the abovementioned sublethal effects of pesticides on moths. Among these, reproductive malfunctions are the most frequently reported.

Usually, survival, fecundity (number of eggs), fertility (hatching) and integrity of egg structure are reduced in a dose-dependent manner, although there are cases in which the insecticides do not obviously affect fertility. For example, the chitin synthesis inhibitors, novaluronon and diflubenzuron, and a nicotinoid insecticide, imidacloprid, do not affect the fecundity of *Cydia pomonella, Platynota idaeusalis* and *Tryporyza incertulas*, respectively.

According to the literature, frequently observed decreased fertility may be caused by various factors, including suppression of sperm and/or egg production, impairment of those germ cells that are produced, sublethal genetic abnormalities within these cells, or lethality in the embryonic stage.

In the insecticide-exposed populations, subsequent generations may lay small eggs, and larval and pupal development may be affected too. Newly hatched larvae of moths may show morphologic defects and feeding disabilities, while adult moths show unusual choices for egg laying sites, compared to control (unexposed) moths or moths exposed to other insecticides. Such aberrant effects are especially important in the case of alkaloids and other natural components showing biological activity against moths. Some research groups report a particularly interesting phenomenon, reduction in egg numbers laid by untreated females mated to treated males or vice versa. This can be an early stage of so-called "assortative mating," as in cases in which conspecific insects do not mate freely if males and females are raised on different diets.

In this review we discuss, in light of their known chemical and biological activities, sublethal effects of pesticides in different insects, and we speculate on the possible ecological significance of these effects.

Why Are Sublethal Effects so Important?

The English word "pest" originates from the latin word "*pestis*"; *caedo* means "I kill." Therefore, the word "pesticides" includes substances generally used to kill pests. Readers can find a wealth of manuscripts showing the lethal effects i.e., the ovicidal, larvicidal, pupacidal or adulticidal actions, of these substances on insects . However, species and populations often show various levels of tolerance to the same chemicals, and we need to understand sublethal effects as well as lethal ones. Moth larvae are among the most important pests. When exposed to insecticides in the field, some individuals/populations/species can survive, and the sublethal effects of these insecticides may be revealed only during the imaginal stages. Even though a population survives, it can disturb whole foodwebs by decreased availability of "postexposed" generations as food for predators and/or by decreased herbivorous activity of later generations of larvae.. Alternatively, weakened fitness of the populations may lead to their total extinction. Furthermore, due to a wide variety and intensity of effects, large varieties of *pesticides* (not only insecticides) affect insects; these sublethal effects are not sufficiently known and we still need more and more data concerning ranges of their effects on various groups of animals. Exposure to sublethal concentrations may affect the imaginal stage, causing reduced adult emergence (Sáenz-de-Cabezón Irigaray et al., 2005; Jenson et al., 2009), as well as a variety of such effects as malformations of wings (Adamski and Ziemnicki, 2004), behavioral changes (Knight and Flexner, 2007) and ultrastructural abnormalities (Adamski et al. 2005 a, c, 2009a). Malformations within hormonal and reproductive systems may weaken exposed and post-exposed generations (Adamski et al., 2005b).Therefore, careful studies of sublethal effects are very important. Sublethal effects can be divided into three categories, according to the stages at which they appear:

1. sublethal effects revealed by exposed larvae: decreased growth, prolonged duration of larval stage, etc.
2. sublethal effects revealed by pupae and imagos after larval exposure: altered mortality and length and weight of pupae, imaginal malformations, reduced longevity, decreased fecundity (numbers of eggs laid) and fertility (number of eggs hatching), etc.
3. sublethal effects revealed in next generations (F_1 – F_x generations, all stages), after larval (P generation) exposure: limited hatching, disturbed growth, decreased fecundity, and so on.

Simultaneously, sublethal effects can be tested and observed at various levels of biological organization, from the suborganismal level (biochemical and ultrastructural changes, malformations and malfunctioning of tissues and organs), through single organisms (malformations of various stages, behavioral alterations, shorter longevity), to the level of populations (decreased numbers of animals in exposed/postexposed generations, weakened resistance to environmental stress, etc.). In this chapter we would like to focus on possible effects of pesticides used during pre-adult development on the adult stages of animals and on their reproductive abilities, as seen in effects on postexposed F1 and subsequent generations.

What Kind of Sublethal Effects Can We Observe and What Are Their Causes?

To our knowledge, effects of sublethal concentrations of some pesticides on moths were first reported by Kuribayashi (1981) and referred to reproductive abnormalities observed in *Bombyx mori* individuals exposed to organophosphorous and carbamate pesticides. Among the reported abnormalities are decreased fecundity, decreased fertilization, increased mortality of embryos, and decreased hatching. The author described these changes as a novel, ovicidal action. Based on the literature (Borkovel, 1964), the author suggested four manners in which the pesticides may cause the abovementioned effects:

- suppression of sperm and/or egg production,
- impairment of produced germ cells,
- genetic, sublethal abnormalities within these cells,
- lethality to embryos.

Any and all of these leave new generations of moths that are weaker and smaller. In consequence, pesticide usage disturbs the whole trophic chain.

From that time, the number of reports has increased. We have gathered a number of papers concerning effects of pesticides on postexposed stages and generations of insects. These describe effects of various types of substances: a huge group of insect growth regulators, moulting hormone agonists, chitin synthesis inhibitors, ecdysteroid agonists, juvenile hormone mimics, "classical" insecticides: organophosphates, carbamates, nicotinoids, pyrethroids, spinosyn insecticides, diamides, avermectins, plant extracts and other groups (Table 1).

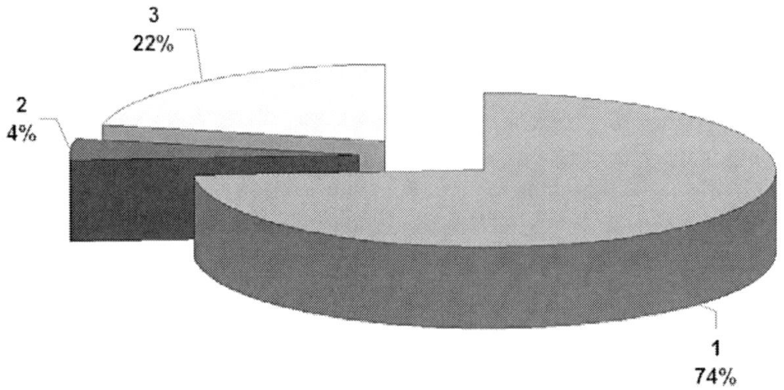

Figure 1. Percentage of reports describing various effects of pesticides on moths' fecundity and fertility: 1 – decreased, 2 – increased, 3 – no significant effect.

Spodoptera, Plutella, Cydia and *Plodia* are among the most frequently tested genera. We can form four groups from the observed effects: A) reproductive effects: altered fecundity and fertility, reduced egg laying, increased egg mortality (i.e. decreased hatching success), B) morphological alterations: changed weight of pupae, malformed structures, malformed eggs, C) biochemical, ultrastructural or anatomical malfunctions: altered enzyme activity or protein content, malformations of organelles, changed size of gonads and D) altered development: postponed development, altered lifespan and duration of stages etc.. (This classification is used in Table 1)

Interestingly, the reports describe negative effects on eggs (decreased hatching) and larvae (increased mortality), whereas effects on fecundity vary: the majority of papers report decreased egg laying, but there are also reports describing no such effects, or even increased egg laying, within pesticide-exposed populations (Figure 1).

Effect of Insecticides on Fecundity and Fertility – To Reproduce or Not to Reproduce? – That Is the Question

Of course the action of xenobiotics differs, depending on the tested chemicals, even though their target tissues and modes of action are judged to be the same (Adamski et al., 2005a and Adamski, 2007). The effects depend on the exposed species and their life stages or on methods of application (Smagghe et al., 2003; Pineda et al., 2007; Pineda et al. 2009). For example, the action of methoxyfenozide - an insecticide that mimics the action of the molting hormone and causes an incomplete and premature molt - depended of the time of application. When adults were topically exposed, fecundity was affected in a dose-dependant manner (Pineda et al., 2007; Pineda et al. 2009), whereas fertility could not be calculated for some of the concentrations, due to abnormal deposition of eggs in clumps (Pineda et al. 2009).

Nevertheless, even the lowest concentration of methoxyfenozide caused a drastic drop in hatching success. Also, a lower percentage of the progeny of insects exposed to this insecticide reached the pupal stage, although the ratio of successful imaginal molts in these cases did not differ from that of the control. This suggests directional selection caused by the pesticide: the weakest animals die, the less susceptible ones survive. Therefore, sublethal effects may be important factors that lead to development of resistant populations.

Reduced fecundity and fertility after exposure to methoxyfenozide were also reported by other authors (Sáenz-de-Cabezón Irigaray et al., 2005). Bylemans and co-workers (2003) pointed at decreased egg deposition and increased egg sterility when *Cydia pomonella* females and males had been exposed, respectively, to this insecticide. Borchert et al. (2005) did not observe any sublethal effect of methoxyfenozide on fecundity and fertility of *Grapholita molesta* in laboratory tests, but during field studies, they found significantly fewer eggs and larvae on methoxyfenozide-treated trees than on control ones. Fruits were also less damaged. Borchert and his co-workers suggested that the difference had been caused more probably by direct mortality than by sublethal effects. This confirms that species may differ in their response to pesticides and may have different levels of susceptibility to these environmental stressors.

Tebufenozide, another moulting hormone agonist, caused effects similar to methoxyfenozide. In field trials some authors observed reduced fecundity and fertility (described as decreased numbers of eggs, reduced hatching success and lower numbers of larvae produced per female) after both 1 and 10-day-long exposures (Knight, 2000) and after chronic exposure of larvae (Biddinger et al., 2006), or topical treatment of pupae (Khebbeb et al., 2008). Biddinger and Hull (1996) also reported reduced fecundity in matings where larvae of both sexes were treated, while Rodreguez et al., (2001) reported lack of oviposition and extinction of exposed populations, together with ovicidal effects, after some exposed generations. Interestingly, oviposition and egg hatching also decreased when only males had been exposed to tebufenozide (Seth et al., 2004). These authors pointed out that such sublethal effects could be caused by relatively low concentrations of insecticides. Therefore, for example, accidental exposure may significantly influence populations and seriously unbalance communities. However, Cadogan et al. (2002) reported lack of significant changes in adult emergence and number of eggs laid by *Choristoneura fumiferana* females exposed to this insecticide during larval development, and Rodriguez and co-authors (2001) did not observe changed fecundity or egg viability for *Diatraea saccharalis* moths exposed to relatively low (EC_5, EC_{15} and EC_{30}) concentrations of tebufenizide.

Other data show decreased fertility and fecundity of moths exposed to other moulting hormone agonists - RH-5849 and the nonsteroidal ecdysteroid agonist 1,2-dibenzoyl-1-tert-butylhydrazine - (Seth et al., 2004), as well as juvenile hormone mimics (Biddinger and Hull, 1999). On the other hand, the chitin synthesis inhibitor, diflubenzuron, did not affect development and reproduction of *Platynota idaeusalis* (Biddinger and Hull, 1999). Another chitin synthesis inhibitor, novaluron, was reported as an agent that does not affect fecundity but does reduce egg hatching of *C. pomonella* (Gökçe et al., 2009). Limited hatching suggests improper embryonic development. If this is true, one can speculate that this insecticide affects not only gonads and their development but can also be transferred to the developing zygote just as are ecdysteroid hormones (Gullan and Cranston, 2010). Buprofezin, the other insecticide of the same class as novaluron, not only did not decrease fecundity, but rather stimulated it in *Tryporyza incertulas* (Wang et al., 2005). However, one must remember that inhibition of chitin synthesis is a reverse process to stimulation of molting.

Effects of endocrine disruptors seem to be species/order-specific, since oviposition and hatching of eggs from adult *Podisus maculiventris* (Heteroptera: Pentatomidae) on novaluron treated plants was significantly reduced (Cutler et al., 2006). In one of the early papers focusing on such sublethal effects, Dorcas et al. (1992) reported various responses of some species of Lepidoptera, Coleoptera, Heteroptera, Diptera and Homoptera to the ecdysteroid agonist, 1,2-dibenzoyl-I-tertbutylhydrazine (RH 5849). The observed sublethal effects concerned delayed postembryonic development, molting disturbances, malformed structures (crochets, in the case of lepidopteran species), and decreased body weight. Thus, pesticides produce a wide range of sublethal effects that may be revealed at various levels and with various strengths.

Moulting hormone agonists may affect the hormonal equilibrium within exposed animals. These can cause malfunctions of metabolism, including of reproductive processes. Pineda at al. (2006) postulated that such effects are due to malfunction in accumulation of proteins in the eggs. Ecdysone itself stimulates transcription and translation within fat body of adult females exposed to it (Bar-Zev and Kaulenas, 1975). Moreover, the ovaries themselves can be sources of ecdysone-like material in insects (Hagedorn et al., 1975) and can regulate vitellogenesis (in: Nation (ed.), 2008). Therefore, ecdysone mimics are likely to affect the development of eggs, their ability to be fertilized and their hatching success, as seen in limited/decreased numbers of properly developed egg cells. This has been suggested by Smagghe and co-workers (Smagghe and Degheele, 1994; Smagghe et al., 1996), who postulated that moulting hormone agonists have sterilizing effects on moth females, and that they affect ovulation. Therefore, these substances may influence the dynamics of the process of egg production and oviposition. To make the effect more significant, the abovementioned work of Seth et al. (2004) shows that the male reproductive system is affected, too. This can be due the regulation of spermatogenesis by ecdysteroids (Dumser, 1980). Hence, decreased hatching success appears probably due to disturbance within both male and female reproductive systems. Even if a female manages to produce healthy egg cells, the probability of their proper fertilization and proper further development is much lower than in a control. On the other hand, such chitin synthesis inhibitors as diflubenzuron are suggested to block transport of vesicles or inhibit translocation of chitin fibrils within cells (Nakagawa and Matsumura, 1994; Cohen, 2001). Therefore, their effect may not be limited to specific target molecules only.

The papers presented in Table 1 point also to another, very important aspect in moths' resistance to insecticides. Pineda et al. (2007) tested the effect of two pesticides, methoxyfenozide and spinosad. Both of these caused reduced fecundity and fertility of moths. Various types of reproductive disturbances were also reported in the cases of many other pesticides belonging to various chemical groups: spinosyn insecticides (Pineda et al., 2007; Wang et al., 2009), pyrethroids (Ho and Goh, 1984; Jallow and Hoy, 2005), organophosphates (Isaacs et al., 2005; Adamski et al., 2009a; Adamski et al., 2009b), diamides (Knight and Flexner, 2007), avermectin insecticides (Yu et al., 2007), inorganic insecticides (Durmus and Büyükgüzel, 2008), as well as various types of botanical extracts (Charleston et al., 2005; Abd El-Aziz and El-Din 2007; Pineda et al., 2009; Adamski et al., 2009b), fungicides (Adamski and Ziemnicki, 2004), and even herbicides (Pourmirza, 2007). However, imidacloprid did cause a reverse effect – stimulated fecundity (Wang et al., 2005).

Moreover, azinphosmethyl (an organothiophosphate insecticide) and diflubenzuron (a chitin synthesis inhibitor) did not cause developmental and reproductive malfunctions in moths (Biddinger and Hull, 1999). One can note that both organophosphate and nicotinoid insecticides lead to hyperactivity within synapses by inhibiting acetylcholinesterase or by binding to receptors on the postsynaptic memberane, respectively. Therefore, the different modes of action are probably due to direct action of these chemicals on non-target tissues and cells, not to activity via nervous system and nervous regulation of metabolic processes.

Decreased fecundity and fertility may result from various factors, acting independently or acting synergistically. First, exposure to pesticides may reduce moths' vigour, which affects their mating activity (for example Sáenz-de-Cabezón and co-workers, 2006, observed inability to perforate grape berries by larvae exposed to luneuron). Since environmental stress alters energy expendatures (Calow, 1991), less energy is used for physical activity, growth, development and reproduction. The difference is shifted to detoxification processes. Next, egg and sperm production, mating, and chorion production demand a lot of substrates. Due to taste, toxic activity etc., the toxicants often discourage insects from eating; therefore the substrates for production of cells and chorion are limited. Third, as mentioned above, pesticides may imbalance hormonal metabolism, which obviously affects development and physiological functions of exposed organisms. Taken together, all these factors decrease the number of individuals in the next generation. However, decreased reproduction lets insects "focus" on detoxification processes, and postpone development and reproduction (e.g. egg laying) (Radwan et al., 1978; Kajiura et al., 1987; Dorcas et al., 1992; Breuer et al., 2005; Adamski et al., 2005a; Durmuş and Büyükgüzel, 2008; Sadeghi et al., 2008; Wang et al., 2009; Jenson et al., 2009) and reproduction. In consequence, the next generation is given off by the healthiest, most resistant individuals and in environments where the concentration of toxicants may be lower. Otherwise, the next generation (which is weaker, produced by weak parents) would have to survive in unfavourable conditions, with limited energy and food supply, often to became competition for their parents. Hence, reduced and postponed reproduction is mutually beneficial for both generations. Such responses are rather non - specific and cannot be treated as precise bioindicators of stress. They rather show that the environment is somehow poisoned, without pointing to any particular substance. Such "uniform" reaction to various stress factors is somehow favourable for insects. It may be a kind of "standard procedure" involved in response to stress and in letting the insects focus on specific, precise detoxification mechanisms at the level of enzymes, synapses or cells. In other words, they focus first on detoxification, stop other activities, and if/when they are more healthy, they can focus on development and reproduction again.

Effect of Pesticides on Number of Offspring – Is the Pairing Important?

One of the reproductive effects that can be observed after exposure of larvae to pesticides is the limited number of eggs laid by adult females. This seems somehow natural when females are treated with xenobiotics (compare Biddinger et al., 2006). What if males are exposed, and interbreed with non-exposed females? Seth et al. (2004) showed that males' mating success decreases significantly with concentration of the moulting hormone agonist,

RH-5849, to which they were exposed, even if females were not affected. Effects like the those mentioned above were reported by Wei et al. (2004): treatment of *Spodoptera litura* with deltamethrin reduced mating success in pairs containing either males or females reared from larvae exposed to this insecticide. Also, Pineda and his co-workers (2007) reported highly decreased numbers of eggs laid after a cross between unexposed females and exposed males; the percentage of decrease was insignificantly lower than in cases when both male and female were exposed. In contrast to this finding, Khebbeb et al. (2008) found significantly decreased fecundity of pairs where one or both partners were exposed to tebufenozide, especially between exposed male and unexposed female. Adamski et al. (2009a) observed decreased numbers of eggs laid by pairings where males were treated and females not, or inversely. Moreover, the number of eggs laid after such a cross was lower than when both sexes were exposed. We know that insecticides disturb pheromone physiology (Yang and Du, 2003; Wei et al., 2004) and ability to locate a pheromone source (Zhou et al., 2005). Therefore, as environmental stressors, xenobiotics may initiate sympatric speciation by the phenomenon described as assortative mating, which may result in disruptive selection of the original population. This was already described for other species exposed to different diets (Malusa et al., 2005) and may result in reproductive isolation of the two strains. Similar effects were observed also for other insect species and for other environmental factors (Hirsch and Tompkins, 1994) or as an effect of various pheromone phenotypes (Zhu et al., 1997). This shows that sublethal effects may cause not only directional selection, where the exposed population evolves towards a more and more resistant one, but also disruptive selection, leading to the appearance of two different strains of the same species.

MORPHOLOGICAL MALFORMATIONS – HOW DO INSECTICIDES AFFECT MORPHOLOGY?

Numerous authors reported morphological alterations after insect exposure to xenobiotics. Generally, larvae, pupae and imagos are smaller and slighter (Biddinger and Hull, 1999; Rodriguez et al., 2001; Biddinger et al., 2006; Adamski et al. 2009a) and display malformation of crochets (Dorcas et al., 1992). Khebbeb et al. (2008) and Yin et al. (2009) described decreased size and number of eggs laid in generations of moths exposed to tebufenozide or postexposed to spinosad, respectively) . Also, pupae from exposed larvae may be smaller (Breuer et al., 2005), and imagos show morphological malformations (Adamski and Ziemnicki, 2004; Breuer et al., 2005; Abd El-Aziz and El-Di, 2007). These changes are in contrast with findings concerning the juvenile hormone analog, fenoxycarb, which caused an increase in larval weight (Mauchamp et al., 1989).

Fila et al. (2002), and Adamski et al. (2005b; 2009a) reported malformations of eggshells of eggs laid by *S. exigua* moths exposed to fenitrotion (Figure.2). Altered structure of the eggshell could possible change levels of elements and compounds. By elemental mapping and analysis of eggs of the butterfly, *Lycaeides melissa samuelis*, Nickles et al. (2002) found that the egg top surface is generally rich in sulfur, phosphorus and chlorine, while the micropylar region (where the sperm must enter) is a "hot spot" for calcium and phosphorus. The authors speculate that the localization of these last two elements may signal the micropyle location to the sperm. If so, any disturbance here, chemical or otherwise, could lead to decreased fertilization which, together with decreased vitality of germ cells and embryos and higher mortality of organisms due improper eggshell structure (see below) would lead to decreased hatching success. Changes in egg structure may also lead to egg desiccation due to water diffusion and to easier predation.

Figure 2. Malformations of micropylar region of an egg laid by *Spodoptera exigua* female dosed with fenitrothion during the larval stage.

These changes, together with the abovementioned sublethal effects, may lead to female sterility, drastically reduced life span (Abd El-Aziz and El-Di, 2007) or inability to produce and lay eggs.

The next generation may also show lower growth rates (Jallow and Hoy, 2005). Such sublethal effects would influence whole exposed populations and their dynamics. More important, any such effect in the natural environment could well be of more critical concern than for protected laboratory strains. In general, weaker, smaller and slower-developing populations would not be expected to resist stress factors as healthy populations would.

Among the most interesting findings of our own research was indeed the inheritance of eggshell malformations (Adamski et al., 2005b) (Fig 2). Malformed structures could be observed for three postexposed generations, although their intensity was progressively lower. Inheritable sublethal effects of pesticides were also observed in rats exposed to the organochlorine insecticide, methoxychlor (Anway et al., 2005). These authors suggest epigenetic or chromosomal alteration in the germ line, which in turn also suggests that sublethal effects may act as 'long-term' harm for the exposed populations. However (again), morphological malformations are not largely chemical-specific. This may be due to a complexity of effects: for example, the malfunctioning of enzymes, blocked at a sublethal level by a given xenobiotic, might show effects at various levels ranging from cytophysiology to behaviour. Therefore, one might expect complex and multilevel effects.

Biochemical, Ultrastructural and Anatomical Effects

Such a wide range of sublethal effects has to have representation at the level of organs and cells. Not only may target tissues and organs be exposed to potent stress, which affects condition of whole populations, but nontarget tissues and organs as well. These changes may be of economical importance; for example, methoprene can cause significant growth of silk glands (and other tissues, too) in *B. mori* larvae (Kajiura and Yamashita, 1989). Increased size of larval tissues of insects exposed to methoprene is in tune with increased overall body size of larvae exposed to another juvenile hormone mimic, fenoxycarb (Mauchamp et al., 1989).

Various abnormalities were observed within reproductive organs of insects exposed to pesticides. The testes of *B. mori* larvae exposed to methoprene revealed increased percentages of G1 somatic cells, spermatogonia and secondary spermatocytes, and fewer somatic and germinal cells (Kajiura et al., 1993). On the other hand, larvae or pupae exposed to ecdysone-agonists exhibited smaller ovaries (Khebbeb et al., 2008) and testes in a dose-dependent manner (Seth et al., 2004; Khebbeb et al., 2008). Diflubenzuron, a chitin synthesis inhibitor resulted in ultrastructural abnormalities of gonads: decreased thickness of the follicular epithelium, size of the basal oocytes and number of oocytes per ovary (Soltani and Soltani-Mazouni, 1992). Seth and co-workers (2004) reported a significant reduction in the number of types of sperm cell bundles and in the dynamics of their physiology. This of course may affect fertility and hatching success. Lower quality of germ cells to some extent may be compensated for by a higher number of delivered sperm. Maybe that was the reason the authors observed increased numbers of matings. Such behavioral change can also be explained by the very probable phenomenon of transfer of toxic substances (hormone agonists) with spermatophores. (Seth and co-workers noted shorter lifespans of control females mated with insecticide-treated males.) In this case mating would be rather a type of detoxification strategy for males, instead of or in addition to transfer of germ cells. Hence, frequent mating would be beneficial to exposed males. Since pesticides may decrease numbers of eggs laid by control females mated with exposed males (Adamski et al., 2009a), we should also look for possible abnormalities in the males. We did not observe significant ultrastructural malformations within sperm cells. However, crypts producing sperm cells did reveal some malformations (Adamski et al., 2005c). We suggested that while sperm cells might have been of a good quality themselves, their numbers (i.e. numbers of spermatophores) had not been satisfactory. Unfortunately, we do not have data concerning longevity and viability of sperm cells. But assuming that the lifespan of germ cells may be negatively affected by exposure to xenobiotics, frequent mating would be both behaviorally and physiologically beneficial. Our research (Adamski et al., 2009c) shows that follicle cells are also targets of pesticide action. We reported altered structure of nuclei and their envelopes, malformed (and probably malfunctioning) mitochondria, lost connections between follicle cells and some changes within cytoplasm (Figure 3). These alterations obviously can lead to the ultrastructural changes observed within the chorion structure. Observations on reproductive systems show that not only target tissues and organs may be exposed to potent stress, but nontarget tissues and organs, too. Since these ultrastructural changes are similar to those observed in other insect species and in various tissues, and in cases of the same and different pesticides (Retnakaran et al., 1997; Adamski, 2007; Adamski et al. 2005a;

Mehlhorn, 2001), they are presumably caused by unspecific chemical interaction of insecticides with biomolecules, as well as by increased oxidative stress due to enhanced levels of reactive oxygen species (Adamski et al., 2003; Içen et al., 2005; Büyükgüzel, 2009). Oxygen free radicals interact with proteins, nucleic acids, hydrocarbons and other molecules. Therefore, these highly reactive molecules may cause various unspecific effects. Various biochemical alterations may take place within exposed gonads. Khebbeb et al. reported a tebufenozide concentration - dependant decrease in carbohydrate, protein and lipid content in the testes of male mediterranean flour moths, *Ephestia Kuehniella,* exposed to this insecticide. Interestingly, females showed decreased content of carbohydrates and lipids but not of proteins. Therefore, males and their reproductive cells seem to be more vulnerable to these insecticides, which may be why crosses with exposed males give fewer offspring Similar results were reported by Soltani and Soltani-Mazouni (1992): exposure to diflubenzuron caused reduced protein content in ovaries of *C. pomonella.* On the other hand, Wang et al. (2005) observed increased content of proteins in ovaries of *T. incertulas* exposed to buprofezin in the diet. In this case also, the effect on fecundity was reversed – this chitin synthesis inhibitor increased fecundity and larval weight.

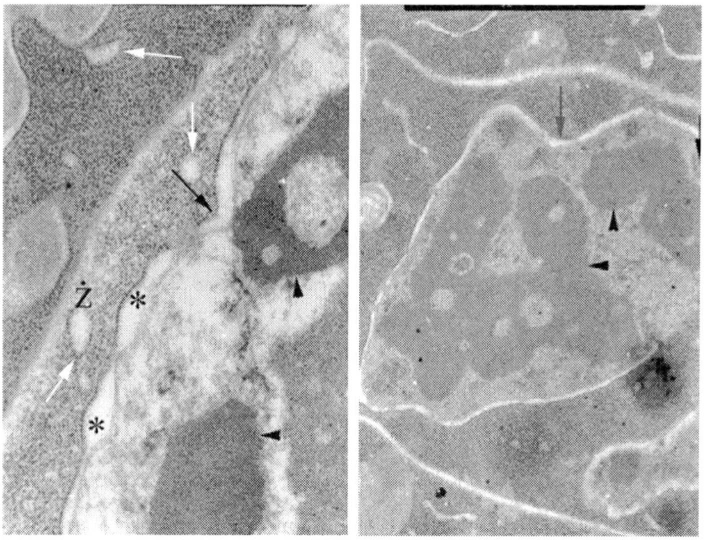

Figure 3. Ultrastructural malformations of follicular cells of imagos exposed to fenitrothion during larval instars. Note electron dense cytoplasm, swollen RER (white arrows), invaginations of nuclear envelope (black arrows) and its swelling (asterisks), and chromatin condensation (arrowheads).

This shows that although pesticides cause some universal or at least similar effects, their action, especially as hormone-mimic-substances, cannot be treated in common. Taken together, the variability of substances that cause malformations and the number of affected levels of biological organization prove that the research on sublethal effects is very important. Chemical interactions between pesticides and moths' cellular content (enzymes, structural proteins, hormones, other macromolecules) lead to various, often nonspecific reactions. These then influence cellular metabolism and tissue and organ physiology and structure, and in consequence they weaken organisms. This affects single organisms and then, since populations are groups of organisms, whole exposed populations and eventually ecosystems.

Although we may not observe massive mortality after exposure of insects to certain chemicals, the activity of these chemicals may manifest itself in weakened, reduced postexposed generations. To sum up, we must agree with Haynes (1988) that sublethal effects on reproduction may be caused by any class of pesticide. Therefore, one must be aware of the importance of potential extended effects before we introduce any xenobiotic to the environment. The variability of sublethal, apparently subtle effects proves that the use of these substances should be constantly monitored and that varied and multilevel research on any possible effects needs to be carried on.

Table 1. Observed sublethal effects of pesticides on moth fecundity and fertility

Pesticide	Species	Observed effects	Reported in:
Methoxyfenozide (moulting hormone agonist)	*Spodoptera littoralis*	A	Pineda et al., 2009
	Spodoptera littoralis	A	Pineda et al., 2007
	Cydia pomonella	A	Bylemans et al., 2003
	Spodoptera exigua	A	Osorio et al., 2008
	Endopiza viteana	A	Isaacs et al., 2005
	Lobesia botrana	A, D	Sáenz-de-Cabezón Irigaray et al., 2005
	Grapholita molesta	A, D.	Borchert et al., 2005
Tebufenozide (moulting hormone agonist)	*Endopiza viteana*	A	Isaacs et al., 2005
	Platynota idaeusalis	A, B	Biddinger and Hull, 1999
	Platynota idaeusalis	A, B, D	Biddinger et al., 2006
	Choristoneura fumiferana	A	Cadogan et al., 2002
	Spodoptera litura	A, B, C, D	Seth et al., 2004
	Diatraea saccharalis	A, B	Rodriguez et al., 2001
	Ephestia kuehniella	A, B, C	Khebbeb et al., 2008
	Cydia pomonella	A	Knight 2000
RH-5849 (1,2-dibenzoyl-1-tert-butylhydrazine, moulting hormone agonist)	*Spodoptera litura*	A, B, C, D	Seth et al., 2004
RH-5992 (non-steroidal ecdysone agonist	*Choristoneura fumiferana*	C	Retnakaran et al., 1997
Fenoxycarb (**juvenile hormone mimics)**	*Platynota idaeusalis*	A, B	Biddinger and Hull, 1999
	Heliothis virescens	B, D	Mauchamp et al., 1989
Methoprene (**juvenile hormone mimics)**	*Plodia interpunctella*	D	Jenson et al., 2009
	Bombyx mori	B, C, D	Kajiura et al., 1987
	Bombyx mori	C	Kajiura and Yamashita., 1989
	Bombyx mori	C	Kajiura et al., 1993
Novaluron (chitin synthesis inhibitor)	*Cydia pomonella*	A	Gökçe et al., 2009
Buprofezin (chitin synthesis inhibitor)	*Tryporyza incertulas*	A, B, C	Wang et al., 2005
Diflubenzuron (chitin synthesis inhibitor)	*Platynota idaeusalis*	A, D	Biddinger and Hull 1999
	Spodoptera littoralis	A, D	Radwan et al., 1978
	Cydia pomonella	C, D	Soltani and Soltani-Mazouni 1992
Azinphosmethyl (organothiophosphate insecticide)	*Endopiza viteana*	A, D	Isaacs et al., 2005
	Platynota idaeusalis	A, D	Biddinger and Hull, 1999

Pesticide	Species	*Observed effects*	Reported in:
Fenitrothion (organothiophosphate insecticide)	*Spodoptera exigua*	A, B, D	Adamski et al., 2009a
	Spodoptera exigua	B	Fila et al., 2002
	Spodoptera exigua	B	Adamski et al. 2005b
	Spodoptera exigua	A	Adamski et al., 2009b
	Spodoptera exigua	B, C	Adamski 2007
Malathion (organothiophosphate insecticide)	*Galleria mellonella*	C	Büyükgüzel, E., 2009
Methyl parathion and ethyl parathion (organophosphorus insecticides)	*Galleria mellonella*	C	Içen et al., 2005
Imidacloprid (**nicotinoid insecticide**)	*Tryporyza incertulas*	A	Wang et al., 2005
Spinosad (spinosyn insecticide)	*Spodoptera littoralis*	A	Pineda et al., 2007
	Helicoverpa armigera	A, B, D	Wang et al., 2009
	Plutella xylostella	B	Yin et al., 2009
Deltamethrin (pyrethroid insecticide)	*Plutella xylostella*	A	Ho and Goh, 1984
Permethrin (pyrethroid insecticide)	*Plutella xylostella*	A, B, D	Jallow and Hoy, 2005
Chlorantranilipole (**diamide insecticide**)	*Cydia pomonella*	A	Knight and Flexner, 2007
Emamectin (**avermectin insecticide**)	*Plutella xylostella*	A, D	Yu et al., 2007
Abamectin	*Platynota idaeusalis*	D	Biddinger and Hull, 1999
Lufenuron	*Lobesia botrana*	D	Sáenz-de-Cabezón et al., 2006
Sodium tetraborate (inorganic insecticide)	*Galleria mellonella*	D	Durmuş and Büyükgüzel, 2008
Boric acid	*Galleria mellonella*	B	Hyršl et al. 2008
Azadirachtin (**botanical insecticides**)	*Spodoptera littoralis*	A, D	Pineda et al., 2009
Syringa extracts	*Plutella xylostella*	A	Charleston et al., 2005
Neem extracts	*Plutella xylostella*	A	Charleston et al., 2005
Solanum extract	*Spodoptera exigua*	A	Adamski et al., 2009b
Anabasis setifera, Cleom amblyocarpa, Maytenus senegalensis, Solenostemma argel extracts	*Spodoptera littoralis*	A, B	Abd El-Aziz and El-Din 2007
Sophora alopecuroids extract	*Plutella xylostella*	A, D	Yu et al., 2007
Acrolein (unclassified herbicide)	*Plodia interpunctella*	A	Pourmirza, 2007
Lectins (*Allium sativum*)	*Spodoptera littoralis*	B, C	Sadeghi A et al., 2008
Mancozeb (dithiocarbamate fungicide)	*Spodoptera exigua*	A	Adamski et al., 2009b
	Spodoptera exigua	B, D	Adamski and Ziemnicki, 2004
Acetone	*Plodia interpunctella*	A	Pourmirza et al, 2007
(E)-5-(2-bromovinyl)-2'-deoxyuridine (BVDU) (antiherpetic agent, nucleoside analogue)	*Spodoptera frugiperda*	B	Breuer et al., 2005

Groups of effects are described as letters: A – effect on fecundity, fertility, B – morphological alterations, C – biochemical, ultrastructural and anatomical alterations, D – altered development. Classification according to Compendium of Pesticide Common Names: http://www.alanwood.net/pesticides/index.html

References

Abd El-Aziz SE., El-Din AA. (2007) Insecticidal activity of some wild plant extracts against cotton leafworm, Spodoptera littoralis (Boisd.) (Lepidoptera: Noctuidae). *Pak J. Biol. Sci.* 10: 2192-2197.

Adamski Z., Ziemnicki K., Fila K., Žikić R., Štajn A. (2003) Effects of long-term exposure to fenitrothion on *Spodoptera exigua* and *Tenebrio molitor* larval development and antioxidant enzyme activity. *Bioll Lett.* 40: 43-52.

Adamski Z., Banaszkiewicz M., Ziemnicki K. (2005a). Ultrastructural and developmental alterations in larvae of *Tenebrio molitor* L. (Insecta, Coleoptera) induced by sublethal concentrations of fenitrothion. *J. Biol. Res.* 3:15–22.

Adamski Z. (2007) Exposure to carbaryl leads to ultrastructural changes and alters activity of antioxidant enzymes in *Spodoptera exigua* (Lepidoptera: Noctuidae). *Invertebr. Biol.* 126: 191–201.

Adamski Z., Machalska K., Chorostkowska K., Niewadzi M., Ziemnicki K., Hirsch HVB. (2009a) Effects of sublethal concentrations of fenitrothion on beet armyworm (Lepidoptera: Noctuidae) development and reproduction. *Pestic. Biochem. Physiol.* 94: 73–78.

Adamski Z., Korolczuk D., Purgiel M., Grewling A., Ratajczak M., Bankiet M., Ziemnicki K. (2009c) Effect of fenitrothion on *Spodoptera exigua* larval development and ultrastructure of follicle cells. *Biologia* 64: 197—202.

Adamski Z., Niewadzi M., Ziemnicki K. (2005b) Inheritance of chorionic malformations and insecticide resistance by *Spodoptera exigua. J. Appl. Entomol.* 129: 526–533.

Adamski Z., Purgiel M., Nogaj D., Grewling A., Ziemnicki K. (2005c). Sublethal concentrations of fenitrothion do not cause significant ultrastructural malformations of sperm cells in *Spodoptera exigua* moths. *Pestycydy* (Pesticides). 1-2: 23-29.

Adamski Z., Ziemnicki K. (2004) Side-effects of mancozeb on *Spodoptera exigua* (Hübn.) *larvae J. Appl. Entomol.* 128: 212–217.

Adamski Z., Ziemnicki K., Marciniak P., Halamunda J., Nawrocka M., Lelario F., Scrano L., Bufo SA. (2009b) Effect of various xenobiotics on hatching success of *Spodoptera exigua* eggs as compared to a natural plant extract. *J. Toxicol. Environ. Health*, Part A. 72: 1132–1134.

Anway MD., Cupp AS., Uzumcu M., Skinne MK. (2005) Epigenetic transgenerational actions of endocrine disruptors and male fertility. *Science* 308: 1466–1469.

Bar-Zev A., Kaulenas MS. (1975) The effect of β-ecdysone on *Gromphadorhina* adult female fat body at the transcriptional and translational levels. *Comp. Biochem. Physiol.* B. 51: 355-361.

Biddinger D., Hull L., Huang H., McPheron B., Loyer M. (2006) Sublethal effects of chronic exposure to tebufenozide on the development, survival, and reproduction of the tufted apple bud moth (Lepidoptera: Tortricidae). *J Econ Entomol.*, 99:834-842.

Biddinger DJ., Hull LA., Mcpheron BA. (1996) Cross-resistance and synergism in azinphosmethyl resistant and susceptible strains of tufted apple bud moth (Lepidoptera: Tortricidae) to various insect growth regulators and abamectin. *J. Econ. Entomol.*, 89: 274-287.

Biddinger DJ.; Hull LA. (1999) Sublethal effects of selected insecticides on growth and reproduction of a laboratory susceptible strain of tufted apple bud moth (Lepidoptera: Tortricidae). *J. Econ. Entomol.* 92: 314-324.

Borchert DM., Walgenbach JF., Kennedy GG. (2005) Assessment of sublethal effects of methoxyfenozide on oriental fruit moth (Lepidoptera: Tortricidae). *J. Econ. Entomol.*, 98:765-771

Borkovel AB. (1964) Insect chemosterilants: Their chemistry and application. *Residue Rev.* 6: 87-103.

Breuer M., De Loof A., Balzarini J., Huybrechts R. (2005) Insecticidal activity of the pyrimidine nucleoside analogue (E)-5-(2-bromovinyl)-2'-deoxyuridine (BVDU). *Pest Manag. Sci.* 61: 737-741.

Büyükgüzel E. (2009) Evidence of oxidative and antioxidative responses by *Galleria mellonella* larvae to malathion. *J. Econ. Entomol.* 102: 152-159.

Bylemans D., De Maeyer L, Auwerkerken A., De Craen H., Wijsmuller JW., Peeters D. (2003) Methoxyfenozide, a reliable IPM compatible compound against Lepidoptera in pome fruit and vegetables with sterilising, ovicidal and larvicidal efficacy on codling moth. *Commun Agric. Appl. Biol. Sci.*, 68:189-202.

Cadogan BL., Scharbach RD., Krause RE., Knowles KR. (2002) Evaluation of tebufenozide carry-over and residual effects on spruce budworm (Lepidoptera: Tortricidae). *J. Econ. Entomol.*, 95:578-586.

Calow P. (1991) Physiological costs of combating chemical toxicants: ecological implications. *Comp. Biochem. Physiol C.* 100: 3-6.

Charleston DS., Kfir R., Vet LE., Dicke M. (2005) Behavioural responses of diamondback moth *Plutella xylostella* (Lepidoptera: Plutellidae) to extracts derived from *Melia azedarach* and *Azadirachta indica. Bull. Entomol. Res.* 95: 457-465.

Cohen E. (2001) Chitin synthesis and inhibition: a revisit. *Pest Manag. Sci.* 57: 946-950.

Compendium of Pesticide Common Names: http://www.alanwood.net/pesticides/index.html. Last update: Nov. 2010.

Cutler GC., Scott-Dupree CD., Tolman JH., Harris CR. (2006) Toxicity of novaluron to the non-target predatory bug *Podisus maculiventris* (Heteroptera: Pentatomidae). *Biol. Control.* 38: 196-204.

Dorcas B., Polar L., Tag E-D., Mahmoud H., Eross K., Wing KD. (1992) Developmental disturbances in different insect orders caused by an edy steroid agonise, RH 5849. *J. Econ. Entomol.*, 85: 2107-2112.

Dumser JB. (1980) In vitro effects of ecdysterone on the spermatogonial cell cycle in *Locusta. Int. J. Invert. Rep.* 2: 165-174

Durmuş Y., Büyükgüzel K. (2008) Biological and immune response of *Galleria mellonella* (Lepidoptera: Pyralidae) to sodium tetraborate. *J. Econ. Entomol.* 101: 777-783.

Fila K., Adamski Z., Ziemnicki K. (2002) Exposure to fenitrothion causes malfunctions of *Spodoptera exigua* Hübn. (Lep., Noctuidae) eggs. *J. Appl. Ent.* 126, 114–118.

Gökçe A., Kim SH., Wise JC., Whalon ME. (2009) Reduced egg viability in codling moth *Cydia pomonella* (L.) (Lepidoptera: Tortricidae) following adult exposure to novaluron. *Pest Manag. Sci.* 65:283-287.

Gullan PJ., Cranston PS. (2010) The Insects: An outline of entomology. (IV) Wiley-Blackwell.

Hagedorn HH., O'Connor JD., Fuchs MS., Sage B., Schlaeger DA., Bohm MK. (1975) The ovary as a source of alpha-ecdysone in an adult mosquito. *Proc. Natl. Acad. Sci. USA*. 72: 3255–3259.

Haynes KF. (1988) Sublethal Effects of Neurotoxic Insecticides on Insect Behavior. *Annu.Rev. Entomol.* 33: 149-168.

Hirsch HVB., Tompkins L. (1994) The flexible fly: experience-dependent development of complex behaviors in *Drosophila melanogaster*, *J. Exp. Biol.* 195: 1–18.

Ho SH., Goh PM. (1984) Deltamethrin as a potential ovicidal pyrethroid against *Plutella xylostella* L. *Toxicol Lett.* 22:161-164.

Hyršl, P., Büyükgüzel, E., Büyükgüzel, K. (2008) Boric acid-induced effects on protein profiles of *Galleria mellonella* hemolymph and fat body. *Acta Biol. Hung.* 59: 281-288.

Içen E., Armutçu F., Büyükgüzel K. , Gürel A. (2005) Biochemical stress indicators of greater wax moth exposure to organophosphorus insecticides. *J. Econ. Entomol.* 98: 358-366.

Isaacs R., Mason KS., Maxwell E. (2005) Stage-specific control of grape berry moth, *Endopiza viteana* (Clemens) (Lepidoptera: Tortricidae), by selective and broad-spectrum insecticides. *J. Econ. Entomol.* 98:415-422.

Jallow MF., Hoy CW. (2005) Phenotypic variation in adult behavioral response and offspring fitness in *Plutella xylostella* (Lepidoptera: Plutellidae) in response to permethrin. *J. Econ. Entomol.* 98: 2195-2202.

Jenson EA., Arthur FH., Nechols JR. (2009) Efficacy of methoprene applied at different temperatures and rates on surface substrates to control eggs and fifth instars of *Plodia interpunctella*. *J. Econ. Entomol.* 102:1992-2002.

Kajiura Z., Kadono-Okuda K., Yamashita O. (1987) Induction of dauer larvae by a juvenile hormone analogue and their response to ecdysteroids in the silkworm, *Bombyx mori*. *J. Seric. Sci. Jpn.* 56: 398-406.

Kajiura Z., Nakagaki M., Takei. R. (1993). Spermiogenesis of the testes of juvenile hormone-treated silkworm larvae, *Bombyx mori*. *Comp. Biochem. Physiol.* A: 106: 495-499.

Kajiura Z., Yamashita O. (1989) Super growth of silkglands in the daure larvae of the silkworm, *Bombyx mori*, induced by a juvenile hormone analogue. *J. Seric. Sci. Jpn.* 58: 39-46.

Khebbeb MEH., Gaouaoui R., Bendjeddou F. (2008) Tebufenozide effects on the reproductive potentials of the mediterranean flour moth, *Ephestia kuehniella*. *Afr. J. Biotechnol.* 7: 1166-1170.

Knight AL. (2000) Tebufenozide targeted against codling moth (Lepidoptera: Tortricidae) adults, eggs, and larvae. *J. Econ. Entomol.* 93:1760-1767.

Knight AL., Flexner L. (2007) Disruption of mating in codling moth (Lepidoptera: Tortricidae) by chlorantranilipole, an anthranilic diamide insecticide. *Pest Manag. Sci.* 63: 180-189.

Kuribayashi S. (1981) Studies on the effect of pesticides on the reproduction of the silkworm, *Bombyx mori* L. I. Effects of chemicals administered during the larval stage on egg-laying and hatching. *J. Toxicol. Sci.* 6:169-176.

Malausa T., Bethenod MT., Bontemps A., Bourguet D., Cornuet JM., Ponsard S. (2005) Assortative mating in sympatric host races of the European corn borer, *Science*. 308: 258–260.

Mauchamp B., Malosse C., Saroglia P. (1989) Biological effects and metabolism of fenoxycarb after treatment of the fourth and the fifth instars of the tobacco budworm, *Heliothis virescens* F. *Pesticide Sci.* 26: 283–301.

Mehlhorn H., Hansen O., Mencke N. (2001) Comparative study on the effects of three insecticides (fipronil, imidacloprid, selamectin) on developmental stages of the cat flea (Ctenocephalides felis Bouche 1835): a light and electron microscopic analysis of in vivo and in vitro experiments. *Parasitol. Res.* 87: 198–207.

Nakagawa Y., Matsumura F. (1994) Diflubenzuron affects gamma-thioGTP stimulated Ca2+ transport in vitro in intracellular vesicles from the integument of the newly molted American cockroach, *Periplaneta americana* L. *Insect Biochem. Mol. Biol.* 24: 1009-1015.

Nation JL. (2008) *Insect Physiology and Biochemistry*, Second Edition. CRC Press.

Nickles EP., Ghiradella H., Bakhru H., Haberl A. (2002) Egg of the Karner Blue Butterfly (*Lycaeides Melissa samuelis*): Morphology ana elemental analysis. *J. Morphol.* 251: 140-148.

Osorio A., Martínez AM., Schneider MI., Díaz O., Corrales JL., Avilés MC., Smagghe G., Pineda S. (2008) Monitoring of beet armyworm resistance to spinosad and methoxyfenozide in Mexico. *Pest Manag. Sci.* 64:1001-1007.

Pineda S., Martínez A-M., Figueroa J-I., Schneider M-I., Del Estal P., Viñuela E., Benjamín Gómez B., Smagghe G., Budia F. (2009) Influence of Azadirachtin and Methoxyfenozide on Life Parameters of *Spodoptera littoralis* Lepidoptera: Noctuidae). *J. Econ. Entomol.* 102:1490-1496.

Pineda S., Schneider M-I., Smagghe G., Martínez A-M., Del Estal P., Viñuela E., Valle J., Budia F. (2007) Lethal and sublethal effects of methoxyfenozide and spinosad. *J. Econ. Entomol.* 100:773-780.

Pineda S., Smagghe G., Schneider MI., Del Estal P., Viñuela E., Martínez AM., Budia F. (2006) Toxicity and pharmacokinetics of spinosad and methoxyfenozide to *Spodoptera littoralis* (Lepidoptera: Noctuidae). *Environ. Entomol.* 35: 856-864.

Pourmirza AA. (2007) Ovicidal activity of acrolein vapors to Indian meal moth eggs of various ages. *Pak. J. Biol. Sci.* 10: 2945-2949.

Radwan HSA., Abo-Elghar MR., Ammar IMA. (1978), Reproductive performance of Spodoptera littoralis (Boisd.) treated topically with sublethal doses of an antimoulting IGR (Dimilin). *Zeitschrift für Angewandte Entomol.* 86: 414–419.

Retnakaran A., MacDonald A., Tomkins WL., Davis CN., Brownwright AJ., Palli SR. (1997) Ultrastructural effects of a non-steroidal ecdysone agonist, RH-5992, on the sixth instar larva of the spruce budworm, *Choristoneura fumiferana. J. Insect Physiol.*, 43: 55-68.

Rodriguez LM., Ottea JA., Reagan TE. (2001) Selection, egg viability, and fecundity of the sugarcane borer (Lepidoptera: Crambidae) with tebufenozide. *J. Econ. Entomol.* 94:1553-1557.

Sadeghi A., Smagghe G., Broeders S., Hernalsteens JP., De Greve H., Peumans WJ., Van Damme EJ. (2008) Ectopically expressed leaf and bulb lectins from garlic (*Allium sativum* L.) protect transgenic tobacco plants against cotton leafworm (*Spodoptera littoralis*). *Transgenic Res.* 17: 9-18.

Sáenz-de-Cabezón Irigaray FJ., Marco V., Zalom FG., Pérez-Moreno I. (2005) Effects of methoxyfenozide on *Lobesia botrana* Den and Schiff (Lepidoptera: Tortricidae) egg, larval and adult stages. *Pest Manag. Sci.*, 61:1133-1137.

Sáenz-de-Cabezón FJ., Pérez-Moreno I., Zalom FG., Marco V. (2006) Effects of lufenuron on *Lobesia botrana* (Lepidoptera: Tortricidae) egg, larval, and adult stages. *J. Econ. Entomol.* 99: 427-431.

Seth RK., Kaur JJ., Rao DK., Reynolds SE. (2004) Effects of larval exposure to sublethal concentrations of the ecdysteroid agonists RH-5849 and tebufenozide (RH-5992) on male reproductive physiology in *Spodoptera litura* . *J. Insect. Physiol.*, 50:505-517.

Smagghe G., Pineda S., Carton B., Del Estal P., Budia F., Viñuela E. (2003) Toxicity and kinetics of methoxyfenozide in greenhouse-selected *Spodoptera exigua* (Lepidoptera: Noctuidae). *Pest Manag. Sci* 59: 1203 -1209.

Smagghe G., Degheele D. (1994) Action of a novel nonsteroidal ecdysteroid mimics, tebufenozide (RH-5992), on insects of different orders. *Pestic. Sci.* 42: 85-92.

Smagghe G., Viñuela E., Budia F., Degheele D. (1996) In vivo and in vitro effects of the nonsteroidal ecdysteroid agonist tebufenozide on cuticle formation in *Spodoptera exigua*: An ultrastructural approach. *Archiv. Insect Biochem. Physiol.* 33: 121-134.

Soltani N. Soltani-Mazouni N. (1992) Diflubenzuron and oogenesis in the codling moth, *Cydia pomonella* (L.). *Pestic. Sci.* 34: 257–261.

Wang AH., Wu JC., Yu YS., Liu JL., Yue JF., Wang MY. (2005) Selective insecticide-induced stimulation on fecundity and biochemical changes in *Tryporyza incertulas* (Lepidoptera: Pyralidae). *J. Econ. Entomol.* 98:1144-1149.

Wang D., Gong P., Li M., Qiu X., Wang K. (2009) Sublethal effects of spinosad on survival, growth and reproduction of *Helicoverpa armigera* (Lepidoptera: Noctuidae). *Pest Manag. Sci.* 65:223-227.

Wei H., Huang Y., Du J. (2004) Sex pheromones and reproductive behavior of *Spodoptera litura* (Fabricius) moths reared from larvae treated with four insecticides. *J. Chem. Ecol.* 30: 1457-1466.

Yang Z., Du JW. (2003) Effects of Sublethal Deltamethrin on the Chemical Communication System and PBAN Activity of Asian Corn Borer, *Ostrinia furnacalis* (Güenee). *J. Chem. Ecology*. 29: 1611-1619.

Yin XH., Wu QJ., Li XF., Zhang YJ., Xu BY. (2009) Demographic changes in multigeneration *Plutella xylostella* (Lepidoptera: Plutellidae) after exposure to sublethal concentrations of spinosad. *J. Econ. Entomol.* 102:357-365.

Yu TC., Luo WC., Ding J., Yan L., Xiao T., Niu HT. (2007) Effects of applying *Sophora alopecuroids* extracts and emamectin on the growth, development, and fecundity of diamondback moth *Plutella xylostella*. Ying Yong Sheng Tai Xue Bao. 18: 2791-2794.

Zhou H., Du J., Huang Y. (2005) Effects of Sublethal Doses of Malathion on Responses to Sex Pheromones by Male Asian Corn Borer Moths, *Ostrinia furnacalis* (Guenée). *J. Chem. Ecology*. 31: 1645-1656.

Zhu J., Chastain BB., Spohn BG., Haynes KF. (1997) Assortative mating in two pheromone strains of the cabbage looper moth, *Trichoplusia ni*. J. Insect Behav. 10: 805-817.

Reviewed by prof. Kemal Büyükgüzel, Department of Biology, Faculty of Arts and Science, Karaelmas University, Zonguldak, Turkey

In: Moths: Types, Ecological Significance and Control... ISBN: 978-1-61470-626-7
Editor: Luis Cauterruccio, pp. 187-205

Chapter 7

GENETICS OF INTERACTIONS AMONG MOTHS, THEIR HOST PLANTS AND ENEMIES IN CRIMEAN OAK FORESTS, AND ITS PERSPECTIVE FOR THEIR CONTROL

Andrei P. Simchuk[1,*], Volodymyr V. Oberemok[2] and Anatoly V. Ivashov[1]
[1] Department of Ecology, V.I. Vernadsky Taurida National University
Vernadsky Ave., 4,Simferopol, Ukraine 95007.
[2] Department of Biochemistry, V.I. Vernadsky Taurida National University.

ABSTRACT

Any organism interacts with an environment in accordance to its genetic constitution, raises unique requirements to the environment and in its own manner interacts with representatives of other species. Thus, intra-population genetic variation should strongly influence integrative ecological pattern of the population in the community or ecosystem. This thesis is a basic principle of our investigation on ecological ties among moths and other species, with which they interact in forest ecosystems. Oak leaf roller and, gypsy moths strongly contribute to a pathological situation in the Crimean forests. Application of allozymes, random and specific DNA markers has shown that the consequences of interaction among the moths, their host plants, parasitoids and micro-pathogens strongly depend on their genetics, raising ties among gene pools of their populations. Evolution of the ties during long co-existence of the moths with host plants and enemies forms natural mechanisms of their regulation. These mechanisms have a great perspective in respect to their application in agricultural environments, changing aim of pest control from pest elimination to control of its evolution towards decrease in its damaging capacity.

Long co-existence of moths and pathogenic viruses forms feedbacks between their genomes. Nuclear polyhedrosis viruses, for instance, can regulate apoptosis processes in host insect cells to increase the efficiency of the host resources exploitation. Application

* E-mail: ecology@crimea.edu

of relatively short, artificially created DNA fragments of antiapoptosis genes of nuclear polyhedrosis virus on its host insect, gypsy moth, increases in its death rate. These fragments may serve as DNA insecticides. We have found that external application of solution with two short single strand iap-3 gene fragments of *Lymantria dispar* multiple nuclear polyhedrosis virus causes reliably high mortality of gypsy moth caterpillars. DNA insecticides may become alternative means of control methods for phyllophagous insects, including the oak leaf roller, and gypsy moth.

Keywords: *Tortrix viridana* L., *Lymantria dispar* L., niche partitioning, host-parasitoid interaction, pest control, community genetics, iap genes, DNA insecticides.

INTRODUCTION

Any organism interacts with an environment in accordance to its genetic constitution, raises unique requirements to the environment and in its own manner interacts with representatives of other species. Thus, intra-population genetic variation should strongly influence integrative ecological pattern of the population in the community or ecosystem. This thesis is a basic principle of our investigation on ecological ties among moths and other species, with which they interact in forest ecosystems.

Trees are main elements of forest ecosystem. Each of the trees reproduces an environment for many organisms, living in, on and around it. This central organism, a tree and all other organisms, living in the sphere of its influence, constitute together an elementary community or ecosystem if to take into the consideration soil, air and other elements, which are included into the sphere. In East European ecological tradition such the elementary ecosystems are named individual consortia (Beclemishev, 1951). Population and species consortia integrate individual consortia of population and species representatives correspondingly.

Each tree has unique genotype, and, thus, reproduces unique environment for its inhabitants. In this reason, the result of interaction between moths and their host tree depend both from the tree genotype and from genetic structure of the moth population. As interspecies interaction is an ecology prerogative, we have reasonable background to incorporate genetics in ecological explanations.

Recent investigations concerning interplay between genetics and ecology raised a question on foundation of a new field of enquiry "community genetics" (Neuhauser et al., 2003; Whitham et al., 2003; Bailey et al., 2006), usefulness of which was discussed in the Special Feature (Ecology, 2003, vol. 84, no. 3). Community genetics, by our opinion, should address its attention to the gene pools of the populations, which interact in a community, discovering the situation when alteration in gene pool of one population evokes corresponding alterations in another one. Such the situation may arise only in that case when the fitness of a specimen depends on the kind of another species representative, with which it interacts in a community.

Among others, community genetics perspective may bring new ideas and solutions to the problem of pest control. Comparing interplay between genetic and ecological processes in native and disturbed environments we may better understand mechanisms, making ecosystems sustainable. These insights may be then adapted for agricultural ecosystems and bring ecologically sound approaches to pest control in artificial environments.

Crimean forests are mainly represented by young-wood areas and forest patches, what produces a favourable environment for pest outbreaks. Such the environment, spreading around the southern slopes of the Crimean mountains and watersheds, is especially profitable for the pests adapted to dry and warm conditions. Besides, Crimean forests are exposed to strong recreational activities, which add to the weakening of the forest ecosystems. Gypsy moth (*Lymantria dispar* L.) and oak leaf roller moth (*Tortrix viridana* L.) strongly contribute to the pathological situation in the Crimean forests (Apostolov et al., 1998). In this respect we address main attention to these moths, sometimes incorporating into the investigation representatives of other species, which inhabit oaks.

Relation between Oak Genotype and Structure of Moth Micro-Community

As inter-species differences in genetics and, thus, in ecological requirements are much greater than those intra-species, structure of moth community should be sensitive to genetic variation in host plant. Many tree futures, controlled by many genes, add to the unique combination of conditions, which the tree reproduces for its inhabitants. Nevertheless, we are looking for the evidence that difference in one or another oak gene may influence structure of the moth community, which inhabit it.

Study of oak (*Quercus pubescens* Willd) genetic heterogeneity with application of RAPD PCR procedure allows detecting a range of DNA fragments, varying among sixteen model oaks (figure 1a). The set of the markers could specifically characterize each of the model trees. Model trees exhibit wide polymorphism on the obtained RAPD-markers. Individual spectrum could contain from 4 to 12 amplified DNA fragments.

Nei's (1987) coefficient of genetic similarity, calculated from pairwise comparisons of oak RAPD spectrums, was related to similarity in the *T. viridana* densities (R = 0,38, d.f. = 118, P<0,01). Thus, genetic variation in host tree influences variation in pest density. Besides, each the DNA fragment may have impact on the herbivore micro-populations. If to divide oaks in accordance with presence or absence of one or another DNA fraction in their genotypes, then we may estimate the moth densities on oaks from each pair of the groups, and recognize if the DNA fraction influences densities and ratios of the moths.

Oaks with OPA 14-11 DNA fraction (about 120 bp in length) in their spectrums of amplified fragments were less infested by *T. viridana* and *L. dispar* than those without the fraction (figure 1 b). OPA 14-4 DNA fraction (about 400 bp in length) in the same manner influences densities of *Agroteria nemoralis* Sc. and *Toeniocampa gothica* L. (figure 1 b). These DNA fragments may mark genes, influencing oak futures, critical for the moths. For instance, oaks with OPA 14-4 DNA fragment in their genotypes accumulate more zinc and copper in their leaves, which, besides, are covered with more trichoids, than those without the fragment (figure 1 c). Contents of toxic metals and protective trichoids may prevent leaves from damage by herbivores. The data obtained show that oak genes have effects on the levels higher than population. Such the effects were named "extended phenotype" (Whitham et al., 2003).

a) Oak genotypes

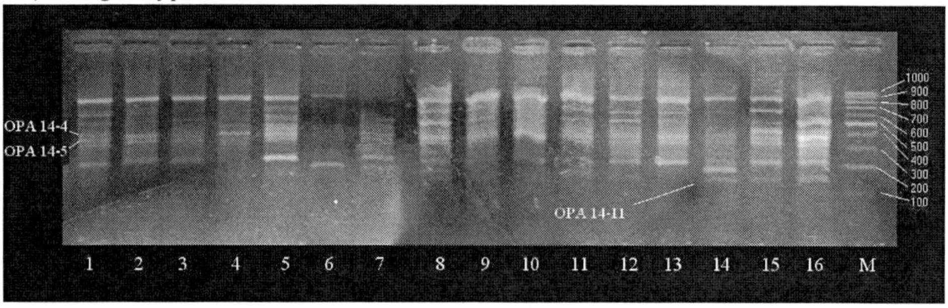

b) Density of insect larvae (individuals per 100 apical points)

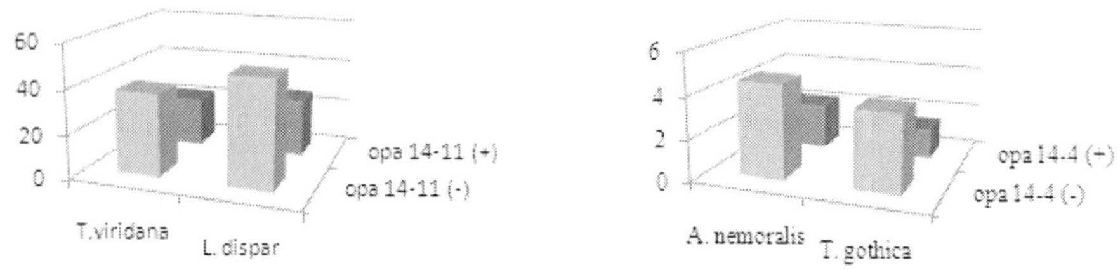

c) Properties of oak leaves

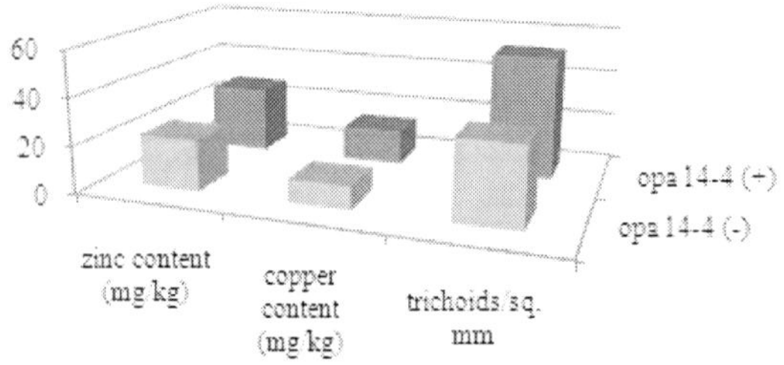

Figure 1. Oak genotypes (a), some properties of leaves from the oaks with and without OPA 14-4 DNA fraction in their RAPD-PCR spectrums (c), *T. viridana, L. dispar, A. nemoralis* and *T. gothica* densities on the oaks (b); statistic comparisons: *T. viridana* ($t=2.69$; $p=0.017$), *L. dispar* ($t=3.65$; $p=0.003$), *A. nemoralis* ($t=2.20$; $p=0.05$), *T. gothica* ($t=2.40$; $p=0.044$); zinc ($t=2.89$; $p=0.02$), copper ($t=2.49$; $p=0.03$), trichoids ($t=4.71$; $p<0.001$), M – weight markers (bp).

Methods: Leaves were sampled from16 trees of *Quercus pubescens* Willd, growing in 15 km to Northeast from Yalta (South Coast of the Crimean peninsula). Simultaneously, densities of the most common insect herbivores, *Tortrix viridana* L., *Lymantria dispar* L., *Toeniocampa gothica* L., *Agrotera nemoralis* Sc., were documented in the model trees. The insect densities were expressed in numbers of found individuals per 100 apical points. RAPD-PCR analysis was performed using DNA from each of the model trees. DNA was extracted from 0.5 cm^2 of fresh leaf tissue in accordance with standard method (Sambrook, Fritisch and Maniatis, 1989) and with OPA-14 primer, TCTGTGCTGG (Operon Technologies, USA). Data on the heavy metal contents were obtained after dry mineralization procedure with following detection of copper, zinc, cadmium and lead using atomic absorption method (Methodic instructions..., 1989). Numbers of trichoids per mm^2 of leaf surface were calculated under the binocular microscope MBS-9.

Among others, extended phenotype (*sensu* Whitham et al., 2003) may cause predictable composition of arthropod community in eucalypt hybrids (Dungey et al., 2000), willow (Hochwender and Fritz, 2004), evening primrose (Johnson and Agraval, 2005) *etc.* At the same time, we find no any author, pointing that extended phenotype depend on genotypes of all the interacting representatives. Herbivore composition depends not only from genetic variation in host plant, but also from genetic futures of the herbivores, because they choose the host plants due to their specific genetic constitution, including genes, which determinate this choice and its variation. In this relation, the question arises what organism expresses the extended phenotype? It obviously should be peculiar to the whole community, including both the host plant and the herbivores. In this case the term "phenotype" is not correct, because it should belong to an organism. Misuse of the term is a first step to further incorrect conclusions. As classic phenotype is heritable, this was incorrectly transferred also to "extended phenotype", and, hence, to community level.

FOOD NICHE PARTITIONING AMONG HERBIVORES ON SINGLE OAK TREE

Many investigations on the host plant – herbivore interaction have shown that host plant produces highly heterogeneous environment for herbivores, and each of them exploits this environment in one or another way (Semerikov and Glotov, 1980). As the same host plant produces resource for many herbivore species both monophagous and polyphagous, this causes competition among them for limited resource, and may result in their trophic niche partitioning. Single tree (individual consortium) with many species, exploiting its resource, serves as very suitable object for study on trophic niches of herbivores. From another hand, the herbivore niche partitioning in individual oak consortium may serve as a clear model showing species "packing up" into the elementary ecosystem.

Figure 2. presents the data on the frequencies of four species, including some of the genotypes, in different zones of two-dimensional leaf width – stiffness distribution. Zones of the distribution were picked out after Yu. P. Altukhov (1989). As it is seen from figure 2 a, *T. viridana* and *T. gothica* from one hand, and *N. quercus-baccarum* from another hand had opposite trends of their frequencies in the zones of two-dimensional leaf width – stiffness distribution. The first two species predominated in the M- and Mw zones while frequencies of the last species were high in the M+ and Ms zones. Such the trends mean that these three species were much more susceptible to the leaf stiffness than to the leaf size. Another was peculiar to new *L. dispar* larvae. Both the leaf features influenced the species occurrence in the zones of host plant leaf distribution (Figure 2 a).

T. viridana and *N. quercus-baccarum* show intra-population partitioning of their trophic niches (Fig. 2 b, c), with many genotypes dividing population trophic niche into narrow micro-niches. Each genotype had its own micro-niche, to which it was most adapted, i.e. larvae of given genotype might eat only leaves of certain properties and did not eat other leaves (Simchuk et.al, 1999). In this relation, we may conclude that not the whole population competes with populations of other species sharing for the same or close resources, but only separate genotypes do that. In other words, individuals of some genotype, eating leaves with corresponding properties, have much higher probability to conflict with representatives of another species preferring leaves with the same properties, than individuals of other genotypes.For instance, the *T. viridana* individuals of "soft-leaf-eating" genotypes (see above and Simchuk et al., 1999) have higher probability to compete with *L. dispar* and *T. gothica* larvae than with *N. quercus-baccarum* in comparison with individuals of "stiff-leaf-eating" genotypes and *vice versa.*

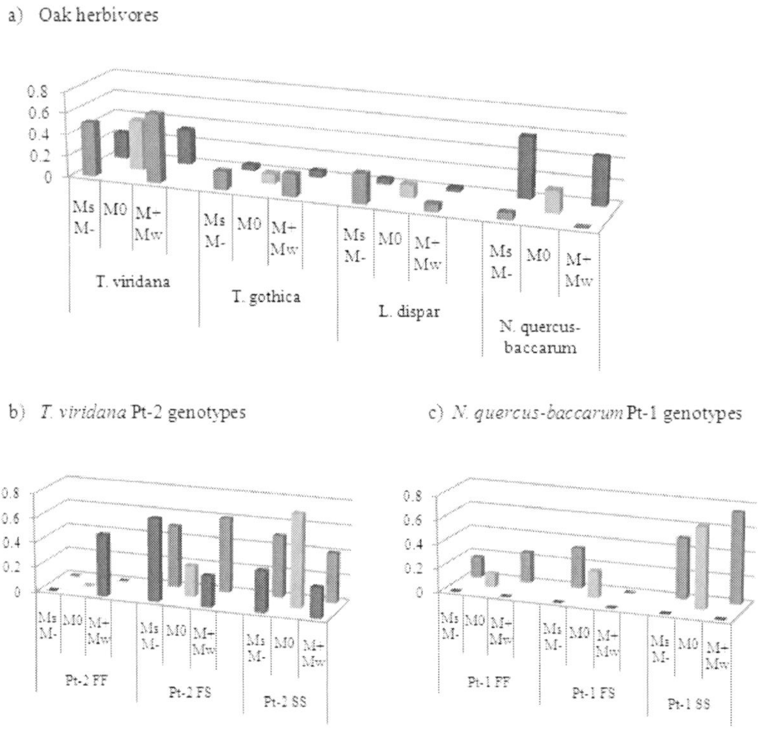

Figure 2. Frequencies of the oak herbivores (a), *T. viridana* Pt-2 (b) and *N. quercus-baccarum* Pt-1 (c) genotypes in zones (defined after Altukhov, 1989) of leaf width-stiffness two-dimensional distribution in single oak tree; M- - leaves with small values of both the features, M0 – moderate leaves, M+ - leaves with high values of both the features, Mw – disproportional leaves with high values of width and low values of stiffness, Ms - disproportional leaves with low values of width and high values of stiffness; Values of the test for frequency homogeneity (after Altukhov, 1989): herbivores, χ^2=94.2, d.f.=15, P<0.001; *T. viridana* Pt-2 genotypes, χ^2=43.4, d.f.=8, P<0.001; *N. quercus-baccarum* Pt-1 genotypes, χ^2=40.1, d.f.=8, P<0.001.

Methods: representatives of all the found herbivores (214 in total) were collected from single tree of *Q. pubescens*, growing in a testing ground named 'Lavrovoe' (chaparral on southern coast of the Crimean peninsula near Yalta), simultaneously with the leaves, on which they fed. Each larva placed in separate vial together with proper leaf. The same day that material was delivered in laboratory.Species of the larvae were determined, and each leaf was measured (maximal width) and taken to define its stiffness using device, consisting of punch on a spring. Leaf stiffness was expressed in symbolic units (distance in mm, which must be passed by a spring, pushing the punch to make a hole in a leaf tissue). Representatives of following most common species were found in the tree: *Tortrix viridana* L. (Lepidoptera; Tortricidae) (101 larvae of 4-th – 5-th instars), *Toeniocampa gothica* L. (Lepidoptera; Noctuidae) (11 larvae of 4-th – 5-th instars), *Lymantria dispar* L. (Lepidoptera; Lymantriidae) (21 larvae of 1-st – 2-nd instars), *Neuroterus quercus-baccarum* L (Hymeniptera; Cynipidae) (85 adults in galls) and some rare species.

As we have found earlier, distribution of gypsy moth specimens along the leaf stiffness gradient significantly differed from analogous distribution of oak leaf roller FS-genotype class and coincided with distributions of its other genotypes. Distribution of the oak leaf roller SS-genotype class along the leaf stiffness gradient significantly differed from analogous distribution of *N. quercus-baccarum* FS-genotype class and coincided with distributions of its FF and SS genotype classes (Simchuk and Ivashov, 2005).

Comparing the Czechanowski niche overlap index (Feinsinger et al., 1981) values, we have found (Simchuk and Ivashov, 2005), that lowest mean value of the index marks niche overlap between genotype classes from different species (0.324±0.046). Growing mean index was found when analyzing niche overlap between genotype classes of given species and other species (0.420±0.029). Highest values of the index describe niche overlap between different species (0.578±0.049) or between genotype classes in the same species (0.585±0.078). As other oak herbivores, *N. quercus-baccarum* (fig. 2 c) for instance, also have different genotypes preferring leaves of different properties as a food, we obtain much more complex picture of competition in which interaction occur between concrete genotypes of different species. Competitive relations between genotypes of different species may be considered as system-creating ties joining genofonds of interacting populations into new extra-populative genetic system, genetic system of the community. For such the genetic system of community M.A. Holubets (1982) has proposed the term "genoplast". This term includes a couple of genotype-genotype interaction between representatives of different species in a community.

Food Niche Partitioning among Parasitoids Infesting Oak Leaf Roller Pupae

As we have shown above, herbivores differentially exploit food resources, produced by the plants, which they inhabit. At the same time, herbivores, in their turn, are the food resources for other organisms, parasitoids in particular. Studying pupal parasitoids of oak leaf roller (*T. viridana* L.), we have found that they preferably attack small host pupae (Figure 3a). Two-dimensional distribution of the parasitized pupae is very close to distribution of pupae, which died from other causes (Figure 3a). As mortality usually is a consequence of 'straggle for existence', we may conclude that parasitoids firstly attack unfit host individuals. On the whole, the data obtained show that thorax widths of the oak leaf roller pupae better reflect their fitness than thorax lengths, because highest viability was peculiar to disproportional pupae with 'wide' and 'short' thoraxes, while pupae with opposite thorax characteristics died more often than others (Figure 3a).

No all the individuals in a population have a chance to reproduction. Only about 30% of individuals may contribute to the next generation (Altukhov, 1989), while others are excluded due to non-random mating, when only fit individuals are involved (Simmons, 1987). As parasitoids exploit unfit hosts, they, thus, are not able to influence sufficiently numbers of the host offspring. Among the parasitized pupae, five parasitoid species differ in their choice of hosts (Figure 3b). *I. maculator* predominated in disproportional pupae with 'long' and 'narrow' thoraxes. Frequencies of *I. alternans* and *B. intermedia* were higher in disproportional 'short' and 'wide' pupae than in others. *Ph. invisor* and *M. aureus* preferred small and large pupae correspondingly. Observed niche partitioning obviously is a consequence of inter-species competition for food resource. Nevertheless, in spite of competition, no any of the parasitoids occupied free niche with the fit hosts as a resource. Parasitoids, thus, balance in equilibrium between the competition and host ability to defense, which grows with host fitness.

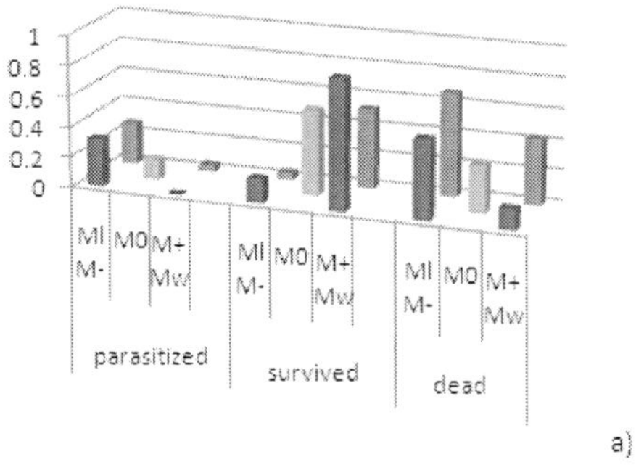

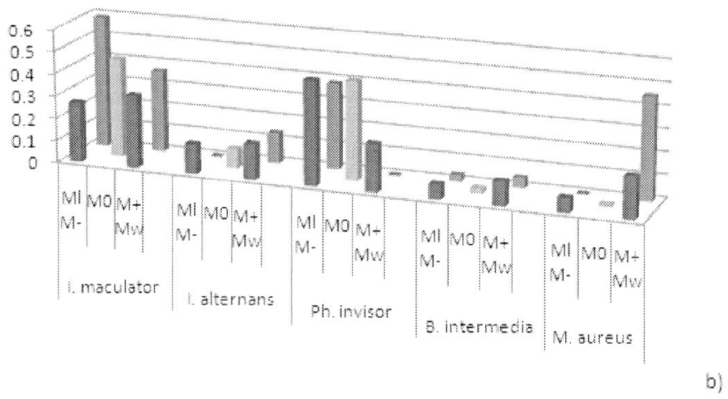

Figure 3. Frequencies of parasitized, survived and dead oak leaf roller pupae (a) and frequencies of five parasitoid species among parasitized pupae (b) in zones (defined after Altukhov, 1989) of two-dimensional pupae distribution on their thorax widths (x-axis) and lengths (z-axis); M- - pupae with small values of both the features, M0 – pupae of moderate sizes, M+ - pupae with high values of both the features, Mw – disproportional pupae with high values of thorax width and low values of thorax length, Ml - disproportional pupae with low values of thorax width and high values of thorax length; Values of the test for frequency homogeneity (after Altukhov, 1989): (a) $\chi2=303.4$, d.f.=8, P<<0.001; (b) $\chi2=172.3$, d.f.=16, P<<0.001

Method: oak leaf roller pupae (918 in total) were collected from pubescent oaks growing in testing ground 'Lavrovoe' (chaparral on southern coast of the Crimean peninsula near Yalta). Each pupa was measured under binocular microscope (thorax width and length), placed in separate vial and stored under the laboratory conditions. The vials were then daily inspected in respect to emergence of adults or parasitoids. All the parasitoids (207 in total), emerged in the laboratory, belong to Hymenoptera: Itoplectis maculator F. (85 individuals), I. alternans F. (19 individuals), Phaeogenes invisor Thunb. (69 individuals), Brachimeria intermedia Nees (9 individuals), Monodontomerus aureus Walk (19 individuals) and some rare species, which were excluded from the analysis.

a) *I. maculator* tibia phenotypes (under Ivashov, 2001)

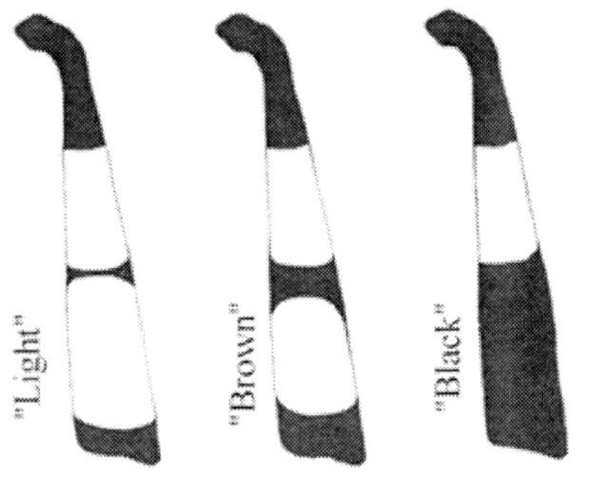

b) frequencies of *I. maculator* tibia phenotypes c) frequencies of *Ph. Invisor* coxa phenotypes

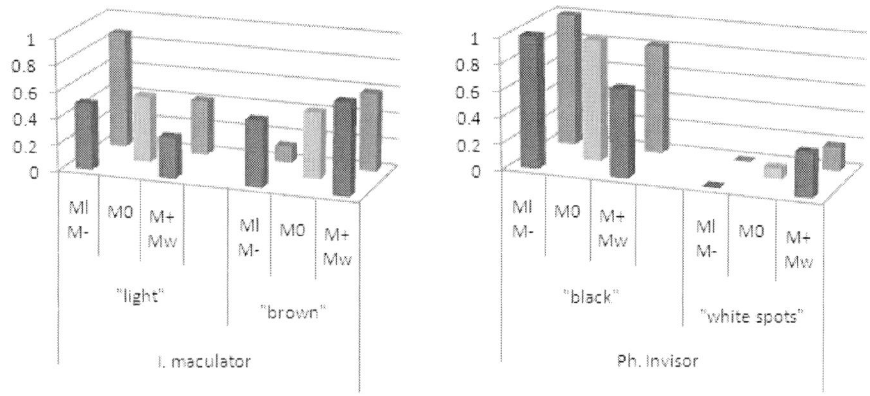

Figure 4. Frequencies of *I. maculator* F. tibia phenotypes (a, b) and *Ph. invisor* Thunb. coxa phenotypes among parasitized individuals in zones (defined after Altukhov, 1989) of two-dimensional pupae distribution on their thorax widths (x-axis) and lengths (z-axis); M- - pupae with small values of both the features, M0 – pupae of moderate sizes, M+ - pupae with high values of both the features, Mw – disproportional pupae with high values of thorax width and low values of thorax length, Ml - disproportional pupae with low values of thorax width and high values of thorax length; Values of the test for frequency homogeneity (after Altukhov, 1989): (a) χ^2=37.6, d.f.=4, P<0.001; (b) χ^2=36.4, d.f.=4, P<0.001.

Method: *I. maculator* and *Ph. Invisor* individuals, emerged from oak leaf roller pupae (see Figure 3), were divided into groups in accordance to their phenotypes, defined after A.V. Ivashov (2001). Individuals of *Ph. Invisor* exhibit two alternative phenotypes of coxa coloring with variants: black and black with white spots (Ivashov, 2001).

Two most common *T. viridana* parasitoids, *I. maculator* and *Ph. invisor* exhibit intra-species morphological variation with alternative phenotypes (Ivashov, 2001) of tibia coloring in *I. maculator* (Figure 4a) and coxa coloring in *Ph. invisor* (black coxa and black coxa with white spots, fater Ivashiv, 2001). Parasitoid phenotypes also show niche partitioning (Figure 4b and c). In both species pairs of alternative phenotypes have opposite priorities in host pupa choice. At the same time, trends in distributions of phenotype frequencies were close in phenotype pairs: 'light' tibia in *I. maculator* – 'black' coxa in *Ph. invisor* and 'brown' tibia in

I. maculator – 'white spots' coxa in *Ph. invisor*. This means that inter-species competition in these pairs should be higher than across them.

Comparing the Czechanowski niche overlap index values, we found the same picture as in the case with herbivore niche partitioning: lowest niche overlap between phenotypes from different species (0.356±0.023), growing niche overlap between phenotypes of given species and other species (0.407±0.024), highest niche overlap between different species (0.427±0.055) or between phenotypes in the same species (0.455±0.024). Variation of the indexes among the groups was significant (one-way ANOVA $F=4.83$, d.f. = 4, 232, $P < 0.01$). Thus, entire picture of the niche partitioning in the *T. viridana* parasitoids as well as in herbivores, damaging oaks, has hierarchic construction. Parasitoids attack preferably unfit hosts and divide them among themselves with each species having its own niche, which then divided among intra-species groups such as genotype classes or phenotypes.

Genetic Aspects of Interaction among Oaks, Oak Leaf Roller and Its Parasitoid, *Itoplectis Maculator* F

We study here a question what role inter-population genetic heterogeneity plays in tri-trophic interaction among oaks, their herbivores and parasitoids, infesting these herbivores. When two individuals of different species interact, their genotypes, coding for their futures and requirements, should influence result of this interaction. Thus, fitness of an individual depends, among others, on the genotypes of individuals, with which it interacts in a community. Direct estimation of specimen fitness is a difficult task if possible at all. One of useful ways lies in measurement of quantitative signs related with average fitness, such as body sizes in insects (Watt et al., 1986; Simchuk et al., 1999). Studying variation of fitness-related signs in dependence on genetic variation in interacting species we may find inter-level genetic ties.

The data obtained show that variation in the *T. viridana* pupal weights depends not only from their phenotypes, but also from genotypes of the oaks, which they inhabit (Figure 5a). Variation in parasitoid sizes depends from their phenotypes, phenotypes of their hosts (Figure 5b) and genotypes of the oaks (Figure 5c). Moreover, the relations have some common regularity. 'Light' phenotype parasitoids reach maximum sizes when developing in the oak leaf roller pupae with the 'high' cremaster phenotype (Figure 5b). Thus, the parasitoid phenotype is most adapted to the host phenotype. At the same time, these hosts are unfit on oaks without OPA 14-5 DNA fragment in their genotypes (Figure 5a). As the result, the parasitoids, attacking small, unfit hosts, obtain advantage on these oaks and, thus, reach maximum sizes (Figure 5c). Analogous causal chain may be created for 'brown' parasite and 'lower' oak leaf roller phenotypes on oaks with the DNA fragment. Thus, genotypes of representatives of the three trophic levels are related. This means that alterations in genetic structure at one of the levels should influence genetic structures at other dependent levels. This example reflects the role, which inter-species genetic variation plays in ecological interactions among the species in a community.

Found associations represent ties among the gene pools of populations, interacting in a community. As we have elements, gene pools, and ties among them, we once more may say about some system, "genetic" system of a community. M.A. Holubets (1982, 2000) heuristically described this system as ecosystem "memory", regulatory, informative subsystem driving ecosystem functioning, and called it genoplast.

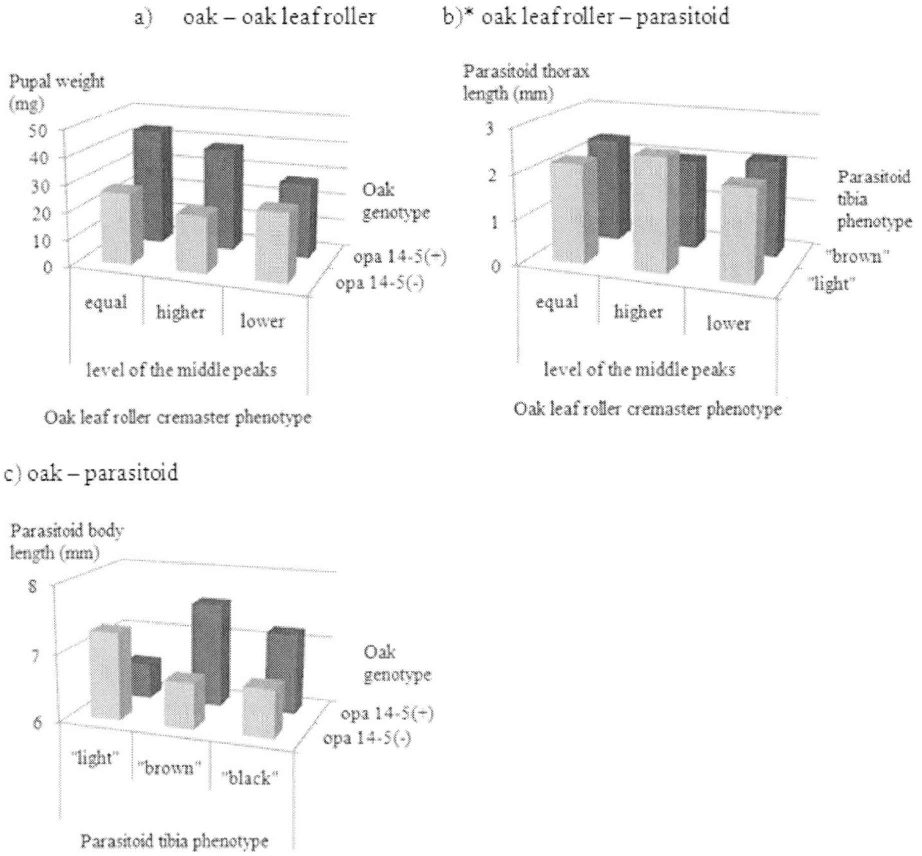

Figure 5. Genetic aspects of interaction among oaks (*Quercus pubescens* Willd.) (fig. 1a), oak leaf roller (*T. viridana* L.) and its parasitoid (*Itoplectis maculator* F.); oak leaf roller pupal weights in relation to the pupae phenotypes and the genotypes of the oaks which they inhabit (a) (two-way ANOVA interaction $F=3.21$, d.f.=2;85, $p<0.05$); parasitoid sizes in relation to their phenotypes (fig. 4a) and the phenotypes of oak leaf roller pupae which they parasitize (b) (two-way ANOVA interaction $F=3.40$, d.f.=2;38, $p<0.05$); parasitoid sizes in relation to their phenotypes and the genotypes of the oaks which their hosts inhabit (c) (two-way ANOVA interaction $F=4.26$, d.f.=2;40, $p<0.05$); *statistics for body length was little lower than the 0.95 confidence level; 'black' phenotype was no presented in all the variants and, thus, excluded from the analysis.

Methods: Oak leaf roller pupae were sampled from 16 oak trees, marked with DNA fragments after amplification with OPA 14 primer (fig. 1a). Each pupa was placed in separate vial and stored under the laboratory conditions after the measurement of its weight with precision of ± 0.5 mg. The vials were then daily inspected in respect to emergence of adults or parasitoids. Genetic variations in the oak leaf roller and its parasitoid were studied using alternative morphological phenotypes as genetic markers (Yablokov, 1980). Oak leaf roller pupae differ in levels of middle cremaster peaks in comparison to lateral peaks (Ivashov, 2001). Parasitoids differ in tibia phenotypes (fig. 4a). Thorax and whole body lengths were measured in each of the morphologically assessed parasitoids under the binocular microscope MBS-9 with precision ± 0.025 mm. The results obtained were statistically processed using two-way ANOVA.

As genoplast arises on the base of selective processes within each of the interacting populations, its explanation requires no introduction of such the categories as "ecosystem selection" or "ecosystem heritability" (Whitham et al., 2003; Bailey et al., 2006). Of course, as heritability is peculiar to each of the individuals included in a community, there should be some succession among the past, present and future states of the community. This ecosystem succession is equivalent to the heritability at the level of individuals, and may mimicry for it in the case, when species participants of a community strongly differ in their life cycle durations. Genoplast is the next hierarchy level after population gene pools. It arises from interactions among the representatives of different species, and its succession integrates individual heritabilities in the same manner as pressure of a gas integrates velocities of the molecules it consists of.

Community Genetics Perspective for Pest Control

Forest pests as a rule show relatively rare-repeated outbreaks. From one hand, strong repeated defoliation may influence tree viability. From another hand, artificial interference with natural self-regulation could break sustainable community (Southwood, 1977). We have shown here that forest ecosystems are very complex with precise adjustment all their components. World-wide experience has shown that, when human takes a role of a regulator (even temporary), he as a rule disturbs ecosystem self-regulatory mechanisms (Marshakov, 1985; Thompson et al., 1979). In this respect pest control activities are often groundless and a decision on them should follow preliminary prognostic simulation predicting their economic effects (Gottschalk et al., 1998). With that, community-genetic approaches may be useful for pest control in simple artificial ecosystems such as agro-ecosystems.

Agricultural practice includes several procedures of pest control based on application of chemical and biological pesticides, or introduction of the genetically modified crops. Although these methods differ each from other in active substances and procedures, all of them have the same strategy with aim to kill effectively the pests. Because mortality usually is a consequence of `struggle for existence', and individuals differ in their susceptibility to the applied agents, these agents stimulate pests to evolve resistance to them. Thus, under the present agricultural practice the pests become more and more resistant and aggressive. Consequently, we need to find new and new killing substances and agents to win the pests, our competitors for food. This way is no interminable and it finally will result in desperate situation when new pesticide will not be found.

In natural environments species evolve towards the weakness of the negative interspecies interactions. Plants produce more biomass than they need for their normal life. Herbivores eat leaves with certain properties avoiding extremely harmful consequences for the plants (Zucker, 1982; Claridge, 1986; Simchuk et al., 1999). Competing species evolve towards partitioning of their niches (Winemiller, Pianka, 1990; Albrecht, Gotelli, 2001). Predators, parasitoids and diseases consume preferably those individuals of exploited species which have low probability of successful reproduction (Pianka, 1978). In natural ecosystems interacting species as a rule reach an equilibrium allowing all of them to coexist. Applied pest control strategies destroy natural mechanisms regulating interspecies interactions. In this respect, we develop a strategy based on the natural regulating mechanisms (Simchuk, 2008).

Modern molecular-genetic procedures allow finding genetic markers related with plant adaptation to local environments (Jasienski et al., 1997). Using such the markers we are able to separate seeds in respect to their future adaptive capacity. Growing in nearest locations, low and high adapted crop populations form natural protective system. Herbivores, exploiting weak low-adapted plants as an accessible food resource, should have an advantage in the intra-specific competition, and their genes should spread over the herbivore population. At the same time, herbivores exploiting high-adapted plants should spend exceeding resources to overcome their protective systems. Consequently, they should lose in the intra-specific competition and thus become an accessible resource for predators, parasitoids and diseases. Preserving the pest population from year to year, we therefore may keep the pests evolving towards weakening of their damaging capacity. Proposed system allows partitioning of human and pest food niches, avoiding application of chemicals, polluting the environment, as well as introduction of the genetically modified crops (Simchuk, 2008).

Influence of a Single Chain DNA Fragments from *Lymantria Dispar* L. Multiple Nucleocapsid Polyhedrosis Virus on Viability of Host Larvae

The increase of efficiency in forest defense from gypsy moth is economically very important. Baculoviruses have enormous potential in ecologically based microbiological control of the quantity of pests including the gypsy moth (Evans, Harrap, 2004). The basic problem of the use of baculoviruses is to increase the speed of infectious process and death of the host (Vlak, 1992; Moscardi, 1999).

Biopreparations based on viral DNA fragments (DNA insecticides) may become alternative means of struggle against phyllophagous insects, including the gypsy moth, since viral DNA has information about the management of the host metabolism. Apparently, viral DNA alone (out of mature viral particle) has the ability to take part in cellular biochemical reactions.

The creation of effective biological preparations based on DNA fragments is quite promising due to important role viral DNA plays in host cells, and the great variability of even small fragments of the molecule. The possible mechanism of their action may be DNA interference or process similar to it.

From the pioneer work on DNA interference (Kawai-Toyooka, 2004) where the authors used this term first, it is known that the application of double strand DNA fragments of the organism causes strict specific inhibition of genes that coincide with the used nucleotide sequences.

As a result, the synthesis of corresponding protein is blocked. The first work on this phenomenon was made at plant *Adiantum capillus-veneris*. Similar research works for the virus-host system we have not found in the scientific literature. We must add that there are no similar research works on single strand DNA fragments and their practical use as DNA insecticides.

The ability of viral short DNA fragments to influence the cells of host organism can be due to the following circumstances:

a) viral DNA in its structure has coded information about the control of the host cell;
b) DNA has the affinity to large number of proteins and, thus, affects their functional activity.

Our idea was to investigate the influence of a single chain DNA fragments of *Lymantria dispar* multiple nucleocapsid polyhedrosis virus iap-3 gene on viability of gypsy moth caterpillars. Baculoviruses have two classes of antiapoptosis genes – p35 and iap genes that can block apoptosis through induction of various signals in a phylogenetically wide range of organisms (Bertin et al., 1996; Miller, 1997). It is known, for example, that antiapoptosis genes of baculoviruses are homologous to antiapoptosis genes of round worms, insects, humans (Zhimulev, 2007). Phylogenetic analysis of the iap genes allowed putting forward the hypothesis that these genes were obtained by viruses from their hosts at least twice during evolution (Hughes, 2002).

If antiapoptosis genes of virus are homologous to the host antiapoptosis genes, then application of such fragments of viral antiapoptosis genes is supposed to cause changes (interfere) in the biochemical reactions of the gypsy moth cells (for example, reactions similar to RNA and DNA interference). According to the data found in scientific literature we were the first who used single chain DNA fragments of *Lymantria dispar* multiple nucleocapsid polyhedrosis virus iap-3 gene to show insecticidal effect of DNA fragments.

One of the possible ways of penetration of the DNA fragments into the body of a caterpillar is an external one. It is known that the presence of developed epicuticule limits to some extent the permeability of a caterpillar`s covers against most insecticides. Clororganical and other contact insecticides get through the epicuticule easily into the organism of an insect through the most permeable areas of the covers (Tyschenko, 1986). Single strand DNA molecules have both hydrophilic (sugar-phosphate backbone) and relatively hydrophobic (bases) parts that allow them to penetrate through the polar and non-polar obstacles of insect tissues.

External Application of Viral Iap-3 Gene DNA Fragments

On the average, during the 11 day period in the experimental groups 53% of individuals perished, that is on the average 30% more than in the control groups ($\chi^2 = 23.72$; df = 1; P <0.01) (table 1)*. Thus, viral DNA fragments acted as signaling molecules (for death) in the host cells of the gypsy moth. The death rate of caterpillars was observed through a short interval of time (2-3 days), while during experimental infection of insects with nucleopolyhedroviruses the latent period of illness lasts from 5 up to 10-12 days (Gouli, Rybina, 1988). It implies on the faster effect of DNA fragments in comparison with viral biopreparations. Reliable correlation in the dynamics of caterpillars' death rate in the experimental groups was found (table 2)*. In the control groups dynamics of caterpillars' death rate did not correlate (exception is the pair of egg masses 1 and 3). It means that the caterpillars from experimental groups mostly died synchronously. Thus, in the control groups there was no factor which would bring caterpillars on to synchronization of their death rate, but in the experimental groups this factor existed – the viral DNA fragments.

The strong correlation between the dynamics of death rate of caterpillars from the experimental groups indicates the existence of a unique cell mechanism of the gypsy moths reaction on the viral DNA fragments.

Table 1. Death rate of I-II instars caterpillars in experimental and control groups from three egg masses after an 11 day period

Egg masses	Control group	Experimental group	Criteria χ^2
1	10%	50%	11.43*
2	10 %	33%	5.03**
3	50%	72%	3.55***

*P < 0.01; **P < 0.025; ***P <0,1

*We used $\chi 2$-criteria of Pearson to count up the insecticidal effect of viral iap-3 gene fragments. The 1st and the 2nd instar caterpillars from three egg masses were used for the research (for each egg mass it is 20 1st instar caterpillars and 20 2nd instar caterpillars per each experimental and control group). Water solution with 2 single strand DNA fragments (100 pmol/μl) was applied outwardly (0.1 μl/caterpillar). After applying the solution we waited till the moment when the drop was completely absorbed by the caterpillar. Caterpillars from the control group were treated with distilled water (0.1 μl/caterpillar). Insects were grown on artificial forage based on dried oak leaves. DNA fragments were designed by us according to the viral genome sequence found in ICTVdb and synthesized by metabion international AG (Germany) with HPLC clearance. The sequences of these 2 single chain DNA fragments were the following: a) 5′- GCC GGC GGA ACT GGC CCA -3′ (134843-134860; antisense chain); b) 5′-CGA CGT GGT GGC ACG GCG -3′ (135159-135142; sense chain) (Kuzio et al., 1999). The used DNA fragments are the part of iap-3 gene which belongs to genes that inhibit apoptosis of host cells (Kuzio et al., 1999).

PCR method was used to find whether applicated DNA fragments of viral iap-3 gene is homologous to host antiapoptosis genes. Viral iap-3 gene DNA fragments were used as specific primers for PCR. In the amplification products of DNA isolated from the gypsy moth eggs from 7 to 8 fragments for each individual spectrum were found (Figure 6). The results indicate the presence of parts in the gypsy moth genome which to some extent coincide with fragments of viral iap-3 gene. Thus, the findings indicate a possibility of influence of viral iap-3 gene DNA fragments on gypsy moth cells by the mechanism of DNA interference or by processes similar to it.

The found effect of the viral DNA fragments on its host can be used for the creation of selective fast-acting insecticides to protect plants from the gypsy moth and other phyllophagous insects (Oberemok, 2008; Oberemok, 2009). Such phenomenon seems to be adaptively important for a system host-virus, maybe as one of natural pathways of interaction between the gypsy moth and *Lymantria dispar* multiple nucleopolyhedrovirus and requires further investigation.

Table 2. Value of correlation coefficients (r) between the dynamics of death rate of 1st and 2nd instars caterpillars from three egg masses for an 11 day period

Egg masses	Untreated	Treated
2,3	0.97*	0.42
1,3	0.95*	0.94*
1,2	0.92	0.45

*P<0.01

*We used correlation analysis (Word Excel 2003) to evaluate the influence of DNA insecticides on each pair of experimental and control groups to find out if there any factor that bring caterpillars on to synchronization of their increase of death during 11 day period. The dynamics of death rate in the experimental groups was largely alike.

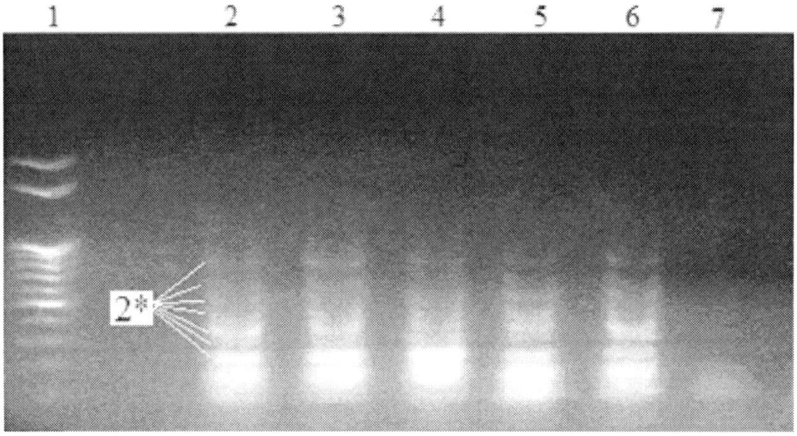

Figure 6. Electrophoregram of DNA amplification products (with iap3 gene DNA fragments as primers) of gypsy moth: 1 – marker of DNA length from 100 to 1000 bp with a step of 100 bp and from 1000 to 3000 bp with a step of 1000 bp (from bottom to top); 2-6 – electrophoretic DNA spectra of individuals; 7 – control; 2* – individual set of amplicons.

DNA Insecticides Perspective for Pest Control

The creation of effective biological preparations based on viral DNA fragments (DNA-insecticides) for pest control methods is quite promising. However, there are many literature data about reproduction ability of viral DNA or RNA in non-host cells. It is characteristic for some viruses of mammals, phages, baculoviruses (Kok et al., 1966; Gouli, Rybina, 1988). For short DNA fragments of the virus this phenomenon is not possible due to lack of genetic material for reproduction of the virus.

On the other hand, it is possible to harm with DNA fragments other organisms through DNA interference phenomenon or mechanisms similar to it. For more accurate conclusions the given theoretical assumptions require to be investigated in practice.

Another important side of DNA insecticides that requires improving is the economical profit. Frankly speaking, today DNA insecticides are not cheap enough to be used on vast agricultural territories. 2 ml of DNA insecticides used in our experiments cost 115 dollars. It is much more expensive than to use viral biopreparations. Technological improvements in that area of biochemistry (synthesis of artificial oligonucleotides) would make DNA insecticides more affordable.

Thus, cheapness and safety (eco-friendness) are the main characteristics of the DNA insecticides that require to be studied attentively.

REFERENCES

Albrecht, M.,Gotelli, N.J., 2001. Spatial and temporal niche partitioning in grassland ants. *Oecologia*, 126, 134–141.

Altukhov, Ju.P., 1989. *Genetic processes in populations*. Moscow. Nauka Publishers.

Apostolov, L.G., Ivashov, A.V., Kolodochka, A.A., 1998. Retrospective analysis and optimization of protective measures in forest and park plantings of the Crimean South coast (in Russian). Ucheniye zapiski Simferopolskogo universiteta, 46, 11-19.

Bailey, J.K., Wooley, S.C., Lindroth, R.L., Whitham, T.G., 2006. Importance of species interactions to community heritability: a genetic basis to trophic-level interactions. *Ecology Letters*, 9, 78–85.

Beklemishev, V.N., 1951. On the classification of bio-geo-cenotic (symphysiological) ties. *Bull. MOIP, Otd. Biol*, 65. P. 3-30.

Bertin, J., Mendrysa, S.M., LaCount, D.J., Gaur, S., Krebs, J.F., Armstrong, R.C., Tomaselli, K.J., Friesen, P.D., 1996. Apoptic supression by baculovirus P35 involves cleavage by and inhibition of a virus induced CED – 3/ICE-like protease. *J. Virol.* 70, 6251–6259.

Claridge, M.F., 1986. The distribution of a typhlocibine leafhopper, Ribautiana ulmi (Homoptera: Cicadellidae) on a specimen wych elm tree. *Ecol. Entomol.*, 11, 31-39.

Dungey, H.S. et al., 2000. Plant genetics affects arthropod community richness and composition: evidence from a synthetic eucalypt hybrid population. *Evolution* 54, 1938–1946.

Evans, H.F., Harrap, K.A., 1982. Persistence of insect viruses. In: Minson A.C., Darby G.K. (Eds) Virus Persistence. SGM Symposium. Cambridge, Cambridge University Press. pp. 57–96.

Feinsinger, P., Spears, E.E., Poole, R.W., 1981. A simple measure of niche breadth. *Ecology*, 62, 27–32.

Gottschalk, K.W., Colbert, J.J., Feicht, D.L., 1998. The mortality risk of oak due to gypsy moth. *European Journal of Forest Pathology*. 28, 121-132.

Gouli, V.V., Rybina, S.Y., 1988. Virusnye bolezni nasekomykh i ikh diagnostika. Shtiintsa, Kishinev.

Hochwender, C.G., Fritzl, R.S., 2004. Plant genetic differences influence herbivore community structure: evidence from a hybrid willow system. *Oecologia*, 138, 547 – 557.

Holubets, M.A., 1982. *Actual questions of Ecology*. Kyiv. Naukova Dumka.

Holubets, M.A., 2000. *Ecosystemology. Lviv*. Polli Publishers.

Hughes, A.L., 2002. Evolution of inhibitors of apoptosis in baculoviruses and their insecthosts. *Infect. Genet. Evol.* 2, 3–10.

Jasienski, M., F.J. Ayala, F.A. Bazzaz, 1997. Phenotypic plasticity and similarity of DNA among genotypes of an annual plant. *Heredity*, 78, 176–181.

Johnson, M.T.J., Agrawal, A.A., 2005. Plant genotype and environment interact to shape a diverse arthropod community on evening primrose (Oenothera biennis). *Ecology*, 86(4), 874–885.

Kawai-Toyooka, H., Kuramoto, C., Orui, K., Motoyama, K., Kikuchi, K., Kanegae, T., Wada, M. 2004. DNA interference: a simple and efficient gene-silencing system for high-throughput functional analysis in the fern Adiantum. *Plant Cell Physiol.* – 45, 1648–1657.

Kok, I. P., Skuratovskaya I. N., Strokovskaya L. I., 1980. Molekulyarniye osnovi reproduktsii bakulovirusov. Kyiv, Naukova dumka.

Kuzio, J., Pearson, M.N., Harwood, S.H., Funk, C.J., Evans, J.T., Slavicek, J., and Rohrmann, G.F., 1999. Sequence and analysis of the genome of a baculovirus pathogenic for Lymantria dispar. *Virology*. 253, 17–34.

Miller, L. K. 1997. Baculovirus interaction with host apoptotic pathways. *J. Cell. Physiol.*173, 178–182.

Marshakov, V.G., 1985. Control of pest numbers in system: host is parasite (Insecta) on the base of forecast of result of species interaction (in Russian). PhD dissertation, Leningrad.

Moscardi, F., 1999. Assessment of the application of baculoviruses for control of lepidoptera.*Annual review of entomology*. 44, 257–289.

Nei, M., 1987. *Molecular Evolutionary Genetics*. New York. Columbia University Press.

Neuhauser, C., Andow, D.A., Heimpel, G.E., May, G., Shaw, R.G., Wagenius, S., 2003. Community genetics: expanding the synthesis of ecology and genetics. *Ecology*, 83, 545–558.

Oberemok, V.V., 2008. Method of elimination of phyllophagous insects from order Lepidoptera. Patent of Ukraine № 36445, Publ. 27.10.2008. Bull. 20.

Oberemok, V.V., 2009. Influence of a single chain DNA fragments of virus iap gene LdnVgp140 on viability of Lymantria dispar caterpillars. *Ukr. Biochem. J.* 81(4) special edition), 171.

Pianka, E.R., 1978 Evolutionary Ecology (2nd ed.). New York : Harper and Row. Publishers
Semerikov, L.F., Glotov, N.V., 1980. Variability of durmast oaks in Dagestan. *Ecologia* (Rus), 4, 25-36.

Simchuk, A.P., 2008. A method of pest control in agriculture. Patent of Ukraine № 31813. Pub. 25.04.2008. Bul. 8.

Simchuk, A.P., Ivashov, A.V., Companiytsev, V.A., 1999. Genetic patterns as possible factors causing population cycles in oak leafroller moth, Tortrix viridana L. *Forest Ecology and Management*, 113, 35-49.

Simchuk, A.P., Ivashov, A.V., 2005. Ecological-genetic aspects of trophic preference partitioning in a micro-assemlage of oak herbivores. *Zhurnal Obschey Biologii*, 66, 491-499.

Simmons L.W., 1987. Female choice contributes to offspring fitness in the field cricket, Grillus bimaculatus (De Geer). *Behav. Ecol. and Sociobiol.*, 21, 313-321.

Southwood T.R.E., 1977. The relevance of population dynamic theory to pest status. Origins Pest Parasite, Disease and Weed Probl. - *18 th Symp. Brit. Ecol. Soc., Bangor*, 1976. Oxford, 35-54.

Thompson, W.A., Holling, C.S., Kira, I.N., Huang, C.C., Vertinsky, I., 1979. Evaluation of alternative forest system management policies. *Journal of Environmental Economy and Management*, 6, 51-68.

Tyschenko, V.P., 1986. *Phiziologiya nasekomyh.* Vyssh. Shk, Moscow.

Vlak, J.M., 1992. Genetic engineering of baculoviruses for insect control. In: Oakeshott J., Zhimulev, I.F., 2007. *General and Molecular Genetics*. Gos. Univ. Izd.

Watt W.B., Carter P.A., Donahue K., 1986. Females' choice of "good genotype" as mates is promoted by an insect mating system. *Science*, 233, 1187–1190.

Whitham, T.G., Young V., Martinsen G.D., Gehring C.A., Schweitzer J.A., Shuster S.M., Wimp G.M., Fischer D.G., Bailey J.K., Lindroth R.L., Woolbright S. and Kuske C.R., 2003. Community genetics: a consequence of extended phenotype. *Ecology*, 83, 559–573.

Winemiller, K.O., Pianka, E.R., 1990. Organization in natural assemblages of desert lizards and tropical fishes. *Ecol Monogr.*, 60, 27–56.

Zucker, W.V., 1982. How aphids chose leaves: the role of phenolics in host selection by a galling aphid. *Ecology*. 63, 972-981.

In: Moths: Types, Ecological Significance and Control… ISBN: 978-1-61470-626-7
Editor: Luis Cauterruccio, pp. 207-242

Chapter 8

BIOLOGY, ADAPTATION AND ULTRA-STRUCTURE OF TWO SILK MOTH SPECIES OF NORTH- EAST INDIA

Sudip Dey
Electron Microscope Division, Sophisticated Analytical Instrument Facility,
North Eastern Hill University, Shillong.

INTRODUCTION

North-eastern part of India is considered to be a hot spot of bio-diversity. As far as the lepidopteran species are concerned, the region sustains a large variety of butterfly and moth. Among the moths inhabiting the area, a large number of silk moth species including *Antheraea perny, Samia Cynthia ricini, Antheraea roylei, Antheraea pruthi, Antheraea assamensis* etc. are prominent. Besides these, many species of Atlas moths, which produce 'Fagura' silk, have been recorded from North-east India.

The diversity of moth species in the region is due to its unique topography, climatic conditions, vegetations etc.

Different states of North-east India including Manipur are gifted with natural abundance of the oak plants, which are the primary food of oak Tsar silk moth larvae. The silk moth, *Antheraea proylei* Jolly (A hybrid between *Antheraea perny* and *Antheraea roylei*) , which is reared extensively in Manipur and other states of North-east India promises to better the economy of the region with 'Oak Tasar.'

The silk moth, *Antheraea proylei* reared in different parts of Manipur with different ecological conditions shows wide range of variation in its fecundity and oviposition behaviour. It is known that fecundity of insect is influenced by different factors such as humidity, light, temperature, feeding, initial weight of female etc. In locust, the number of eggs laid by a female increases up to 70%, while it declines again at 90% humidity (Hamilton, 1950). On the other hand, fecundity of the weevil, *Sitophilus oryzae* increases steadily up to 70% relative humidity but does not show any decline in the number of eggs laid when the humidity is increased up to 90 %(Birch, 1945). However, there are certain insect species, on which too high humidity does not have any adverse effect on fecundity, development and survival. Thus there is no definite range which is favourable or unfavourable to all insects, and each species has its characteristic behaviour. Again, it has been emphasized that the effect of a particular humidity is influenced by prevailing temperature (Atwal and Bains, 1974). Thus fecundity of a particular insect is affected by various conditions of temperature and humidity.

In some insects, fecundity is influenced by feeding and initial weight of female (Farrel, 1973, 1975). Solomon and Neil (1974) observed that host plant species and generation time affected fecundity of the carpenter worm reared on some oak plants.

With this background, effect of different environmental and physiological factors on fecundity and oviposition behaviour of the silk moth has been studied.

Besides fecundity and oviposition, the other important aspects of silk moth larval eco physiology are the food utilization, food assimilation, increase in larval body substances etc.It is well known that the quality of silk produced by a silk moth is greatly influenced by larval food utilization. Hence knowledge on eco physiology of food utilization appears to be extremely important for assessing the quality of silk produced by the silk moth. Detailed information on constants for mass, caloric and nitrogen relations are available for larval stages of many silk moth species. However, the aforementioned values are not worked out in detail for many silk moth species under Indian conditions. Keeping these in view, a study was performed on utilization of oak (*Quercus acuitissima*) leaves by the final larval stage of the silk moth, *Antheraea proylei*. The study was confined to the final larval stage because in lepidopteran larvae more than 70% of the total food is consumed in the last larval stage (Waldbauer, 1968).

The larval stages of the silk moth species, *Antheraea proylei* encounters a problem of fungal infestation of its host plant leaf. Soon after sprouting, the tender leaves are found to be infested by some aphids, which cause serious injury to the plant through sucking sap (Chakravorty and Das, 1981). It is well known that aphid infestation has some adverse effects on the chlorophyll content of plants (Lal, 1973). Infestation of aphids causes secretion of honey dew and creates favourable condition for growth of the fungus (Chakravorty and Das, 1981). Thus the leaves of the oak plants infested by aphids become unsuitable for rearing of the Tasar silk moth larvae.

In an attempt to find some biological control measures for aphids in oak plants, an aphidophagous coccinellid, *Coccinella septempunctata* L has been used after studying its biology.

In addition to *Antheraea proylei,* the present article describes studies on the 'Muga' silk moth, *Antheraea assamensis,* which is endemic to Assam and adjoining region. *Antheraea assamensis,* popularly known as 'Muga' silk moth is a semi-domesticated multi voltine, polyphagous insect having six broods in a year (Seitz, 1933).The insect is indigenous to Assam and other North-eastern states of India. The environmental conditions during different seasons greatly influence the growth and development of the'Muga' silk moth larvae, pupal weight, cocoon weight, shell weight, fecundity etc. and reelability of the silk (Thangavelu and Sahu, 1986; Yadav andGoswami, 1999).

Out of the six broods of *Antheraea assamensis*, two are popular for commercial rearings (Thangavelu and Sahu, 1986; Yadav and Goswami, 1999), which are conducted during May-June and October-November. The two seasons, Spring (May-June) and Autumn (October-November) are preferred for commercial rearing, because the climatic conditions during the period are favourable for growth and development of 'Muga' silk moth. However, since the 'Muga' silk moth could not be domesticated completely, the commercial rearing is carried out mostly under outdoor conditions subjecting the delicate larvae to environmental constraints such as rain, hail storm and a number of predators such as bird. Further, due to compulsion of outdoor rearing, dependence on favourable climatic conditions for growth and development of the larvae is quite high and hence only two commercial rearings are possible, although the insect has six broods in a year.

Several attempts have been made in the past to rear the larvae under indoor condition to overcome the problems associated with outdoor rearing (Thangavelu and Sahu, 1986). However the success of large scale indoor rearing was not very encouraging due to a number of problems. It is obvious that before attempting indoor rearing of 'Muga' silk moth larvae, a detailed knowledge about its feeding behaviour, physiology of feeding and digestion, specialization in structural features of associated organs etc.are essential. It is surprising to note that reports on basic studies on 'Muga' silk moth adult and larvae are extremely low (Hazarika and Bordoloi1998; Dey et al., 1997; Dey, 1999), and probably this lacuna in our knowledge is one of the reasons for less success in any experiment with the moth including its indoor rearing. A recent study carried out by us showed some unique structural features of the mid gut of the silk moth larvae which is likely to have a direct impact on indoor rearing (Dey et al., 1999).

Preliminary studies on the feeding behaviour of the silk moth larvae suggest that the rate of feeding, food consumption, larval growth etc., particularly at the final instar is very low under indoor rearing conditions.

The adaptational diversity of an insect is best expressed by the chemical as well as structural specialization of the chitinous cuticle covering the body (Neville, 1975), which is different in different species and different body parts of the same individual.

Specialized cuticular structures known as sensilla are reported to occur in different body parts of insects and are known to perform a variety of functions (Bland, 1983; George and Nagy, 1984; Alm and Hall, 1986; Crouau and Crouau-Roy, 1991). Scanning Electron Microscopy (SEM), because of its large depth of field and high resolving power has become indispensible for studying the detailed morphology of these cuticular structures. Several authors have used SEM for studying the distribution pattern, specialization in position, directional function, structural features of the socket, the nature of association of the sensilla base with the body cuticle and the finer detail of the surface (Dey, 1995; Dey et al., 1995; Dey and Biswas, 1996).

A review of literature on the subject reveals that SEM has been used extensively for studying the morphological details and surface ultra structural features of sensilla in different groups of insects including Diptera, Coleoptera, Siphonaptera and Lepidoptera. Although reports on a number of important studies on lepidopteran sensilla exist in the literature (Calvert and Hanson, 1974; Chun and Schoonhoven, 1973; Roessingh et al., 1991), the diversity in the habit and habitat of the group suggests the need for further studies on sensory system in different species.

In 'Muga' silk moth, *A. assamensis*, it has been observed that the larvae, particularly at the advanced developmental stages maintain a vertical posture during feeding under out door natural condition. The same however, can not be maintained in tray feeding under conventional indoor rearing. In an attempt to investigate whether or not there exist any sensory basis for preferred vertical feeding posture of *A. assamensis* larvae, a detailed ultra structural study has been undertaken in comparison to other domesticated and semi-domesticated silk moth larvae.

Besides the difficulties experienced in feeding and digestion in the larval stages of *A.assamensis*, some more problems eg. Low auto coupling and uneven sex ratio exist. Under the conventional rearing, the auto coupling efficiency of the moth seems to be very poor. It has been observed that in commercial rearing, during coupling, a single female is approached by three to six males of which one is successful to couple. The low auto coupling frequency of the moth is irrespective of seasonal variation in temperature and humidity, although its extent may vary in different seasons.

As far as the un-even sex ratio is concerned, it has been estimated that if in a commercial 'Muga' silk farm, the total emergence of the moth is 80,000, then the number of female moth emerged is only 20,000, the remaining 60,000 being the males. As a result, although there could have been 40,000 couplings if the sex ratio was1:1, presently due to un-even sex ratio (3:1), only 20,000 couplings are possible and the remaining 40,000 male moths emerged, are simply wastage. The low auto coupling and un-even sex ratio eventually results in low silk out put. Further, because of the existing low auto coupling frequency, manual coupling by hand has become necessary. The manual coupling necessitates additional man power, which involves extra expenditure. Besides this, hand coupling quite often injures the female causing adverse affects on fecundity.

Hence, it appears that there is a need for finding out the reasons for low auto coupling frequency and un-even sex ratio in the 'Muga' silk moth, *A. assamensis*, which is certainly a major problem in 'Muga' silk industry.

It is known that wings of male and female plays a very important role in governing the coupling behaviour of lepidopterans (Gibbs, 1980). The wing movement and appropriate signals to the coupling partners constitute an important component of prenuptial behaviour in butterfly and moth. In this context, it is to be noted that the ultra structural features of lepidopteran wings determine some of the physical, electrical and photo biological properties relevant to certain characteristic behaviour (Dey et al., 1998; Meyer-Rochow and Eguchi, 1983).

Keeping these in view, an ultra structural study on the wing of *A. assamensis* was undertaken to find out if there is any structural basis for low auto coupling frequency and un-even sex ratio in the moth.

MATERIALS AND METHODS

1. Antheraea Proylei

(i). Food Utilization

The larvae of A. proylei were brought from the field of Regional Tasar Research Station, Imphal. The larvae were reared in bamboo trays under indoor environment. When the larvae entered the fifth stage, they were weighed and were provided with known quantity of fresh oak leaves (*Quercus acuitissima*) every day. The same quantity of leaves was kept in another bamboo tray *as* control to know the weight lost due to evaporation. The amount of food consumed and faecal matters excreted were measured following the standard gravimetric method (Waldbauer, 1968; Reddy and Alfred, 1979). The study was continued until spinning. The amount of food consumed (C), food egested (F) and production due to body growth (P) were estimated using the formula of Petrusewicz and Manfadyen (1970) i.e,

C=P+R+F, where R=Respiration. Food energy assimilation (A) =C-F and Respiration (R) =C-(F+P)

(ii). Oviposition and Fecundity

Cocoons of *A.proylei* were collected from Breeding and Genetics section of Regional Tasar Research Station, Imphal, India. They were sorted out according to sex and were placed in separate compartments of the cocoon cage. After emergence, the adults were removed from the chambers and the females were weighed.

A single female with known body weight was allowed to couple with a healthy male. The duration of coupling for moths of all weight groups were maintained the same (4h). After decoupling, the female moths were kept in small bamboo baskets in dark for egg laying. The number of eggs laid and longevity of the female moths of different weight groups were recorded. A total of 384 moths were thus examined.

To know the effect of food plants on fecundity and other rearing performance, larvae from the same disease-free layings were fed with the leaves of two different species of oak plants, *Quercus acuitissima* and *Q. dealbata.* Another experiment was carried out to compare the fecundity of moths, whose larval stages were reared on noticeably vigorous trees with those reared on less vigorous trees.

Effects of temperature and humidity on coupling behaviour of the moth were studied by applying cool air blast (20-22°C) by means of air cooler. Fecundity studies were also conducted on moths emerged from diapausing pupae after light treatment and were compared with those emerged from non-diapausing pupae.

After the natural death of moths, the mature, unlaid eggs were taken out from the ovariole by dissecting the abdomen, and, the number was counted. These, together with the egg deposited by the moth constituted the total egg complement. Observations were also made on the influence of light, temperature and humidity on the oviposition behaviour of the moth.

To study the oviposition response to the bark characters of the host trees, mated females were released on to young trees of *Q.acuitissima* and *Q.dealbata.* Response to some non-host species and some artificial substrates were also studied.

To study the effect of crowding on egg laying, mated females were kept singly, and in a group of 2, 3, 4 and 5 respectively in small bamboo baskets (18cm x 11cm x5cm). 25 replicates were made for the experiment and the females selected were from the initial weight group of 4-4.5 gms.

(iii). Biological Control of Host Plant Pest

The pest of the host plant causing injury to the plant and growth of the sooty mould has been identified as an aphid. In an attempt to make biological control of the aphid, an aphidophagous coccinellid, *Coccinella septempunctata* has been studied in term of its biology and life history. The number of eggs laid by a single female, the time taken for hatching, the percentage of hatching etc.were recorded. The shape and colour of the egg were also noted. The size and colour of the first instar grub was also recorded and time taken for the first moult was noted. Similarly the length, colour and duration of the second, third, fourth and fifth instar grubs were recorded. Besides this, the number of aphids consumed by a single grub in its different developmental stages was also recorded.

2. Antheraea Assamensis

(A). Cephalic Sensilla Governing Preferred Feeding Posture (Dey et al, 2002)

The head of the larva was excised and was fixed in 2.5% glutaraldehyde in 0.1M Sodium cacodylate buffer (pH 7.2-7.4) for 2h at 4°C. The samples were washed thoroughly (1h) in the same buffer, post fixed in 1% Osmium tetroxide and were dehydrated in graded series of acetone (30, 50, 70, 80, 90, 95 and 100%). The dehydrated samples were dried in a Samdri-PVT-3 (Tousimis) Critical Point dryer or by TMS drying (Dey et al., 1989).

The dry samples were secured to brass stubs (10mmdiax 10 mm height) with double sided sticky tape. The samples were coated with gold in a JFC-1100(Jeol) Ion sputter and were viewed in the secondary electron emission mode of a JSM-35 CF (Jeol) Scanning Electron Microscope operated at 15 Kv at the working distance of 15 mm. Five samples from five different fifth instar larvae were examined. Sensilla were identified based on the work of Schneider (1964), Mc Iver (1975) and Schmidt and Smith (1987).For comparison of the surface micro structure of head of *A. assamensis* with other domesticated and semi-domesticated silk worms, samples from the 5th instar larvae of *Bombyx mori, Samia Cynthia ricini* (domesticated) and *A. proylei* (semi-domesticated) were prepared and viewed in Scanning electron microscope following the same method as used for A. assamensis.

(a). Feeding Behaviour under Outdoor Condition

Various larval stages (Ist, IInd, IIIrd, IVth, & Vth instars)of Muga silkworm were brushed on '**Som**' plants (*Machilus bombycina,* King) at about 7.00 am to observe their behaviour before feeding, during feeding, termination of feeding and restart. The budgeted time for each step was recorded. The orientation of the larvae, kineses in relation to different abiotic factors such as temperature, humidity and light was also recorded. Due care was taken to observe how does the worm orient itself on the leaf, it clasps the twigs as well as how does the abdominal appendages work during feeding.

(b). Recording of the Abiotic Factors

i. **Temperature** - Ambient temperature during feeding was recorded using the thermohygrometer, TEMPTEC Indoor-Outdoor Digital Thermohygrometer.
ii. **Humidity** - Humidity of the field during feeding of the worm was also recorded with the help of thermohygrometer, TEMPTEC Indoor-Outdoor Digital Thermohygro-meter.
iii. **Light** - The light intensity during feeding and its effect on the time and rate of feeding as well as the action of the worm was also studied. Light intensify was measured with the help of Digital Luxmeter, TES 1332, (Taiwan).

(c). Posture of the Worm

The preferred posture of the worm during feeding was studied by observing the behaviour of about 10 worms in each bush in each instar. i.e., Ist - Vth. The observations were repeated 10 times. The study was also carried out under indoor environment to investigate the feeding behaviour of the worm in horizontal and vertical posture.

(d). Effect of Posture on Food Consumption

The larvae of *Antheraea assamensis* (Ist to Vth instars) were taken from the Central Silk Board, Muga Seed Development Project (MSDP), Nongpoh (Meghalaya) and were reared inside the laboratory with the leaves of *Machilus bombycina*. The worms were divided into two groups (a) Group with Vertical Feeding posture, (b) Group with Horizontal Feeding posture. For rearing the worm under vertical posture the twigs of **Som** (*Machilus bombycina*) were placed in bottles containing water and the worms were released for feeding.

For horizontal posture, the worms were kept in bamboo cane trays with the twigs and leaves of host plant. For both the sets, frequent changes of host plant leaves were made to maintain its freshness. Dry weight of leaves before consumption was recorded each time when fresh leaves were provided to the worms. Control sets of leaves in the bottles containing water and also in bamboo cane trays were kept and their weight was taken to know the weight lost due to evaporation of water in a particular period. Exactly the same amount of leaves were provided to the worms and also kept in the control set. After recording the weight lost in the control sets of leaves, necessary adjustment were made for calculating the amount of food consumed. The standard Gravimetric Method (Woldbauer, 1968) was used for the daily food consumption and faecal product measurements. Faecal pallets were counted and dried to constant weight at 105 ^{0} C and dried weights were taken. All the gravimetric measurements were made in XP – 600 (Denever) electronic balance. Direct estimations were made on food consumption (C), food egestion (F) and production due to body growth (P). The formula of Petrusewicz and Manfadyen (1970), was used.

(e). Effect of Posture on Moulting and Spinning

To determine the effect of vertical or horizontal posture of the worm (during feeding) on the onset of moulting, all the instars of *Antheraea assamensis* were kept in (i) vertical posture by providing the twigs containing leaves in a water - filled bottle, (ii) by maintaining horizontal posture by feeding the worms in conventional tray rearing. Worms used were from the same disease free laying (DFL). Temperature and humidity of the indoor environment was recorded with a thermohygrometer, TEMPTEC Indoor-Outdoor Digital Thermohygrometer. The study was carried out in March - April, June - July and October – November during the year 2000-2003. Fifty worms of each instar were used for the study in each of the aforementioned seasons. The time taken for Ist, IInd, IIIrd, and IVth moult and for spinning was recorded in each season. The average of the data was calculated and statistical analysis was made using standard procedure.

STATISTICAL ANALYSIS

The statistical analysis of the data was made as follows:

I. Test of Significance (t-Test) for Food consumption in vertical and horizontal posture for all the Instars of Muga silkworm were computed by using Right Trail Test Method (2 Trail Test) (Mood, et.al., 1974)

$$t = \frac{\bar{x}_v - \bar{x}_H}{s\sqrt{\frac{1}{n_1} + \frac{1}{n_2}}}$$

Where,

$\bar{x}_V$ = food consumption in vertical postures,
$\bar{x}_H$ = food consumption in horizontal postures,
n_1, n_2 = number of observation,
s = standard deviation of the samples,

$$s = \sqrt{\frac{1}{n_1+n_2-2}\left[\Sigma(x_v - \bar{x}_v)^2 + \Sigma(x_H - \bar{x}_H)^2\right]}$$

$$d.f. = n_1 + n_2 - 2$$

Where, d. f. is degree of freedom
The t-test was computed from the table 6.

II. Test of significance in onset of moulting with reference to horizontal and vertical posture during feeding for all the instars of muga silk worm were computed by using the Right Trail Test. Table 7 was used as reference table. The formula used was as follows-

$$t = \frac{\bar{x}_{HM} - \bar{x}_{VM}}{s\sqrt{\frac{1}{n_1} + \frac{1}{n_2}}} \qquad d.f. = n_1 + n_2 - 2$$

Where,

$\bar{x}_{HM}$- Moulting data in horizontal posture
$\bar{x}_{VM}$- Moulting date in vertical posture
s- Standard deviation of samples
n_1, n_2 - number of observations

III. Test of significance of correlation coefficient for food consumption and moulting with reference to vertical (table 7) and horizontal (table 7) posture were also computed for all the instars of muga silkworm with the help of following formula:

$$r_{x_i y_i} = \frac{\frac{1}{n}\Sigma x_i y_i - \bar{x}\,\bar{y}}{\sqrt{\left(\frac{1}{n}\Sigma x_i^2 - \bar{x}^2\right)\left(\frac{1}{n}\Sigma y_i^2 - \bar{y}^2\right)}}$$

Where,

$x_i y_i$ - Correlation coefficient
x_i - Food ingested (g) in vertical posture
.y_i-Moulting and spinning in the vertical posture hours
n- Numbers of observations

$$t = \frac{r}{\sqrt{1-r^2}}\sqrt{n-2}$$

Where,

r- Correlation coefficient
n- Numbers of observations
t- Test of Significance for Correlation coefficient

(B). UV-sensitivity and Coupling behaviour (Dey et al, 2011):

MATERIALS

Male and female *Antheraea assamensis* were used for Electron Microscopical, Behavioural and other studies concerning UV sensitivity and coupling behaviour.

The samples were collected from Central Silk Board (Muga Silk Seed Organization) P3 Unit, Nongpoh, Meghalaya.

METHODS

Scanning Electron Microscopy

The different types of stripes and spots of wing parts of male and female moths were washed with acetone to remove dust and other foreign substances. The samples after drying were secured to brass stubs (10mm dia x 30mm high) and were coated in a JFC-1100 (JEOL) ion sputter.

Observations were made in a JSM-6360(JEOL) Scanning Electron Microscope using secondary electron emission mode at an accelerating voltage of 20 Kv.

Ultraviolet Photography

Ultraviolet photography of the whole moth was taken using a Pentex K 1000 camera, a Transilluminator (Phamacia) and a 312nm UV-filter.

Behavioural Studies

Coupling Behavior under Natural Conditions

The general behavior of the male and female moths during coupling was studied by observing the movement patterns of wings, the preferred timing of coupling, number of males approaching a female etc. under natural condition.

Effect of Artificial Source of UV Light on Coupling

To understand whether or not the auto-coupling behavior of the moth is influenced by ultraviolet light, male and female moths showing inability in auto-coupling were exposed to UV lamps for a short period (about 1 minute)

To examine the role of UV in auto-coupling, possible sites of UV reflections (as determined from SEM & UV-photography) were covered with some paints, and the moths with painted wings were exposed to the artificial UV-source.

Effect of Outdoor Natural Source of UV on Auto-Coupling

To determine the effects of natural UV on auto-coupling of *Antheraea assamensis*, male and female moths were exposed to outdoor twilight prior to coupling. Groups of male and female moths, ranging from 60 to 100 each were kept in cages, and were exposed to the outdoor twilight conditions, the only extensive natural source of UV. A set of control with the same number of males and females under indoor conditions was also used to compare the percentage of auto-coupling in the indoor and outdoor condition.

The temperature (maximum and minimum) and humidity (maximum and minimum) were recorded to know whether or not these factors have any relation with the percentage of coupling under indoor or outdoor conditions.

RESULTS

1. Antheraea Proylei

(i). Food Utilization (Dey, 1983)

The amount of food consumed, food assimilated, increase in the larval body substances and food oxidised during the final larval stage of A. proylei are presented in Table 1. There has been a gradual increase in the amount of food ingested, food incorporated as body substances and the food oxidised for energy (respiration). Out of the total food consumed, 78.94% was found to be assimilated. The larvae spent 79.72% of the assimi-lated food on respiration and the rest on production of body substances (Table1).

Table 1. Consumption and utilization of food by the last instar larva of *A. proylei* (Values are based on an average performance of 5 individuals with 10 replicates and in terms of wet weight)

Days	Food ingested (gms)	Food assimilated (gms)	Increase in the larval body substances	Food oxidized for energy (gms) (Respiration)
1st Dey	2.90	2.55	0.88	1.67
2nd Day	3.14	2.74	1.00	1.74
3rd Day	4.73	4.03	1.00	3.04
4th Dey	5.30	4.04	1.00	3.04
5th Day	5.51	4.11	1.00	3.11
6th Day	6.31	4.91	1.13	3.78
7th Day	7.19	5.73	1.16	4.57
8th Day	7.88	6.38	0.66	5.72
9th Day	9.00	7.40	1.19	6.21
10th Day	4.35	2.75	0.20	2.53
11th Day	2.67	1.85	0.20	1.65

(ii). Fecundity and Oviposition (Dey, 1987)

(a). Relation between Initial Body Weight of Female and Longevity and Fecundity

The effect of initial body weight of female on longevity is shown in Table 2. It was observed that life span was always more in heavier female than in lighter ones (table 2).

Table 2. Relation between initial body weight and longevity in adult female *Antheraea proylei.* (Number of moths in each class in parenthesis)

Year	Mean initial body weight (gm) ± SE for each individual class*		
	1	2	3
1981	1.4 ± 0.2 (54)	2.8 ± 0.3 (72)	4 ± 0.4 (64)
1982	1.3 ± 0.3 (54)	3.0 ± 0.2 (75)	4 ± 0.3 (65)

* survival classes: 1, up to 2 – 3 days; 2, between 2 – 3 days and 6 – 7 days and 3, over 6 – 7 days after emergence.

Relation between initial body weight of female and fecundity & oviposition period is shown in Table 3. It was found that initial weight of the female influences the fecundity significantly. Moderately heavy females of 4.24 gms laid approximately75% more eggs than moderately light females of 2.54 gms. Linear regression analysis of the body weight and fecundity showed highly significant correlation (r=0.9). Further, moths of weight group 1.7gm laid eggs up to third day and their fecundity was the least. Those with mean initial weight of about 2.54gm used to lay eggs beyond the third or the fourth days and their fecundity was also higher. Moths having initial weight of 3.23 gms laid eggs up to the seventh day and their fecundity was found to be higher than that of the weight group of 2.54 gms. Moths with initial weight of 4.24 gms continued to lay eggs up to the eighth day and their fecundity was higher than that of the other weight groups (Table 3).

Table 3. Relation between initial body weight, fecundity, and oviposition of *Antheraea proyle*

Mean initial body weight	No. of eggs laid on 1st	2nd	3rd	4th	5th	6th	7th	8th	9th Day	Total egg / female
1.7	62	7	5	0	0	0	0	0	0	74
2.54	93	12	7	4	4	3	0	0	0	123
3.23	131	12	8	5	5	5	4	0	0	170
4.24	158	27	13	7	5	5	4	3	0	217

* the values are mean of egg-laying patterns of 96 individuals in each group.

(b). Effect of Food Plant and Abiotic Conditions on Larval Feeding and Growth

Feeding in the larvae of *A.proylei* was found to be greatly influenced by environmental conditions and food plants. Less light was found to be preferred at the time of feeding. Very high and very low temperature and humidity reduced the feeding rate. Food consumption and the larval growth was more when the larvae were grown outdoor. The shell weight, initial weight of female and fecundity were more in those who were reared outdoor (Table 4). When the larvae were reared on 'Sahi' (*Quercus dealbata*), the larval period was more than when reared on 'Uyung' (*Quercus acuitissima*). Larval growth was also better on 'Uyung'. Growth rate was not uniform in'Sahi', whereas it was uniform in 'Uyung'. The cocoons were dark brown in colour when the larvae were reared on'Uyung' whereas they were light brown in'Sahi'. The shell weight of cocoon was also higher when the larvae were reared on 'Uyung'as compared to those reared on 'Sahi'. Fecundity of moths whose larvae were reared on 'Uyung' was15% more than those reared on'Sahi'

Table 4. Effect of Outdoor and Indoor rearing on shell weight and fecundity*

Rearing condition	Larval period	Cocoon wt. (g)	Shell wt (g)	Initial wt. of female (g)	Survival	Fecundity
Outdoor	35 days	6 gm	0.86	4.5	8 days	230
Indoor	40 days	4gm	0.60	3.0	5 days	153

*The difference between total fecundity values in outdoor and indoor conditions is significant ($P<0.05$).

(Table 5). In addition, there were some indications that moths obtained from larvae reared on vigorous trees were more fecund than those from less vigorous trees.

Table 5. Effect of food plant species on larval growth, initial weight of female and fecundity (outdoor rearing)*

Food plant	Larval period	Cocoon wt. (g)	Shell wt. (g)	Initial wt. of female (g)	Survival	Fecundity
Sahi (*Quercus delbata*)	40	4.2	0.5	3.1	4 days	170
Uyung (*Quercus acuitissima*)	35	5.0	0.6	3.7	6 days	200

*The difference between total fecundity value is significant ($P<0.05$).

(c). Relation Between Generation Time snd Fecundity:

A.proylei is a uni-voltine race. Its pupa enters diapause during winter. This pupal diapause however, could be terminated by exposing the pupa to artificial light for 4hours during night so that they could get additional light besides the normal day-light. By this light treatment, a second generation could be raised in a year. When the larval growth, shell weight, initial weight of female and fecundity of the second generation were compared with those of the first generation, it was found that the aforementioned characters were inferior in the second generation.

(d). Oviposition Behaviour and Rhythm

Emergence of moths usually starts in the afternoon (2pm) and continues up to 9pm. Peak emergence were observed between 6-7pm. The male moths were found to be more active than the female and preferred a pre-nuptial flight. Coupling was found to take place within 2-3 hours after emergence. For coupling, moths were found to prefer darkness. Relatively low temperature (26-28°C) and high relative humidity (70-80%) were found to be ideal for coupling. Cool air-blast (20-22°C) by means of air cooler for 30 minutes to one hour was quite helpful in achieving high coupling percentage (80-90%). The normal duration of coupling was10-12 hours, but experimentally it was found that four hour was sufficient for coupling. After that period, the pair could be decoupled and mails could be used again for pairing. The moths preferred darkness for egg laying. Except for a brief period in the afternoon of emergence, a night time rhythm was observed. Observation on more than 300 females indicated a nocturnal pattern in the oviposition behaviour. Females began ovipositing at about 4-6pm and continued up to late night. The most active period was found to be between 7.30 pm and 11.30 pm and again at the late hours of night.

(e). Substrate Preference for Oviposition

The roughness of the substrate was found to play important role in egg laying. Females released on the young trees of 'Uyung' with rough bark was found to deposit about 30% more eggs than on 'Sahi', whose bark is comparatively smooth. Females released on to some non-host species of plants with rough bark oviposited in the same fashion as on the host species with rough bark ('Uyung'). When artificial substrates, such as dry wood, dead bark, bamboo basket etc. were tried, females were found to oviposit freely.

(f). Effect of Crowding on Egg- Laying

When the effect of crowding of moths on egg-laying was studied, it was observed that overcrowding of moth affected the fecundity adversely. In the experiment, the best result was obtained with single moth per bamboo basket showing an average fecundity of 220 with very less unlaid eggs. Overcrowding, i.e. 5 moths per bamboo basket showed less fecundity (an average of 130), and large number of unlaid eggs.

(iii). Biological Control of Host Plant Pest

A single female of aphidophagous coccinellid beetle, *Coccinella septempunctata* was found to lay 35-45 oval, yellow eggs, which hatched out within 3-4 days, the percentage of hatching being 76-80%. The first instar grub is 3.5 mm in length and is grey in colour. The first moult takes place within 48 hours. The second instar grub has darker body pigmentation and the upper part is brown in colour. Its length is 4.4 mm. The second moult takes place within 24 hour. The third instar grub is 5 mm in length and is characterised by the appearance of a brown shield- like structure on the head. The duration of the third stage is 48hours, after which the third moult takes place. The mature fourth instar grub is 6.7 mm in length. The fourth moult takes place after 24 hours. The fifth instar continues for about 6 days. The length of mature fifth instar grub is about 7.5 mm. One or two days before pupation, the body becomes flattened and the length is reduced to about 6.6 mm. The pupa is 6.3 mm in length and 4.5 mm in width, the pupal stage lasts for 4 days, after which the adult emerges out.

It has been observed that single first instar grubs can consume10-20 aphids on the first day. As the larval stages progress, the number of aphids consumed by a single grub also increases, reaching to 150-200 aphids per day in the fifth stage.

2. Antheraea Assamensis

(A). Cephalic Sensilla Governing Preferred Feeding Posture (Dey et al. 2002)

Feeding Behaviour

The feeding of the larva was found to be carried out in different steps.

i. The larva started feeding from the edge of the leaf blade near the petioles and moved towards the right. It clasped the petiole with thoracic and abdominal legs. After feeding a portion of the leaf, the worm fed again from the lower side of the leaf blade upward. In the process, a crescent shaped structure was formed.
ii. The worm moved back by sliding its head and thorax backward through the midrib. It then clasped the midrib and ate the portion of the blade at a little distance away from the mid rib. In this case also a crescent shaped structure was formed, thus forming a double crescent. The worm then moved downward through the midrib.
iii. The worm then moved up and made a deep cut by moving its head forward in the first crescent. It then went on feeding towards the margin of the leaf blade through the first crescent. The larva moved downward through the midrib and started feeding the portion of the leaf near the mid rib by holding the blade with thoracic legs. It consumed almost the entire portion of the leaf blade leaving only a small portion of it attached to the mid rib.
iv. The feeding continued near the double crescent and the worm consumed completely the portion of the leaf blade till the tip.
v. When the part of the leaf blade at one side of the midrib was consumed completely, the worm moved down and started feeding in the same fashion until it completely consumed the leaf blade till the tip. Finally, even the midrib was also consumed.

At the time of mild to heavy shower the worm moved up and took rest below the leaf blade by holding the midrib with thoracic and abdominal legs. After the rain was over the worm started feeding again near the midrib by normal fashion. During the windy days when the branches of the plant were moved by the wind with considerable velocity, the worms were found to be disturbed, and feeding was observed to be less. When the wind was over, then the worm started feeding with the earlier speed.

During sunshine, the worm placed itself under the partial shade of the leaves. It turned to the left side and came near the midrib and continued feeding.

During the entire process of feeding, the worm maintained a vertical posture. It could not feed properly at horizontal posture. The worm was found to take rest for 7-10 minutes in horizontal posture during feeding after excretion of faecal pallets.

Abiotic Factors

i. **Temperature** - The worm was found to feed properly and was found to spend maximum time in feeding at a temperature of about 26.9 to 29.4 ^{0} C.
ii. **Humidity** - The preferred humidity during feeding was found to be between 70 - 79%.
iii. **Light** - It was observed that the worm preferred less light during feeding. The maximum time spent in feeding was found to be at around 240-343 lux. However, the other weather conditions such as sunshine, cloud, wind temperature and humidity were also found to affect feeding at a particular light intensity.

Effect of Posture on Food Consumption

Comparisons on amount of food consumed by different instars of the 'muga' silk worm at vertical and horizontal postures under indoor conditions revealed that the worms from the Ist to Vth instars consumed a total amount of 333.46g at vertical postures and 175.03 g at horizontal posture. However, differences in the amount of food consumed at the two different postures were different in different instars. While the difference in food consumption of the worms at vertical and horizontal posture was not very high in first and second instars, remarkable reduction in food consumption at horizontal posture compared to vertical postures was recorded in the later stages such as IIIrd, IVth and Vth.It was recorded that, 28 to 41% less food was consumed at horizontal posture in comparison to vertical posture during Ist to IIIrd stages. In the later stages i.e. IVth and Vth instar, 50% less food was consumed at horizontal posture compared to vertical posture (Table 6).

Effect of Posture on Moulting and Spinning

Studies revealed that moulting was much earlier in the worms fed at vertical posture as compared to those in horizontal postures. First moult took almost 18 hours less time in worms fed at vertical posture in contrast to those fed at horizontal posture. In the case of 2nd moult, the worms at horizontal posture took almost 98 hours more time as compared to those at vertical posture. The 3rd and 4th moult took 24 hours more time at horizontal posture in comparison to the worms at vertical posture (Table7).

The spinning of the worms fed at vertical postures took place almost 38 hours earlier than the worms reared at horizontal posture. Total time from brushing to spinning, at horizontal posture under indoor rearing conditions was almost 8 days more than that at vertical postures under the similar indoor environment (Table7). The data were statistically significant (Table 8).

Table 6. Comparative data on food consumption by the different larval stages of muga silk worm, *Antheraea asssamensis* at vertical and horizontal postures (values are average performance of 5 individuals with 10 replicates in terms of wet weight) [Average Temp: 22-29^{0}C; Average RH: 85-92%]

	Days	Food ingested (g) at vertical posture	Food ingested (g) at horizontal posture
Ist larval stage	1st Day	0.83	0.41
	2nd Day	0.71	0.44
	3rd Day	0.40	0.54*
	Total	1.94	1.39
IInd larval stage	1st Day	1.05	0.46
	2nd Day	1.74	0.54
	3rd Day	0.78	0.89
	4th Day	0.21	0.32 *
	Total	3.78	2.21
IIIrd larval stage	1st Day	1.49	1.01
	2nd Day	1.74	1.42
	3rd Day	3.10	1.59
	4th Day	3.13	1.61
	Total	9.46	5.63
IVth larval stage	1st Day	6.54	3.11
	2nd Day	12.52	5.41
	3rd Day	16.70	6.14
	4th Day	7.33	6.00
	Total	43.09	20.66
Vth larval stage	1st Day	28.95	15.25
	2nd Day	41.02	21.61
	3rd Day	46.60	24.55
	4th Day	51.63	27.20
	5th Day	39.80	20.97
	6th Day	48.59	25.60
	7th Day	18.60	9.80
	Total	275.19	144.98

* Loss of food consumption at vertical posture shown in certain days in some instars is due to moulting of some worms earlier than those in horizontal posture resulting in less number of worms at vertical posture at a particular stage.

Table 7. Comparative data on onset of moulting and spinning by Muga silk worm reared in vertical and horizontal position

Moult	TIME TAKEN FOR MOULTING AT VERTICAL POSTURE	AT HORIZONTAL POSTURE
1st moult	32 ± 2 hrs	50 ± 3 hrs
2nd moult	60 ± 3 hrs	158 ± 5 hrs
3rd moult	157 ± 6 hrs	181 ± 6 hrs
4th moult	144 ± 3 hrs	168 ± 4 hrs
Spinning	288 ± 6 hrs	326 ± 7 hrs
Total hours from brushing to spinning	681 hours (or 28.37 days)	883 hours (or 36.79 days)

Table 8. Statistical analysis

I. For food consumption in vertical and horizontal posture

Instars	Reference table	d.f.	Computed 't' value	Significance level	Remarks
Ist	6	4	1.34	$t > t_4$, 0.3	Significant at 0.3%
IInd	6	6	1.18	$t > t_6$, 0.1	Significant at 0.1%
IIIrd	6	6	2.06	$t >. t_6$, 0.1	Significant at 0.1%
IVth	6	6	2.27	$t > t_6$, 0.1	Significant at 0.1%
Vth	6	12	3.74	$t > t_{12}$, 0.01	Significant at 0.01%

II. Onset of moulting and spinning with reference to horizontal–vertical posture during feeding

Reference table	d.f.	Computed 't' value	Significance level	Remarks
7	8	0.643	$t > t_6$, 0.6	Significance at 0.6%

III. For correlation coefficient for the food consumption and moulting

Parameter	Reference table	Computed $r_{x_i y_i}$	d.f.	Computed 't' for $r_{x_i y_i}$	Significance level	Remarks
Vertical posture of feeding and moulting	6	0.89	3	3.4	$t > t_3$; .05	Significant
Horizontal posture of feeding and moulting	6	0.87	3	3	$t < t_3$; .05	Significant

Scanning Electron Microscopy

Head Surface of the Vth Instar

Dorsal Campaniform Sensilla

i. ***Antheraea assamensis***: The general surface of the head in *A. assamensis* (Vth instar) was found to be covered with round or conical protuberances. The diameter of these protuberances ranged between 14μm to 18μm (Fig. 1a). These sensilla are known as Dorsal campaniform sensilla and are reported to be present in the cuticle of joint regions in antennae, appendages, mouth parts, wings, legs, ovipositors etc (McIver,1985).

ii. ***Antheraea proylei***: The head surface of the semi domesticated Tasar silk worm, *Antheraea proylei* also revealed the similar type of round and conical protuberance distributed throughout the entire head surface. The diameter of the protuberances was found to be about 22μm to 34μm. Besides these, small cuticular spines about 1 μm long and 0.2μm broad were also observed. (Fig 1b).

iii. ***Bombyx mori***: When comparisons were made on the head surface micro-architecture of *Antheraea assamensis* with domesticated silk worm *Bombyx mori* the conical or dome shaped protuberances could not be detected (Fig. 1c).

iv. ***Philosamia ricini***: The head surface of the domesticated silkworm *Philosamia ricini* also did not exhibit any dome shaped structure similar to dorsal Campaniform sensilla that was detected on the head surface of *Antheraea assamensis* (Fig. 1d)

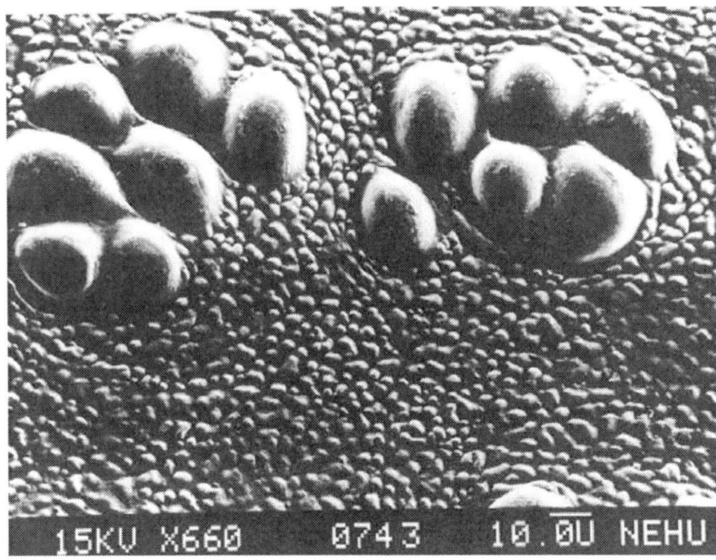

Figure 1a. Enlarged view of groups of dorsal campaniform sensilla of Vth instar *Antheraea assamensis*.

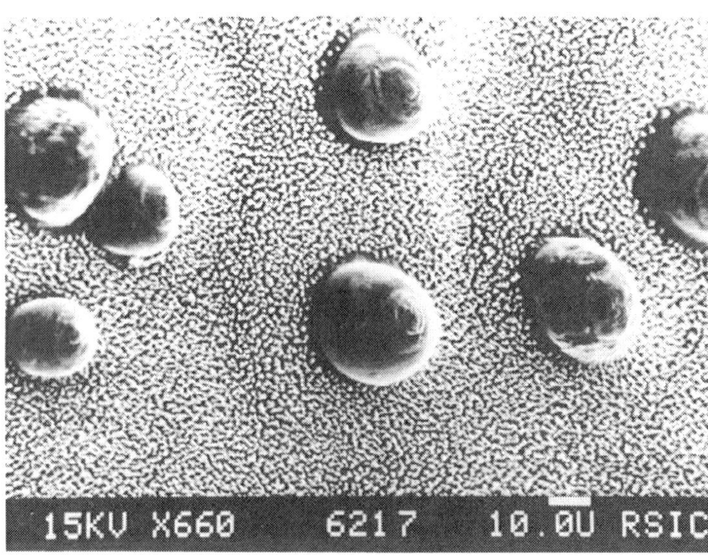

Figure 1b. Head cuticle of Vth instar *Antheraea proylei* showing dome-shaped dorsal campaniform sensilla and innumerable cuticular spines.

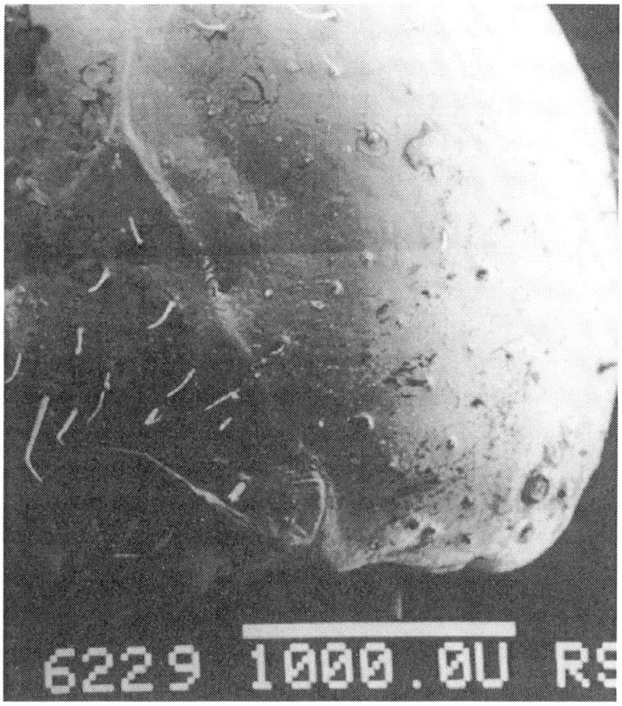

Figure 1.c. Head cuticle of Vth instar *Bombyx mori* showing the absence of dorsal campaniform sensilla and the presence of few sensilla trichoidea.

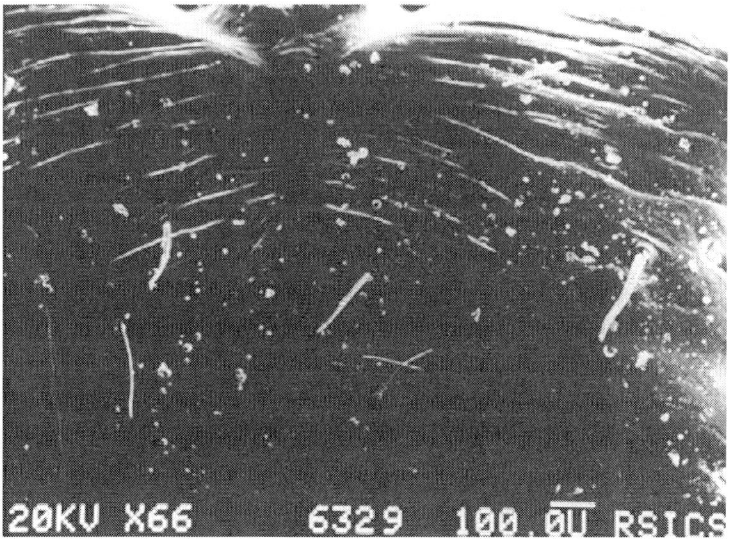

Figure 1d. Head cuticle of Vth instar *Phyllsmia ricini* showing the presence of a few sensilla trichoidea but total absence of dorsal campaniform sensilla.

Distribution pattern and abundance of dorsal campaniform sensilla on the head surface of different instars of *Antheraea assamensis*:

a) **I^st^ instar** - About five cuticular domes of 32μm to 39um diameter were found to be located on the two lateral sides. The sensilla were not grouped and were isolated from one another (Figs. 2 a, b).
b) **IInd instar** - In this stage the numbers of the cuticular domes were more or less same as that of the 1st instar, but the diameter was slightly larger (47μm to 76μm) (Figs. 2 c, d).
c) **IIIrd instar** - Numerous cuticular domes appeared on the lateral parts of the head cuticles. However, they were not raised much above the cuticular surface. The domes were mostly in groups of 4 to 5 individuals, although some were isolated. The diameter of the sensilla ranged between 12μm and 22μm (Figs. 2 e, f).
d) **IVth instar** - Numerous dorsal campaniform sensilla were distributed throughout the head cuticle except the central part. They were mostly in groups of 4 to 5 individuals. Dome shaped appearance of the sensilla was prominent. The diameter of the sensilla was about 13μm to 18μm (Figs. 2 g, h).
e) **V^th^ instar** - Numerous Dorsal campaniform sensilla were found throughout the head cuticle. However the central part was devoid of the cuticular domes. The sensilla were in groups of 4 to 7 individuals. Domes shaped appearance was prominent. The diameter ranged between 14μm and 18μm (Figs. 2 g, h).

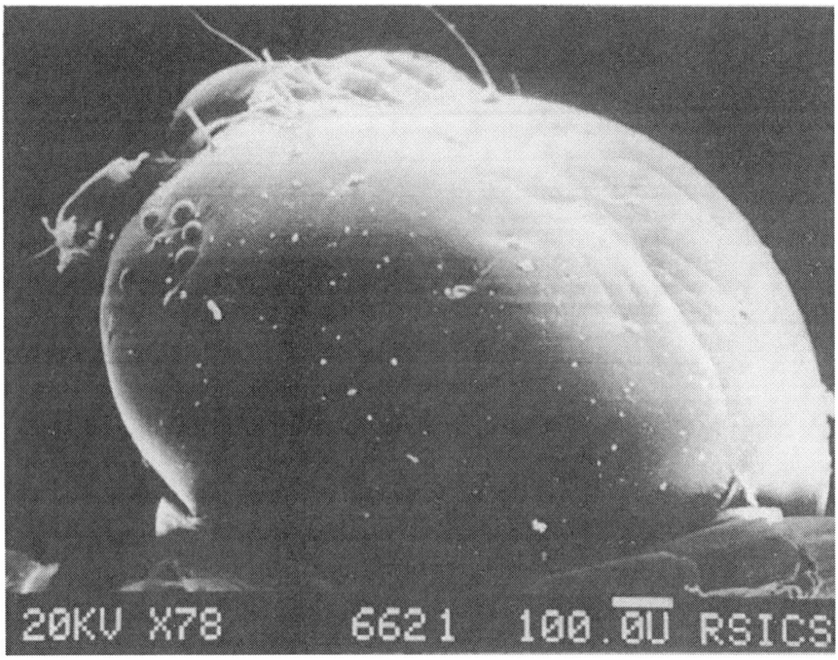

Figure 2a. Gravity receptor I^st^ Instar.

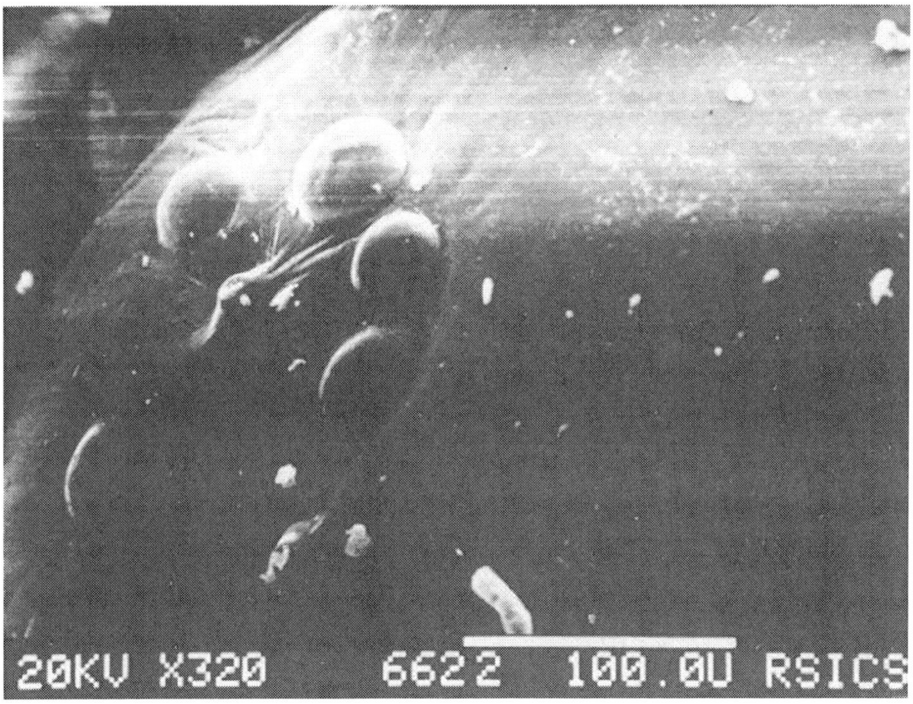

Figure 2b. Enlarged view of Fig.2a.

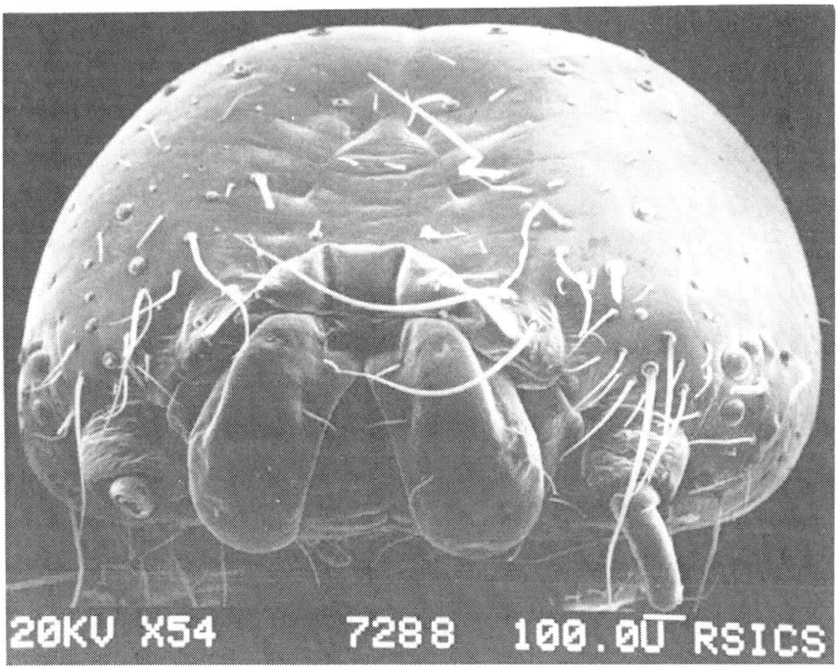

Figure 2c. Gravity recptor IInd Instar.

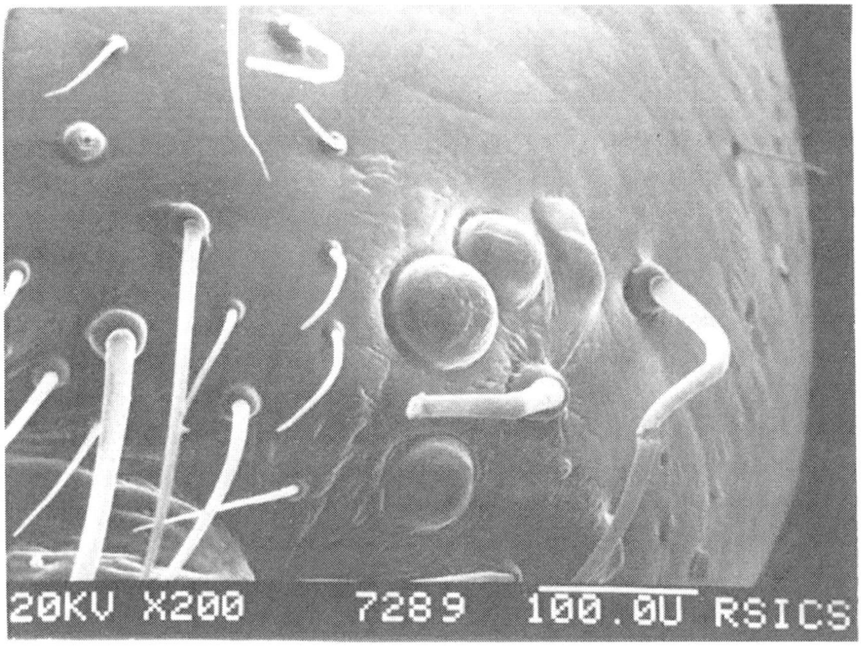

Figure 2 d. Enlarged view of Fig 2 c.

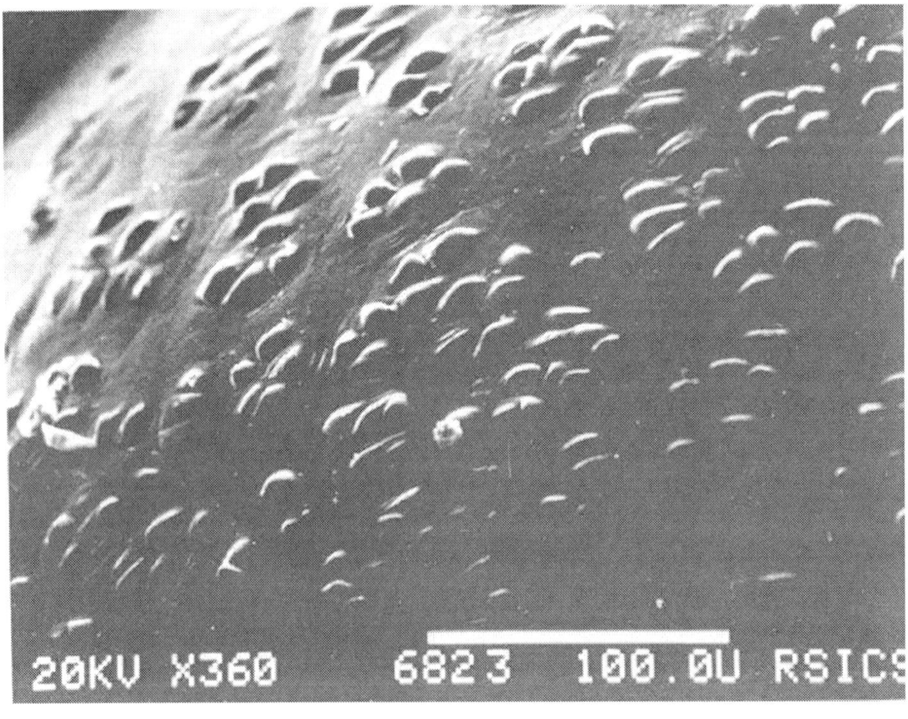

Figure 2e. Gravity receptor IIIrd Instar.

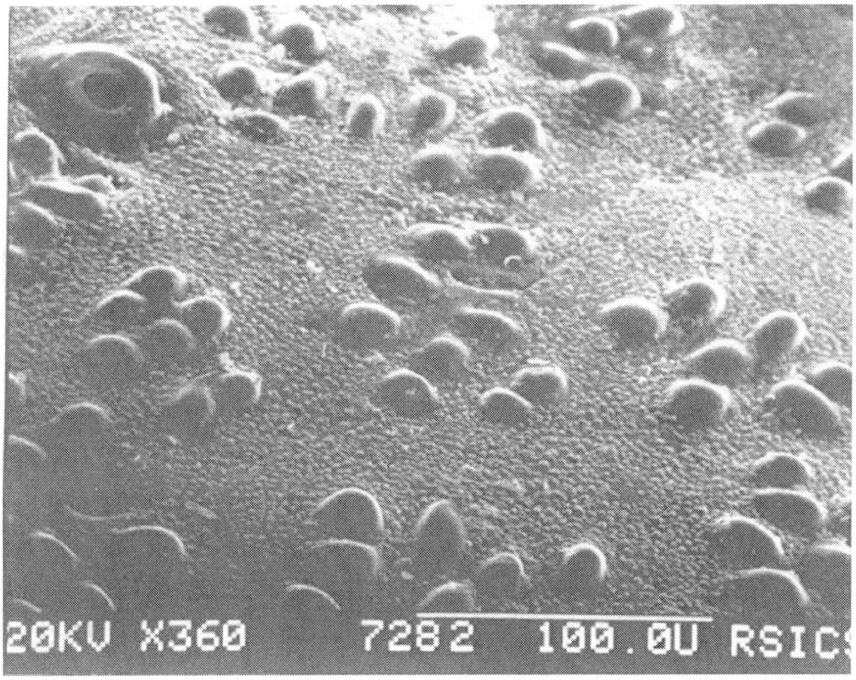

Figure 2f. Enlarged view of Fig.2e.

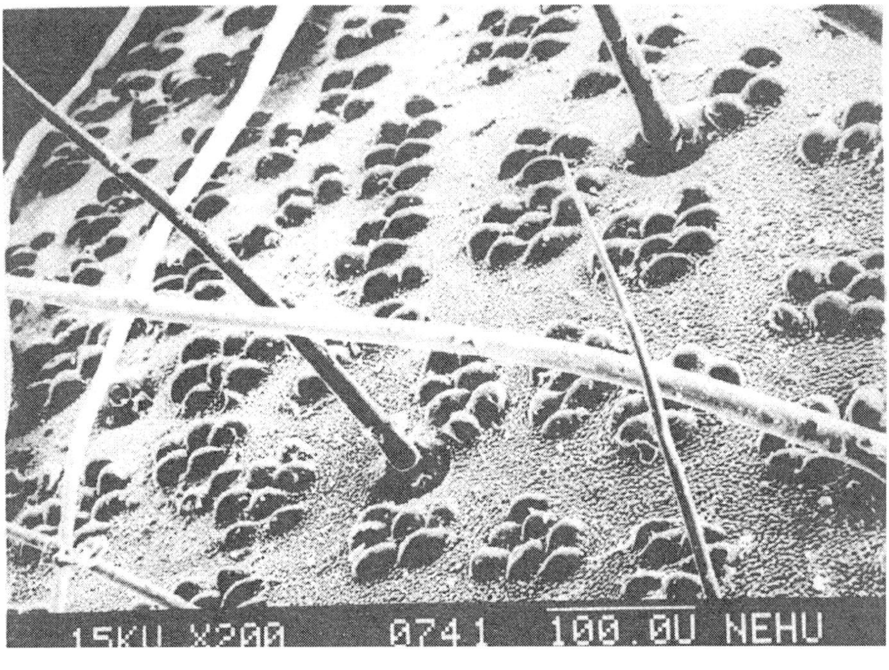

Figure 2g. Gravity receptor IVth & Vth Instar.

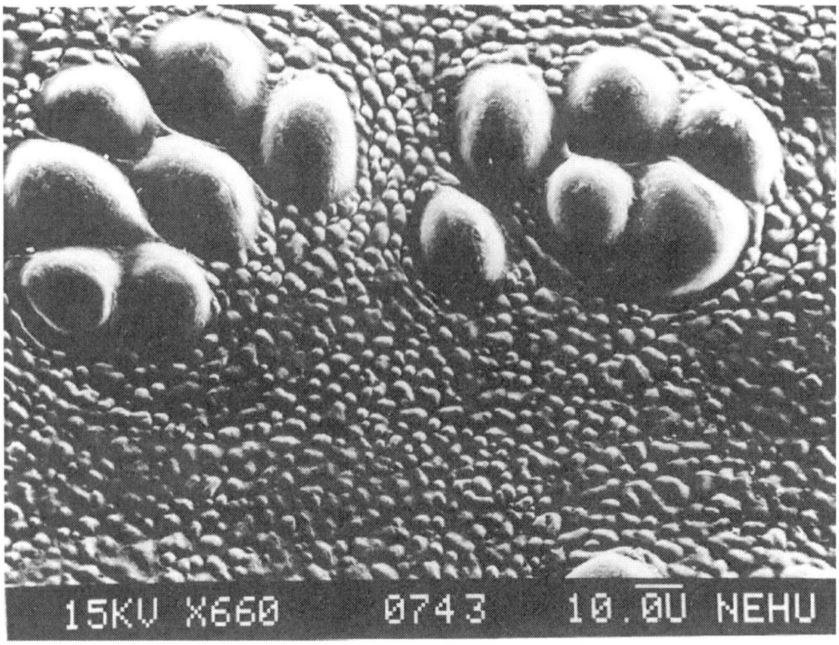

Figure 2h. Enlarged view of Fig.2g.

(B). UV sensitivity and Coupling Behaviour

Scanning Electron Microscopy

Wing Scale in Male and Female

Scanning electron microscopy revealed that the wings of male and female *Antheraea assamensis* have two types of scales:

1. Thin and elongated scale: – These are about 0 .6 mm in length and 25-30µm in width. At the tip the scale is bi-lobed.
2. Broad and flat scales: – These scales are 0.4-0.5 mm in length and 80-100µm in width.

The enlarged view of both elongated and broad wing scales showed diffraction grating pattern and air spaces of similar dimension. The pattern was characteristic of UV-absorbing scales as described by earlier workers. (Ghiradella, et.al., 1972).

General wing part: General part of the wing of male and female contains only one type of scale which was found to be about 180-410µm in length and 100-150µm in breadth. The enlarged view of the scale revealed the arrangement of diffraction grating and air spaces characteristic of UV-absorbing wing scales (Ghiradella, et.al., 1972).

Wing spots: The wing spots of the male and female revealed two types of scales, viz broad. and elongated. The broad scales were found to be 200-225 µm in length and 100-125 µm in breadth. The elongated scales on the other hand were about 250-280µm in length and

40-50μm in breadth. The enlarged view of both the broad and elongated wing scales revealed the presence of 50-60 nm thick micro-ridges and 80 nm thick air spaces between them. This arrangement has been reported to be the characteristic of UV-reflectors (Ghiradella, et.al., 1972).

Wing stripes: The smoky stripes present on the edges of wing of both male and female revealed the presence of two types of scales, viz., the broad and elongated scales. The broad scales were found to be about 300-315μm long and 80-90μm wide. The elongated scales on the other hand were about 415-440μm long and 40-55μm wide. Enlarged view of both the broad and elongated scales revealed the presence of micro ridges of about 50-60 nm thickness along with air spaces of about 80 nm width (Figs 3a, b).

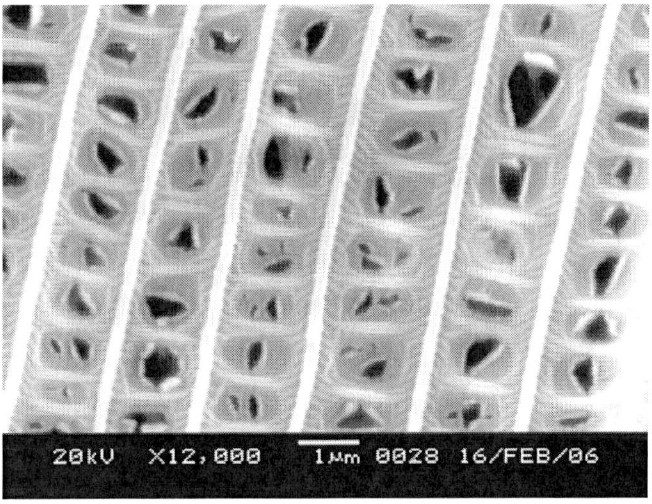

Figure 3a. Further enlargement of scales in smoky stripes.

Figure 3b. An enlarged view of scales in smoky stripes showing fine micro-ridges and air-spaces.

Ultra-violet Photography: Ultra-violet Photography of the whole moth showed white patches on the wing scales representing the spot and smoky stripes in contrast to the scales on general wing scales which appeared dark (Fig. 4 a,b). The white patches suggest UV-reflection while the dark area indicates UV-absorption (Meyer-Rochow, 1991).

Figure 4a. Whole body of *Antheraea assamensis* under normal light.

Figure 4b. Whole body of *Antheraea assamensis* under UV-light.

Behavioural Studies

Coupling Behaviour under Indoor Natural Conditions

The coupling was found to be erratic under indoor conditions, which takes place both in the morning and afternoon. However, preferred coupling time was observed to be just at dawn and dusk i.e. the twilight condition. During coupling, 3-6 males were found to approach a single female. The males as well as the female were found to vibrate their wings constantly until a male succeeds to couple with a female.

Effect of Artificial Source of UV Light on Coupling

When the male and female moths were exposed to artificial source of UV light, a single male approached a female to get coupled in a short time. When the parts of the wing, which exhibit UV-reflecting property (as revealed from SEM and UV-photography), were covered with some paints, the moths did not couple even after exposing them to UV light.

Effect of Outdoor Natural Source of UV on Auto-Coupling

Exposure of male and female moths to outdoor twilight conditions, the only extensive natural source of UV, caused a single male to approach a female. Coupling was found to take place in a short time after the female and male constantly moved their wings. The percentage of auto coupling was much higher (about 98%)(Fig.5a) as compared to that in control(about 52%) (Fig.5b) kept inside the room.

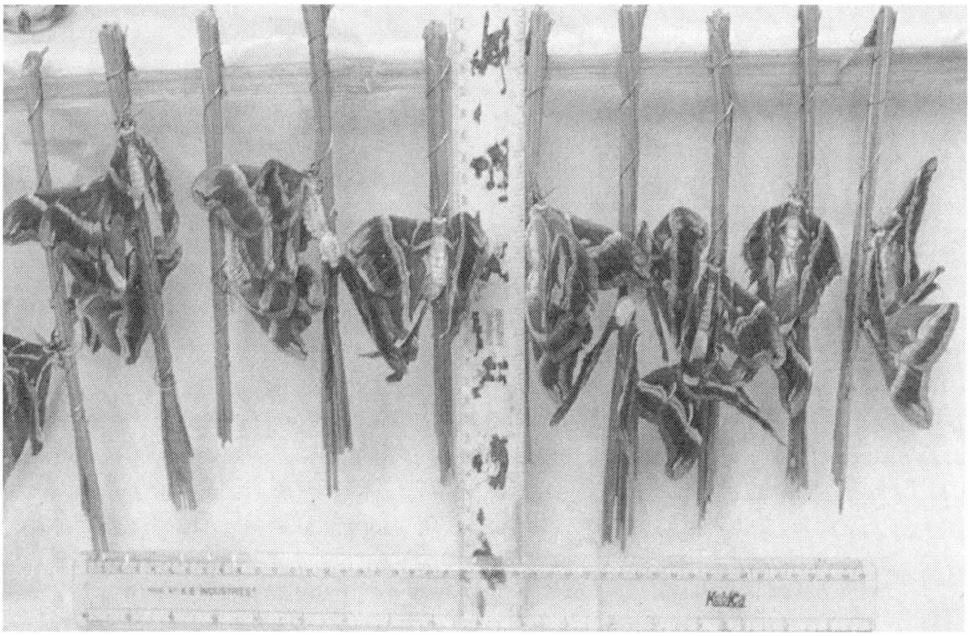

Figure 5a. Coupling under outdoor twilight conditions.

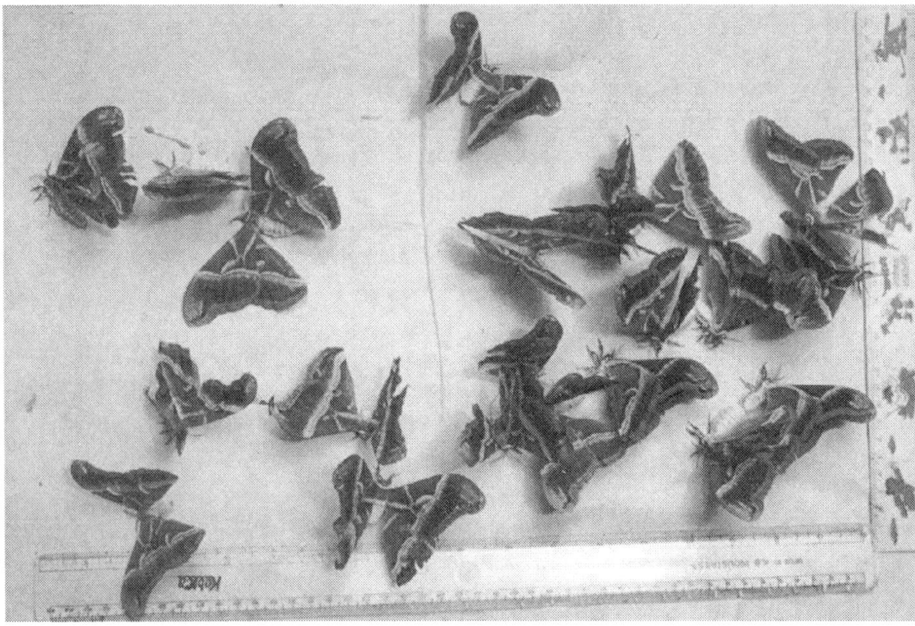

Figure 5b. Inability of auto-coupling inside the room.

DISCUSSION

1. Antheraea Proylei

(i). Food Utilization

The results obtained on food utilization by *A. proylei* larvae show similarity with those reported in other silk moth larvae as well as in larvae of other insects. In *Bombyx mori*, Hiratsuka (1920) reported that 42% of ingested food was assimilated, 52% of assimilated food was metabolised and 48% was used in production of body substances. Stockner (1971) has reported that in the herbivore diptera, *Hedriodiscus truguii,* 62% of the consumed food was assimilated, and out of this assimilated food 44% was spent on respiration, while 56% was converted into body substances. Lawton (1971) found that in the carnivorous Zygopteran, *Pyrrhosoma nymphula,* 88% of the consumed food was assimilated. About 51% of the assimilated food was converted into body substances and 49% was metabolised. Pandian and Mathaban (1974) working with a tropical dragonfly *Displacodes trivialis* (Rambur) reported that 91% of the food consumed was assimilated, 48% of the assimilated food was converted into body substances and the remaining 52% was used on metabolism. Reddy and Alfred (1979) have reported that in the silk moth, *Phylosamia ricini* HUTT, 63.15% of total food ingested was assimilated, out of which 65.76% was used for respiration and 34.25% was converted into body substances. The report of Reddy and Alfred (1979) is in good agreement with our findings in *A.proylei.* In both the findings, a great percentage of the assimilated food was found to be used for respiration, possibly due to the high activity of the larvae during the final instar.

(ii). Fecundity and Oviposition

The observations made on *A. proylei* suggest that fecundity and oviposition of the silk moth are influenced by a variety of environmental and physiological factors. It has been found that there is a direct relationship between the initial weight of the female and fecundity. However the initial weight seems to be influenced significantly by feeding, food plant species and the rearing conditions. Comparatively better fecundity and oviposition of moth reared in outdoor condition indicate that the moths still show their wild nature and may need many more generations to be raised in indoor condition before they could be domesticated successfully.

(iii). Biological Control of Host Plant Pest

Studies on the biology of the aphidophagous coccinellid beetle, *Coccinella septempunctata*, indicating its capability of consuming increasing number of aphids as the larval stages progress, suggest that the coccinellid can be used as a biological control measure against aphids infesting oak plants resulting in growth of sooty mould, which makes the leaf unsuitable for consumption by the silk moth larvae.

2. Antheraea Assamensis

(i). Cephalic Sensilla (Gravity receptor)

Observations made on behaviour of the muga silkworm both under outdoor and indoor conditions are suggestive of some preference to a particular posture during feeding. The maintenance of a vertical posture during feeding under outdoor rearing conditions indicates that this posture is a part of innate behaviour of the insect. The movement of the worm in search of a suitable substratum to maintain a vertical posture during compelled horizontal feeding in conventional tray reared indoor conditions confirms that vertical posture is an inborn tonic reflex of the worm.

The present observations on a positive correlation between vertical posture and high food consumption in the muga silkworm suggest that the innate behaviour of the worm which determines its vertical posture during feeding and causes higher food consumption compared to horizontal posture is extremely important in the rearing process of worms particularly under indoor conditions. It is well known that in *Antheraea assamensis* (Thangavellu and Sahu, 1986) and other species of Antheraea (*Antheraea proyelei*, tassar silk worm) (Dey, 1987), the rearing performance in terms of food consumption, utilization and assimilation are much inferior under the indoor environment compared to outdoor.

It is also reported that moulting and spinning takes longer time in muga (Thangavelu and Sahu, 1986) and Tassar silkworm (Dey, 1987) under indoor rearing conditions compared to outdoor rearing .Observation on the earlier onset of moulting and spinning in worms at vertical posture compared to those at horizontal posture under indoor rearing conditions, suggests that since the food consumption was higher in worms at vertical postures, all its body physiology functioned normally in contrast to the worms feeding at horizontal posture with less food consumption. Further, observations on the identical time taken for moulting & spinning in vertically postured indoor reared worms and the worms reared in their natural outdoor environment suggest that the delay in onset of moulting and spinning under indoor rearing was due to low food consumption at conventional horizontal posture.

In the present study, the cuticular domes detected on the head cuticle of the muga silkworm (*Antheraea assamensis)* are the dorsal campaniform sensilla reported in different locations of the body in various insects. It is well established that the external components of campaniform sensilla appear as round or oval domes on cuticle surface (McIver 1975, 1985). These sensillae are reported in joint regions of mouth parts, antennae, wings, legs, ovipositors etc (McIver, 1985). It is suggested that compression and stretching of the surroundings, result in cuticular stresses to which the dome shaped campaniform sensilla respond (Pearson, 1972; Spinola and Chapman, 1975; Zill and Moran, 1981). The cuticular stresses are reported to arise from external as well as internal forces which include frictional resistance between body parts, muscular contraction, loading, contact with external objects, gravity etc (Hustert et al, 1981; McIver, 1985).These stress forces are reported to concentrate in the cuticle of joint regions and that is why campaniform sensilla are found to be associated with articulations of mouth parts, legs, wings, etc. In this context, the distribution pattern of groups of dorsal campaniform sensilla throughout the head surface appears to be quite unusual. Since the dorsal campaniform sensilla in muga silkworm are located on the head surface and not in a joint region between head and other body parts, it is quite likely that cuticular stress associated with the sensilla does not arise from external forces since external movement of the head surface components is not possible. In that case, the cuticular stress resulting from compression and stretching of the surroundings due to some internal force or gravity seems to be the only plausible explanation.

The presence of dorsal campaniform sensilla throughout the head surface helps the silkworm to maintain their vertical posture as and when required. A correlation between the vertical posture and feeding preference supports this view. The interesting point to be noted here is that the silkworm e.g. *Philosamia ricini* and *Bombyx mori* which generally feed at their horizontal posture, did not exhibit dorsal campaniform sensilla on their head cuticle. The semi-domesticated Tasar silkworm (*Antheraea proylei*), which prefers vertical posture during feeding, on the other hand, exhibited well developed dorsal campaniform sensilla on the head surface.

(ii). UV sensitivity and Coupling Behaviour

Wienzek (1971) found UV- reflection in pierid butterflies, and in other species including the papilionid, *P.helenus*. However, the structural basis of the physically produced UV-reflections did not become clear until Ghiradella et al (1972) measured the dimensions of the series of micro ridges and the air spaces between them. It was suggested by Ghiradella et al (1972) that when the measured average thickness of the lamellae is about 550± 35Å, that of the air spaces between them is 826± 42 Å, then with known refractive index of air (1.0) and infra- red refractive index of lamellar cuticle (1.60), it follows that lamellae and the intervening lamellar air spaces have similar optical thickness. A common optical thickness of 858±51 Å for the two types of layers was then assumed, which meant that the micro ridge structures could function in the fashion of a quarter wavelength interference filter.

Meyer-Rochow and Eguchi (1983) studied the UV and other colour patterns in a wide variety of Lepidoptera and could confirm the observations of Ghiradella, et.al. (1972). However, Meyer-Rochow and Eguchi (1983) found that scales other than those described

by Ghiradella et al (1972) may also function as UV- reflectors. A prerequisite for structurally produced UV reflection seems to be the presence of micro ridges or a combination of fine holes and marginal micro ridges between the regularly spaced scale ribs

(Meyer-Rochow and Eguchi, 1983). The present study on Scanning electron microscopy of wing scales suggest that the UV- reflecting wing scale type described by Ghiradella et al (1972) and scales with a combination of fine holes and marginal micro ridges (Meyer-Rochow and Eguchi , 1983) were both present in the muga silk moth, *Antheraea assamensis.*

As far as the function of UV-reflecting wing scales in Lepidoptera is concerned, it is known that butterfly wing patterns in the visible and ultraviolet spectrum (Eguchi and Meyer-Rochow,1983) play a role in sexual selection (Bernard and Remington, 1991; Meyer-Rochow, 1991), crypsis (Brakefield et al ., 1990), in thermoregulation (Kingsolver and Watt, 1983; Kingsolver, 1995) and in aposematism and mimicry rings (Mallet et al . 1998; Joron and Mallet, 1998; Kapan, 2001). The wing scales that convey those patterns are intricately designed protein-chitin extensions of epidermal wing cells, arranged in rows that run along the anterior-posterior axis of the wing (Nardi and Magee-Adams, 1986).Butterfly wing scales are of two types: those which generate the UV- patterns physically, and those which absorb UV-radiation (Ghiradella, 1974, 1984; Meyer-Rochow and Eguchi, 1983).

The present study on ultra structure of wing scales through Scanning Electron Microscopy revealed that different types of scales in *Antheraea assamensis* have characteristic features of UV- reflecting scales as described by Ghiradella (1974) and Meyer-Rochow and Eguchi (1983). Ultraviolet photography also confirms the presence of UV-reflecting wing scales in the moth, *Antheraea assamensis.*

The observation that female constantly move its wings when 3-4 males approach it, suggests that some signals are being sent as a part of courtship behaviour, which is not being displayed properly under indoor conditions, with very little UV source. This is confirmed by the fact that under twilight outdoor conditions a single male approaches a female and gets coupled immediately.

The presence of UV-reflecting wing scales, difficulties in auto-coupling under indoor conditions, and proper auto-coupling under twilight outdoor conditions, points towards the possibility of UV as the major guiding factor in auto-coupling of *Antheraea assamensis.*

Approaching a female by a single male to get coupled immediately after exposure to artificial source of UV under indoor conditions, and failure in auto-coupling in moths with painted UV reflecting wing parts even after exposure to artificial UV source confirms that auto-coupling in *Antheraea assamensis* is under direct influence of UV-pattern display. The high percentage of auto-coupling (98%) under twilight conditions, irrespective of temperature and humidity suggests that auto-coupling under twilight conditions is not affected by temperature and humidity and it is directly controlled by utilization of UV- reflecting wing scales.

Relatively low percentage of auto-coupling (52%) under indoor conditions suggests that non-utilization of UV-reflecting wing scales by the moth due to less availability of ultra violet light inside the room adversely affects auto-coupling frequency.

It can be concluded that it is the UV-light which is the key factor for auto-coupling, since it results in very high percentage of auto-coupling irrespective of change in temperature and humidity.

Thus the present observation involving Scanning Electron Microscopy, UV-photography and behavioural studies confirms that certain wing parts of the moth *Antheraea assamensis* reflects UV, which influences the coupling behaviour of the moth.

The immediate impact of the present finding on the Muga silk industry is a simpler arrangement of mass coupling, a considerable increase in silk output and requirement of less man power. This is because of the fact that due to a very low auto-coupling frequency, mechanical coupling by hand is practiced in most of the muga silk farm, which involves extra man power. Further, mechanical coupling may sometimes cause injury to the moth resulting in less egg-laying.

The present findings may also have a far-reaching impact on the industry by providing a possible solution to the problem of uneven sex ratio of the moth. It was observed that during coupling under indoor conditions, 3-6 males approach a female which constantly vibrates its wings, obviously to send the UV signals, which cannot reach the male due to the scarcity of UV light inside the room. However, some of them succeed in coupling, probably by recognizing the signals by chance, through whatever little UV enters the room. The observed involvement of chance factor in auto- coupling inside the room explains the uneven sex ratio of 3 male: 1 female in the moth. Difficulties in utilization of UV-reflecting device, generation after generation might have compelled the moth to produce about 3 times more males as compared to females to increase the chance of coupling under indoor condition. This is supported by our observation that unlike the indoor environment with very little UV, the artificial or natural source of UV (twilight sky) could make a single male to approach a female to get coupled in a short time. This follows that reversing the sex ratio back to 1:1 may be possible by providing the moth its natural twilight condition for coupling and by maintaining the generation thus obtained.

It is however, certain that many generations have to be raised from moths auto-coupled under natural twilight conditions to reverse the sex ratio to 1:1.

REFERENCES

Alm, S.R. and Hall, F.R. (1986): Antennal sensory structures of *Conotrachelus nenuphar* (Coleoptera: curculionidae). *Ann. Entomol. Soc. Am*, 79,324 – 333.

Atwal, A.S and Bains, S.S. (1934): Applied Animal Ecology, Kalyani Publishers, New Delhi, pp25-30.

Bernard, G.D. and Remington, C.L. (1991): Color vision in Lycaena butterflies: special tuning of receptor essays in relation to behavioural ecology. *Proc. Natl.Acad. Sci.USA*, 88, 2783- -2787

Birch, L.C (1945): The influence of temperature, humidity and density on the oviposition of the small strain of Calendra oryzae and Rhizopertha dominica Fab. (Coleoptera).Aust.J.Exp.Biol.andmed.Sci, 23, 197-203

Bland, R.G. (1983): Sensilla on the antennae, mouthparts and body of the larva of the Alfalfa weevil, *Hypera postica* (Gyllenhal) (Coleoptera: Curculionidae). *Int. J. Insect Morph. Embryol.* 12(5/6); 261 – 272.

Brakefield, P.M., Gates, J., Keys, D., Kesbeke, F., Wijngaarden, P.J., Monteiro, A., French, V., Carroll, S.B. (1990): A decline of melanism in the peppered moth *Biston betularia* in The Netherlands *P roc. R. Soc. Lond.* B 267, 1953- 1957.

Calvert, W.H. and Hanson, F.E. (1983): The role of sensory structures and preoviposition behaviour in oviposition by the patch butterfly, *Chosyne lacinia. Entomologia Experimentalis et Applicata*, 33, 179 - 187.

Chakravorty, R and Das, B.C. (1981): Curr.Sci, 50,138, Cited in: Dey, S, Report on the occurrence of Coccinella septempunctata L on oak plants in Manipur, Geobios New Report, 4, 59-60, 1985

Crouau, Y., and Crouau Roy, B. (1991): Antennal Sensory organs of a Troglobitic coleptera, *Speonomus hydrophilus Jeannel* (Catopidae): An ultra structural study by chemical fixation and Cryo fixation. *Int. J. Insect Morphol. Embryol*; 20(3): 169 -184.

Dey, Sudip (1983): Food utilization by the last instar larvae of the silk moth, Antheraea proylei in indoor condition, *The Indian Zoologist*, 7,1&2, 85-87.

Dey, Sudip (1987): Ecophysiological studies on fecundity, oviposition behaviour and longevity of the silk moth, Antheraea proylei. *The Indian Zoologist,* 11, 1&2, 13-17

Dey, Sudip (1995): Possible ultrasonic receptor on the bat flies *Mystacinobia zealandica Current Science*, 68, 10,992 -995.

Dey, Sudip (1998): A scanning electron microscopic study on the body cuticle and associated sensilla of an ant (worker) *Tetraponera rufonigra. Proc. Nat. Acad.Sci.India,* 68(B), I, 37 – 44

Dey, Sudip (1999): Scanning Electron Microscopic detection of an unusual giant sensilla on the abdominal cuticle of the fifth instar larvae of the muga silk moth, *Antheraea assama, Micron,* 30/ 4, 339 – 348.

Dey, Sudip, Baul, T.S.B., Roy,B and Dey, D. (1989) : A new rapid method of air – drying for scanning electron microscopy using tetra - methylsilane. *Journal of Microscopy*, 156: 259- 261.

Dey, Sudip, Hooroo, R.N.K. and Wankhar, D. (1995): Scanning elelctron microscopic studies on the external morphology of sensilla on the legs of a butterfly *Graphium sarpedon* (Lepidoptera: Papilliouidae): *Micron* 26, 367-376.

Dey, Sudip, and Biswas N., (1996): Scanning electron microscopic detection of a specialized cuticular structure on the abdominal surface of the maggot of *Blepheria Zebina* (walk) (Diptera: Tachinidae) parasitizing the larva of the muga silkworm, *Antheraea assamensis. Curr. Sci*, 70, 347 -348

Dey, Sudip, Hooroo R.N.K., Biswas N., (1997): Dendritic organization of prominent sensory structures on the body of the first instar larva of the muga silk moth *Antheraea assamensis*: *Cytobios*, 92, 99-110.

Dey, Sudip, Hooroo, R.N.K., and Bhattarcharjee, C.R.(1998): Electron Microscopy and Spectroscopical studies on a coloured patches on wings of a butterfly, *Graphium sarpedon. Pigment Cell Research*, 11, 1-11.

Dey, Sudip, Hooroo, R.N.K., Biswas N., Raghuvarman, A. and Choudhury S. (1999): cellular feature of muga silkworm (*Antheraea assama* Ww; Lepidoptera: Saturniidae) midgut and its possible impact on indoor rearing. *Sericologia* 39 (4); 529 – 535.

Dey, Sudip., Singh, Surendra., Dhar, N.J., Biswas, N., Chakravorty, Rahul ., Das, P.K., Sharma, D.K. and Raghuvarman, A., (2002): Possible relationship between Dorsal Campaniform sensilla in head surface and preferred feeding posture of muga silkworm *Antheraea asasma* during indoor rearing. *Sericologia*, 42(2) 171- 180.

Dey, Sudip, Singh, S, Dey, S, Choudhury, S, Chakraborty, R, Hooroo, R.N.K and Sharma, D.K, (2011): UV reflecting wing scales in the silk moth Antheraea assamensis: Its Biophysical implications, *Microscopy research and Technique*, 71,663-668

Eguchi, E and Meyer – Rochow, V.B. (1983). Ultraviolet photography of forty-three species of Lepidoptera representing ten families. *Annot Zool Japon,* 56, 10 -18.

George, J.A. ., and Naggy, B.A.L (1984):.Morphology, distribution and ultrastructural differences of sensilla trichoidea and basiconica on the antennae of the oriental fruit moth, *Grapholitha molesta* (Busek) (Lepidoptera: Tortricidae).*Int.J.Insect Morphol Embryol,* 13,157 -170.

Farrel, J.A.K (1973): Effect of initial body weight and feeding on the fecundity and longevity of adult Costelytra zealandica (Coleoptera: Scarabaeidae), N.Z.Agr. Res, 18, 53-57

Farrel, J.A.K (1975): Effect of food plant species on the weight loss, survival and fecundity of adult Costelytra zealandica (Coleoptera: Scarabaeidae), N.Z.Agr. Res, 18, 303-305

Ghiradella, H., Aneshansley, D., Eisner, T., Silbergleid, R. E. and Hinton, H.E. (1972): Ultraviolet reflection of a male butterfly: Interference colour caused by thin layer elaboration of wing scales. *Science*, 178, 1214- 1217.

Ghiradella, H. (1974): Development of ultraviolet reflecting butterfly scales: how to make an interference filter. *J. Morphol*, 142, 395 – 409.

Ghiradella, H (1984): Structure of iridescent lepidopteran scales: Variation on several themes *Ann. Entomol. Soc. Am*., 77, 637- 645.

Gibbs, G.W. (1980): New Zealand butterflies ------- *Identification and natural history. Collins, Auckland.*

Hamilton, A.G. (1950): Further studies on relation humidity and temperature to the development of two species of African locusts, Locusta migratoria migratorioides R and F and Schistocerca gregaria Forsk.Trans. R.ent. Soc. Lond., 101, 2-56.

Hazarika, L.Kand Bordoloi, S (1998): Antennal and mouthpart sensilla of the muga silkworm, *Antheraea assama* (Lepidoptera: Saturniidae), *Sericologia, 38*(1), 55 - 63.

Hiratsuka, E. (1920): Bull.Ser.Exp.Sta, Japan, I, 257-315

Hustert R., Pfluger H.J and Brauning P. (1981): Distribution and specific Central projections of mechano – receptors: the campaniform sensilla. *Cell Tissue Res*. 216, 97 – 111.

Joron, M., and Mallet, J.B.L (1998): Diversity in mimicry: Paradox or Paradigm? *Trends Ecol. Evol.* 13, 461 – 466.

Kapan, D.D. (2001): Three- butterfly system provides a field test of Mullerian mimicry, *Nature*, 409, 338 – 340.

Kingsolver, J.G. (1995): .Viability selection on seasonally polyphenic traits: wing melanin patterns in western white butterflies. *Evolution,* 49, 932 – 941.

Kingsolver, J.G. and Watt, W.B. (1983): Thermo-regulatory strategies in Colias butterflies: thermal stress and the limits to adaptation in temporally varying environments. *AM. Nat.* 121, 32-55.

Lal, O.P. (1973): *Z. Angew.Entomol,* 73, 260-262

Lawton, J.H. (1971): Ecological energetic studies on larvae of the damselfly, Pyrrhosoma nymphula (Sulzer) (Odonata: Zygoptera) J.Anim.Ecol.40, 385-423

Mallet, J., McMillan, W.O., Jiggins, C.D. (1998): Mimicry and warning color at the boundary between microevolution and macroevolution. *In*: Endless forms: Species and Speciation. (D.H.A.S. Berlocher, ed) pp. 390 – 402, Oxford University Press.

Mclver, S.B., (1975): Structure of cuticular mechano-receptors of arthropods, *Ann. Rev. Ent*; 20. pp 381 – 397.

Mclver S.B.,(1985): Mechano-reception; *In:* Comprehensive Insect Physiology Biochemistry and Pharmacology vol 6 (Edited by G.A. Kerkut and L.J., Gilbert) pp71 – 132, Pergamon Press, Oxford.

Meyer – Rochow, V.B. (1991): Differences in ultraviolet wing patterns in the New Zealand lycaenid butterflies *Lycaena salustius, L. rauparaha* and *L.feredayi* as a likely isolating mechanism. *J. Soc. N.Z*, 21, 169- 177.

Meyer – Rochow, V.B. and Eguchi, E. (1983): " Flugle farben, wei sie die Falter sehen"- a study of UV and other colour patterns in Lepidoptera.*Annot. Zool. Japon*, 56, 85- 99.

Mood, A.M. Graybill, F.A. & Boes, D.C. (1974): Introduction to the theory of statistics, *M.C. Graw – Hill, International StudentThird Edn.*

Nardi, J.B. and Magee – Adams, S.M. (1986): Formation of scale spacing patterns in a moth wing: 1. Epithelial feet may mediate cell rearrangement. *Dev. Biol.* 116, 278 – 290

Neville, A.C. (1975) *Biology of the Arthropod cuticle.* Springer Verlag, Berlin. New York.

Pandian, T.J. and Mathavan, S. (1974): Patterns of energy utilization in the tropical dragonfly Diplocodes triviales (Rambur) and some other aquatic insects (Anisoptera: Libellulidae), Odonatologica, 4,241-248.

Pearson, K.G.(1972): Central programming and reflex control of walking in cockroach. *J. Exp. Bio.*56,173 - 193.

Petrusewicz, K. and Macfadyen, A. (1970): *Productivity of terrestrial animal.* IBP Hand Book, No 13, Blackwell Sci; Publs., Oxford, 190.

Reddi, M.V. and Alfred, J.R.B. (1979): Utilization of Castor (Ricinus communis) leaves by the last instar larvae of the silk moth, *Phylosamia ricini* HUTT. (Lepidoptera: Saturniidae). *Indian Biologist* XI, 1-2.

Roessingh, P. , Stadler,E. , Schoni,R. , Feeny,P (1991): Swallow tail butterfly *Papilio potyxenes* ;responses to phytochemicals from host and non host plants. *Physiological Entomology*, 16,485 - 495

Schmidt J.M. & Smith J.J.B. (1987): The external morphology of the legs and hair plate system of female *Trichogramma minutum Riley* (Hymenoptera: Trichogrammatidae) *Proc. R. Soc.Lond.* B, 232, 333- 366.

Schneider, D. (1964): Insect antennae. *Annual Review of Entomology.* 9, 103 –122

Seitz, A. (1933): *The Macro Lepidoptera of the world,* Alfred Keren Publishers, Stuttgart, vol.10, p. 909.

Solomon, J.D. and Neel, W.W (1974): Fecundity and oviposition behaviour in the Carpenter worm, Prionoxystus robiniae. *Ann.Entomol.Soc.Am,* 67,238-240

Smith, R.L., (Ed). (1984): Sperm Competition and the Evolution of Animal Mating Systems. Academic Press, New York.

Spinola, S.M., & Chapman, K.M., (1975): Proprioceptive indentation of campaniform sensilla of cockroach legs. *J comp. Physiol.* 95, 257 – 272.

Stockner, J.H. (1971): J. Fish.Res.Bd. Can, 28, 73-94

Thangavelu K. and Sahu A.K. (1986): Further studies on indoor rearing of muga silkworm (*Antheraea assama* Ww.) (Saturniidae: Lepidoptera). *Sericologia*, 26(2): 215 – 224.

Thornhill, R. & Alcock, J. (1983): *The Evolution of Insect Mating* Systems. Harvard University Press, Cambridge, Massachusetts.

Waldbauer, G.P. (1968): The consumption and utilization of food by insects: *Advances in Insect Physiology;* 5: 229 -288

Wienzek, M. (1971): Spektrale Reflextions – messungen an Insekenintegumenten. *Forma et Functio*, 4,340 – 381.

Yadav G.S. and Goswami B.C. (1999): Correlation and regression studies between different quantitative characters of Muga silkworm *Antheraea assamensis* Helfer during two commercial crops. *Bull. Ind. Acad. Seri.* 3(2), 12 – 20

Zill, S.N. and Moran. D.T. (1981a): The exoskeleton and insect proprioception. I Responses of tibial campaniform sensilla to external and muscle – generated force in the American cockroach, *Periplaneta americana. J. exp. Biol*, 91, 1- 24.

In: Moths: Types, Ecological Significance and Control... ISBN: 978-1-61470-626-7
Editor: Luis Cauterruccio, pp. 243-272

Chapter 9

ATTRACT-AND-KILL STRATEGIES FOR CONTROL OF LEPIDOPTERAN PESTS IN GRAPES AND HOPS IN WASHINGTON STATE, U.S.A.

Holly Ferguson, Sally O'Neal and Douglas Walsh*
Washington State University.

ABSTRACT

Traditional control methods for lepidopteran pests in grapes and hops often involve foliar treatments with broad-spectrum organophosphates or pyrethroids. However, as many of these high-risk chemicals face increased regulation, growers have shifted toward using softer insecticide programs including reduced risk insecticides and biopesticides, which are typically more effective against early instar larvae than on later instar larvae. Attract-and-kill technology was explored as a reduced-risk alternative control strategy against lepidopteran pests of grapes and hops. A feeding attractant (1:1 acetic acid:3-methyl-1-butanol (AAMB)) derived from fermented molasses attracts a large number of noctuid species, including some that are pests of grapes and hops. Attract-and-kill experiments using bait stations with AAMB and a pyrethroid insecticide mixed with PTFE grease were conducted in Washington State vineyards during the summers of 2003 to 2005. Bait stations were hung in wine and juice grape vineyards in replicated 2-ha plots at three treatment levels, 0, 25, and 125/ha. Moths were monitored with feeding attractant, pheromone (spotted cutworm only), and light traps throughout the season. Post-deployment, climbing cutworm numbers in feeding attractant traps in bait station-treated plots were significantly reduced in wine grapes in 2003 and in juice grapes in 2005, while numbers of cutworms usually increased in pheromone and light traps. In 2009 and 2010, targeting the hop looper (*Hypena humuli* (Harris), bait stations with feeding attractant and pyrethroid insecticide were deployed in three Washington hop yards in replicated 2-ha plots at three treatment levels, 0, 25, and 150/ha. Moth populations were monitored with AAMB-baited traps throughout the season. Larvae on hop plants were sampled biweekly during the season. In two of the hop yards during July 2010, trap captures of hop looper moths were significantly reduced in bait station-treated plots compared with control plots in July, approximately one month after deployment of

* E-mail address: hferguson@wsu.edu; soneal@wsu.edu; dwalsh@wsu.edu

bait stations. For hop looper and other lepidopteran larvae, numbers collected were variable, and no effects due to bait stations were detected. Bioassays involving field-collected noctuid moths exposed to fresh and field-aged insecticide-treated bait stations showed that field residues of bifenthrin, fenpropathrin, and β-cyfluthrin maintained toxicity against noctuids as far out as 9 weeks post-deployment. Based on these data, the use of AAMB-based bait stations has potential as a reduced-risk method to control pest cutworms in grapes and hop looper in hops.

INTRODUCTION

Lepidopteran larvae can cause significant economic damage in Washington State vineyards and hop yards. Traditional control methods for lepidopteran pests often involve foliar treatments with broad-spectrum insecticides, such as organophosphates and carbamates. However, as the use of insecticides containing these high-risk active ingredients become increasingly regulated, growers have shifted toward using softer insecticide programs that include reduced risk or biopesticide products, which are typically more effective against early instar larvae than on larger, late instar larvae that cause the greatest economic damage. Control of large larvae often requires applications of broad-spectrum insecticides including pyrethroids. While pyrethroid insecticides will provide rapid and inexpensive control of outbreaks of pest lepidopterans, they thwart good intentions to engage in integrated pest management, which may be defined as "a sustainable approach to managing pests by combining biological, physical, and chemical tools in a way that minimizes economic, health, and environmental risks" [1]. When applied as a broadcast, foliar treatment, pyrethroids can destroy beneficial arthropod populations and will often contribute to outbreaks of spider mites and other secondary pests. Attract-and-kill technology (reviewed in [2]) was explored as a reduced-risk alternative management strategy against lepidopteran pests of grapes and hops.

Generally, attract-and-kill insect control systems consist of two basic components, a toxicant, usually a pyrethroid insecticide, and an insect attractant, usually a sex pheromone. The first lepidopteran pest for which an attract-and-kill strategy was implemented was the pink bollworm (*Pectinophora gossypiella* (Saunders) in the early 1980s [3, 4]. Since then, sex pheromone-based attract-and-kill systems have been developed for a number of lepidopteran pests in a diversity of cropping systems, including codling moth, *Cydia pomonella* (L.) [5-7], Oriental fruit moth, *Grapholita molesta* (Busck.) [8], light brown apple moth, *Epiphyas postvittana* (Walker) [9], and the grape root borer, *Vitacea polistiformis* Harris [10]. All of these systems incorporate an attractant to target and trap male moths only, which limits their utility in pest management. Removing the egg-laying females would more directly reduce population growth [11]. Thus, more recently, attractants incorporating floral odorants [12, 13] and feeding attractants [14] have been developed that target both sexes. Floral and feeding attractants take advantage of typical adult noctuid habits of foraging and feeding at sugar sources such as floral and extrafloral nectaries [15]. Camelo et al. [11] developed a floral lure + permethrin bait station system for alfalfa looper control in alfalfa; female moths were reduced 80 to 93% in monitoring traps in plots with 125 bait stations/ha. Presumably the reduction of moths in traps is brought about by moths being killed through contact with the toxicant in the bait station, thereby reducing the numbers of local moths that can respond to trap lures [11]. Feeding attractants derived from the odors of fermented molasses baits are known to attract both sexes of several pest noctuid species [14, 16, 17]. The combination of 3-methyl-1-butanol with acetic acid (AAMB), both isolated and identified from fermented molasses solutions [18], proved to be strongly attractive to certain climbing cutworms, such as the spotted cutworm, *Xestia c-nigrum* (L.) [16], a species known to damage grapevines [19, 20]. Additionally, the hop looper, *Hypena humuli* Harris, is an emerging pest of hops [21], and while this species has not been documented to be attracted to this odor combination, other hypenines have been collected in traps baited with AAMB [22]. This chapter documents the initial development of attract-and-kill management systems based on this feeding attractant for climbing cutworms in grapes and the hop looper in hops. We present evidence of successful reduction of moth numbers in grapes and hops as well as data from experiments in which reduction of moths did not occur. We provide insights into the factors that may contribute to the success or failure of the AAMB-based attract-and-kill system. In addition, we present data from traps baited with this feeding attractant on a nontarget noctuid in grapes, i.e., the true armyworm, *Pseudaletia unipuncta* Haworth, which is a key pest in other crops, and two other potential pest noctuids in hops, i.e., spotted cutworm and Bertha armyworm, *Mamestra configurata* Walker, and one potential pest geometrid, the common gray moth, *Anavitrinella pampinaria* (Guenée) [23].

Bait Station Design

The architecture of the bait station should be dictated initially by its intended use; subsequent optimization would focus on better durability and lower cost. For the initial 2003 experiment in wine grapes, we chose to use the bait station design by E. R. Mitchell (USDA-ARS, Gainesville, FL; deceased) and Landolt [24], which mimicked a flower in appearance, presumably to attract foraging Lepidoptera. It consisted of a badminton shuttlecock base with a centrally-placed polypropylene 15-ml vial containing several cotton balls (Fig. 1a). Ten ml of feeding attractant, comprised of glacial acetic acid and 3-methyl-1-butanol (AAMB), 50:50 ratio by weight [22], was injected into the vial and absorbed into the cotton. A 3.0 mm hole was drilled into the vial's cap, allowing for a constant and controlled release of odorants [16]. The toxicant used in 2003 was 5% technical grade ζ-cypermethrin (donated by FMC Corporation) mixed with polytetrafluoroethylene (PTFE)-based hitch ball lube (Reese Towpower™, TriMas Corporation); approximately 5 g was smeared onto the outside of the shuttlecock. For the 2004 and 2005 grape experiments, we replaced the shuttlecock with a less expensive, more durable compact disc (CD) (Fig. 1b). The center hole of the CD was widened to allow the insertion of the vial containing AAMB. About 7 g of the pyrethroid/grease mixture was applied evenly to the surface of the CD. A field experiment conducted in wine grapes in 2004 revealed no difference in trap catches between plots with CD bait stations versus plots with badminton shuttlecock bait stations. By switching to the CD and a less expensive polyethylene 15 ml vial, the total cost of the bait station was reduced 34%, from \$1.17 to \$0.77 apiece. In order to reduce cost and hazards associated with the pyrethroid ζ-cypermethrin, a different toxicant at a lower rate was used in grapes in 2004 and 2005. We changed the toxicant to 2.5% fenpropathrin (as formulated in Danitol® 2.4 EC (donation from Valent U.S.A. Corporation)) mixed in PTFE-based lube.

Figure 1. Bait station designs. A, badminton shuttlecock used in grapes in 2003 and 2004; B, compact disc (CD) used in grapes in 2004 and 2005; C, PVC slip tee fitting used in hops in 2009 and 2010.

The experiments in grapes proved to be labor-intensive, in particular, the time and labor necessary for the construction of bait stations, for their field deployment, and for any later replenishment of toxicant and/or attractant. In addition, both the shuttlecock device and the CD bait station were considered disposable, that is, durable enough to last one season only in the field. Thus, when the hop experiments were initiated, we sought a bait station design that was permanent, a device that could be attached to a trellis pole or wire and last through several field seasons. By coincidence, John Brown (Washington State University, Department of Entomology) was using a bait station to control carpenterworm moths in hybrid poplars that consisted of a short length of 3.8 cm ID polyvinylchloride (PVC) pipe with a scintillation vial containing feeding attractant inserted into the PVC pipe. Because Brown's bait station was based on PVC, a material generally known to withstand cold temperatures and UV degradation for at least a couple of years, this design appeared to offer the durability we were seeking. We chose a PVC slip tee fitting (3.8 cm X 3.8 cm X 1.27 cm) (Fig. 1c) with inside threading that allowed us to screw the vial containing feeding attractant into the bottom of the tee. A larger 30 ml vial was used that could potentially hold enough feeding attractant to last all season. When one considers that the PVC bait station could last a few years in the field, the higher initial cost of the PVC bait station ($1.96 in 2010) is justifiable. Construction costs for the bait station used in hops were somewhat mitigated by using an aluminum complex USDA H1 food-grade grease (Sta-Lube®, CRC Industries) instead of the more expensive PTFE lube used previously. Finally, a different pyrethroid, β-cyfluthrin, formulated in Baythroid® XL (in-kind donation by Bayer CropScience), was used at a rate of 2.5% active ingredient in food-grade grease. Approximately 5 g of the mixture was daubed onto the inside and outside of the slip tee fitting. This pyrethroid was chosen because in previous bioassay studies conducted by our group, β-cyfluthrin residue on high-density polyethylene netting hung in sunlight and shade conditions remained highly toxic against insects for at least 14 weeks (unpublished data).

Targeting Cutworms in Washington Wine and Juice Grapes

At the time of our grape attract-and-kill experiments, climbing cutworms (Lepidoptera: Noctuidae) were a major economic pest for wine grape growers in Washington State; over half of the 28,000 wine grape acres [25] was being treated for cutworms [26]. Direct damage results when cutworms climb the vine in the springtime to feed on the opening buds; as low as 1 to 5% injured buds is enough to significantly reduce yields and production value [27]. The predominant control program in the early 2000s consisted of broadcast applications of environmentally disruptive insecticides, primarily the organophosphate chlorpyrifos (Lorsban™ 4E, Dow Agrosciences, Indianapolis, IN). We sought to develop a reduced-risk alternative management strategy for cutworms in grapes.

Experiments were conducted in commercial vineyards during the summer and fall of 2003, 2004, and 2005 to test the effectiveness of attract-and-kill bait stations against pest cutworms. In 2003, a randomized complete block design of 2-ha plots with three treatments--0, 25, and 125 bait stations/ha—and four replicates was set up in one wine grape block (*Vitis vinifera* L., cv. 'Semillon') near Paterson, WA. On 25 August 2003, three types of monitoring traps were placed in each plot as follows: two universal bucket traps with AAMB lure, diagonally across from each other; two universal bucket traps with spotted cutworm sex pheromone lure (generous gift from Peter Landolt, USDA-ARS, Wapato, WA), also diagonally across from each other; and one battery-operated light trap located in the center of the plot (basic design in Fig. 2). Shuttlecock bait stations were deployed in the treatment plots on 4 September 2003. Bait stations were hung with cable ties on the trellis wire within the canopy, uniformly distributed within the plot. In 2004, three experiments were carried out simultaneously; all grape blocks were located near Prosser, WA. The first experiment in a wine grape block (*V. vinifera*, cv. 'Merlot') tested the effectiveness of CD bait stations against cutworm moths, using a randomized complete block design with two treatments, 0 and 125 bait stations/ha, and four replicates. A second experiment executed in another wine grape block (*V. vinifera*, cv. 'Chardonnay') tested for a difference in moth trap capture with shuttlecock bait stations versus CD bait stations, using the higher rate of 125 bait stations/ha and three replicates of each treatment. The third 2004 experiment was conducted in juice grapes (*V. labrusca* L., cv. 'Concord') with a design similar to the Merlot experiment. Monitoring traps were placed in the field on 29 July 2004 in the Concord block, on 30 July 2004 in the Chardonnay block, and on 2 August 2004 in the Merlot block. There were two bucket traps baited with AAMB in each plot (Fig. 2) and an additional pheromone and light trap in each of the control plots (in the northwest quadrant and in the center, respectively). Bait stations were hung on trellis wire with cable ties in treated plots on 10 August 2004 in the Concord block, on 11 August 2004 in the Chardonnay block, and on 12 August 2004 in the Merlot block, uniformly distributed within a plot. Feeding attractant was replenished in bait stations and monitoring traps in-field on 31 August 2004, which was three weeks after initial deployment. In 2005, experiments with designs similar to the 2004 Merlot experiment were executed in two wine grape blocks (*V. vinifera*, cvs. 'Semillon' and 'Chardonnay') and the same Concord grape block used in 2004, to test the effectiveness of the higher rate of CD bait stations against cutworm moths. The Semillon block was the same block used in 2003; the Chardonnay block was located near Paterson, WA, approximately 8 km north of the Semillon block. Monitoring traps as described for the 2004 experiments were placed in the field on 2 August 2005 in the Semillon and Chardonnay blocks and on 3 August 2005 in the Concord block. CD bait stations were hung in treatment plots in a uniform distribution on 30 August 2005 in the Semillon block, on 31 August 2005 in the Chardonnay block, and on 7 September 2005 in the Concord block. In all three years, monitoring traps were removed prior to harvest and bait stations were removed post-harvest during the wintertime.

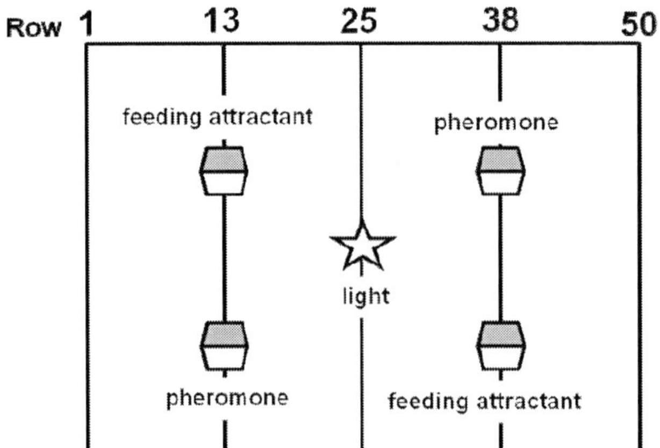

Figure 2. Location of monitoring traps (feeding attractant, pheromone, and black light) in typical 2-ha plot in grapes, 2003.

In 2003 and 2004, trap samples were collected every two to three days. In 2005, trap samples were collected on a weekly basis. Trap samples were frozen, then sorted and counted according to lepidopteran species, when the species could be properly identified. Trap data from similar-length pre-deployment and post-deployment periods were grouped and subjected to ANOVA followed by means separation using Tukey's correction, α=0.05, using SAS [28].

Species Composition of Cutworms in Grapes in 2003

In 2003, there were eight frequently (≥10 specimens) caught noctuid species identified and counted in the AAMB trap captures (total n=3,034). These were: *Euxoa* spp., principally *septentrionalis* (Walker) and *atomaris* (Smith) (n=1,696); the spotted cutworm (n=989); the Western yellow-striped armyworm, *Spodoptera praefica* (Grote) (n=159); the dingy cutworm, *Feltia jaculifera* (Guenée) (n=93); *Agrotis vetusta* (Walker) (n=49); the variegated cutworm, *Peridroma saucia* (Hübner) (n=28); and the black cutworm, *A. ipsilon* (Hufnagel) (n=11). Species composition in light trap captures (total n=2,535) was slightly different; spotted cutworm was the most abundant (n=1,448), followed by *Euxoa* spp. (n=842), dingy cutworm (n=85), western yellow-striped armyworm (n=78), *A. vetusta* (n=50), and variegated cutworm (n=32).

At the time of the study, the cutworm species that were being blamed for bud injury in grapes were the spotted cutworm and the redbacked cutworm, *E. ochrogaster* (Guenée) [29]. Approximately 33 and 57% of the AAMB and light trap captures, respectively, was comprised of spotted cutworm, *X. c-nigrum*, known to feed on grapevines [20]. We did not find redbacked cutworm in any of our 2003 trap captures, but the congeneric species *septentrionalis* and *atomaris* made up over half of the AAMB trap capture and 33% of the light trap capture. Because the literature predating 2003 indicated that these species were known to be primarily ground vegetation feeders, we attributed their abundance in the samples to their strong attraction to AAMB [30] and did not consider these species to be a direct threat to the grapevines.

Bait Station Effects on Cutworms in Grapes

Only the data of species known to feed on grapevines in the springtime were included in the analysis of bait station effects. In 2003, the springtime climber group included the spotted cutworm, *A. vetusta* [20], the variegated cutworm [31], the dingy cutworm (considered a *Feltia* species complex [32]) [33], and the black cutworm [31]. Dingy cutworm has been found on a long list of host plants, including many from the family Asteraceae [32] but not including grape; however, more recently, one specimen was found on a grapevine [33], hence its inclusion in the springtime climber group. Application of shuttlecock-based AAMB bait stations at the high rate of 125/ha significantly reduced AAMB trap captures for this group of cutworms by 36% compared with control plot captures in the post-deployment period (Fig. 3, $P<0.05$). Trap captures were reduced by 22% in the 25/ha plots relative to controls post-deployment, but the difference was not significant. Also, while AAMB captures of springtime climbers were significantly reduced in all plots post-deployment compared with pre-deployment captures (Fig. 3, $P < 0.05$), trap captures in treated plots were reduced by 71 to 85% compared with only 49% reduction in control plots. These results coupled with the declining trend in light trap captures post-deployment probably indicated that by the first part of September in 2003, the springtime climber cutworm moths, comprised primarily of spotted cutworms, were past the maximum amplitude of their summertime flight period. No significant differences were found for the spotted cutworm pheromone trap captures among treatments within either pre- or post-deployment periods or within treatment between pre- and post-deployment periods (Fig. 3).

Our single-vineyard results for climbing cutworms in wine grapes in 2003 are comparable to those obtained by Camelo et al. [11] who conducted similar bait station experiments using a floral odorant combined with permethrin to attract and kill alfalfa looper in commercial alfalfa fields and by Landolt [24] who used AAMB+permethrin bait stations to control Lacanobia fruitworm, *Lacanobia subjuncta* (Grote and Robinson), in apple orchards. Camelo et al. [11] reported 80-93% reduction of female moth captures in floral-lure-baited traps for treated plots from pre-deployment to post-deployment. In addition, no effect due to bait station treatment on pheromone trap captures was observed for alfalfa looper in that study. Perhaps, as was suggested by this group, male moths travel greater distances in their search for mates and hence, recruitment to pheromone traps comes from a larger area than the 2-ha plot. In contrast, search for food sources and hence attraction to feeding or floral lure more likely involves short-range dispersal, well within the confines of the 2-ha plots.

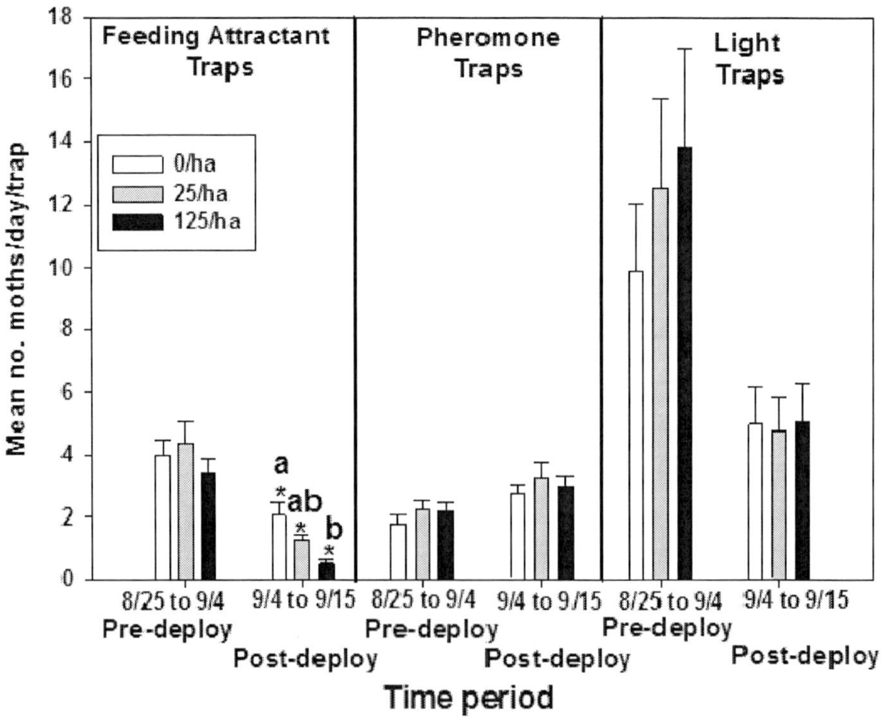

Figure 3. Effect of feeding attractant bait stations on climbing cutworm moth trap captures in a Semillon wine grape block, Paterson, WA, 2003. Asterisks denote significant difference from pre- to post-deployment ($P < 0.05$). Means grouped within time period and block with the same letter above them are not significantly different ($P > 0.05$).

We were encouraged by our 2003 results, which indicated that our AAMB-based attract-and-kill system had great potential for managing grapevine bud-destroying cutworms through reduction of the moth population that gives rise to overwintering larvae that cause damage the following spring. Thus, in 2004, the study was expanded to include two wine grape blocks (but not the 2003 Semillon block) and one Concord grape block, as described above. Improvements to the system included using less expensive materials for the bait station (e.g., using a CD instead of a shuttlecock) and recharging the feeding attractant vials in the field three weeks after initial deployment. In addition, deployment of bait stations was accomplished earlier in August to coincide with the beginning of the second flight period of spotted cutworm. To improve longevity of bait station effectiveness, feeding attractant was replenished in the bait stations and the monitoring traps three weeks after deployment.

Species Composition of Cutworms in Grapes in 2004

Overall moth captures were much lower in the 2004 vineyards compared with the previous year's captures in the Semillon vineyard. In addition, noctuid species composition was notably different in the 2004 sites. Fourteen noctuid species frequently (≥10 specimens) caught in the AAMB and light traps were identified and counted. Species included all of those found in 2003 plus the following: Bertha armyworm, *M. configurata*; clover cutworm, *Discestra trifolii* Hufnagel; true armyworm, *P. unipuncta*; forage looper, *Caenurgina erechtea* (Cramer); *L. subjuncta*; and *Dargida procinctus* (Grote). (In 2003, *P. unipuncta* and *D. procinctus* were noted as incidentals in the pheromone and AAMB trap samples, respectively.) About 66% of the total catch (n=3,070) was comprised of forage looper (n=770), clover cutworm (n=661), and *Euxoa* spp. (principally *septentrionalis* and *atomaris*, n=584); none of these are considered grapevine pests. Only about 13% (n=388) of the moths caught in the 2004 AAMB traps belonged to the five species included in the springtime climber group. Of those, only 147 spotted cutworms were found over the entire season in the feeding attractant traps. In contrast, spotted cutworm was third in abundance (433/3,771) in the 2004 light traps, following *Euxoa* spp. (n=1,482) and clover cutworm (n=802).

Bait Station Effects on Cutworms in Grapes in 2004

Due to the relatively low numbers of springtime climber moths caught in the AAMB traps, no significant effects due to bait stations were found in any of the vineyard blocks (Fig. 4). Trap counts were quite variable; however, there appeared to be a trend toward fewer springtime climber moths post-deployment in the Concord and Merlot blocks. The pheromone trap catches in the control plots showed a significant increase in spotted cutworm male numbers two weeks after deployment of bait stations, particularly in the Chardonnay block where trap catches significantly increased immediately after deployment (Fig. 5). In the control plot light traps, springtime climbing cutworms showed a significant increase in numbers post-deployment in the Concord grape block but not in the wine grape blocks (Fig. 6). The pheromone and light trap data seemed to indicate that our timing of bait station deployment was on target. Springtime climber counts were low pre-deployment in all traps, but remained low in the AAMB traps post-deployment while the local springtime climber moth numbers were increasing as indicated by the pheromone and light trap data. This suggests that the bait stations in the treated plots may have been effective in reducing cutworm populations somewhat but not enough to show a significant difference in AAMB trap captures between control and treated plots or between pre-deployment and post-deployment periods within treatment. For the control plots in the post-deployment period, it is not clear why AAMB trap counts were considerably lower than pheromone and light trap counts, but we suspect that recruitment for light and pheromone traps extends outside of the boundaries of the 2-ha control plot while moths that were caught in AAMB traps came from within the plot. These data might also be viewed as evidence that the feeding attractant odors somehow affected the moths' response to pheromone and/or light. A recent study showed a differential response of spotted cutworm to pheromone and blacklight traps in close proximity depending on the time of year [34], but there have been no similar studies on differential responses to feeding attractant traps when pheromone and/or light traps are close by.

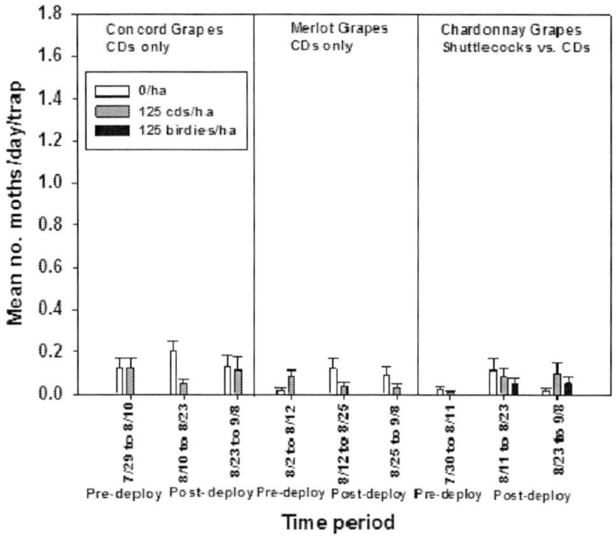

Figure 4. Effect of feeding attractant bait stations on climbing cutworm moth trap captures in Concord juice grape, Merlot wine grape, and Chardonnay wine grape blocks, Prosser, WA, 2004. No significant differences due to bait station effects were found (P > 0.05).

SPECIES COMPOSITION OF CUTWORMS IN GRAPES IN 2005

In 2005, bait station field deployment studies were undertaken in three vineyards: the same Semillon block used in 2003, the same Concord grape block used in 2004, and a Chardonnay wine grape block not previously used for the present study. Even though the 2005 trapping period was shorter by about two weeks compared with 2004 (13 weeks vs. 15 weeks), total moths captured and identified in AAMB traps nearly tripled from 2004 (2005 total n=8,087). All but two of the cutworm species frequently recorded in the 2004 trap captures (black cutworm and Lacanobia fruitworm) were frequently (>10 specimens) caught in the 2005 AAMB samples. True armyworm (n=2,790) was the most abundant cutworm in AAMB samples, followed by forage looper (n=1,678) and *Euxoa* spp. (n=1,668); these species, not considered grapevine pests, comprised 76% of the total AAMB capture over the season. A total of 310 spotted cutworm moths were recorded in the 2005 AAMB trap samples, with more than half of these caught in the Semillon vineyard. Surprisingly, control plot light and pheromone trap captures were greatly reduced in 2005, which made it difficult to assess the background seasonality of the local cutworm populations. Total cutworm moth capture over the whole season in light traps was 788, which included only 20 spotted cutworm moths, while total spotted cutworm males caught in pheromone traps was 658, compared with 1,239 in 2004.

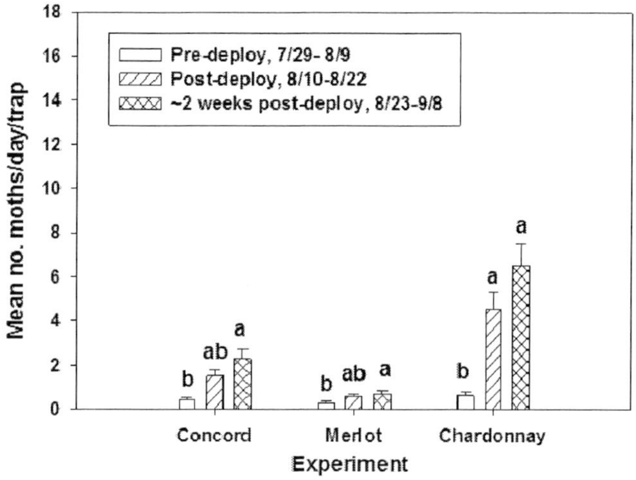

Figure 5. Effect of time period on captures of spotted cutworm male moths in control plot pheromone traps for Concord juice grape, Merlot wine grape, and Chardonnay wine grape blocks, Prosser, WA, 2004. Means grouped within block with the same letter above them are not significantly different ($P > 0.05$).

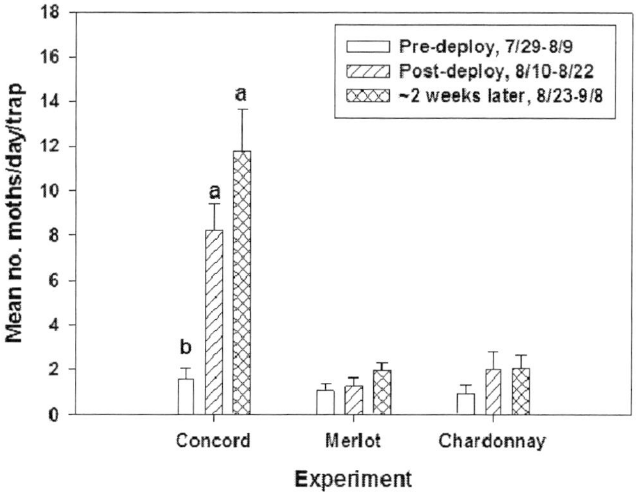

Figure 6. Effect of time period on captures of climbing cutworm moths in control plot light traps for Concord juice grape, Merlot wine grape, and Chardonnay wine grape blocks, Prosser, WA, 2004. Means grouped within block with the same letter above them are not significantly different ($P > 0.05$).

BAIT STATION EFFECTS ON CUTWORMS IN GRAPES IN 2005

Data from springtime climber cutworms were analyzed for bait station effects as done in previous years, separately for each vineyard. For the Concord grape block, AAMB moth trap captures of springtime climbers were significantly reduced in the treated plots post-deployment compared with control plot trap counts (Fig. 7, $P < 0.05$). Though numbers of springtime climber cutworms were lower in the treated plots post-deployment in the wine grape blocks, these differences were not significant. Numbers of spotted cutworm males in control plot pheromone traps were observed to increase post-deployment in both wine grape blocks but not in the Concord block; however, this difference was significant only for the Semillon block (Fig. 8, $P <0.05$). Springtime climber species were observed to significantly increase in numbers post-deployment in the control plot light traps of the Chardonnay block, as was observed in 2004 (Fig. 9, $P<0.05$). A similar increase was not observed for the Semillon block. While springtime climber counts were greater post-deployment in the Concord block light traps, this difference was not significant (Fig. 9). In summary, 2005 pheromone and light trap captures were highly variable with no consistent trends seen for all blocks.

It is not clear why springtime climber cutworm counts in the AAMB traps were much reduced in 2004 and 2005 compared with 2003 counts. It is likely that the variability seen in cutworm numbers year to year was partly due to seasonal weather patterns. Given the general polyphagous nature of the cutworms under investigation in this chapter, it is also probable that cutworm population densities vary widely with geographic location, with each site differing in local and regional agricultural landscape and pest management practices [34]. An example of a site-specific factor would be the prevalence of a flowering cover crop or weed species, which could have a reducing effect on AAMB trap captures if natural floral odors are interfering with the response of foraging moths to the AAMB-based feeding attractant. However, floral odor competition is not likely to be a presiding factor in late-season wine or juice grapes when all vegetation is past the bloom stage. With regard to pest management, however, local and regional changes in pesticide application patterns for control of cutworms in grapes were indeed occurring at the same time we were conducting the attract-and-kill experiments described in this chapter. Author D. Walsh had been developing a novel method to manage pest cutworms by targeting the climbing larvae in the spring. This technique involved the application of a very small amount of pyrethroid spray to the base of the grapevine that served to repel or kill climbing cutworms [35]. This low-input barrier spray technique proved to be quite efficacious against cutworms while reducing the growers' pesticide and application costs. Beginning in 2003, Walsh was recommending an early spring application of the barrier spray, ideally during the first 2 weeks of March, to ensure coverage before the cutworms began to climb. Grape growers throughout southeastern Washington State were quick to adopt this method. We believe that the widespread adoption of the barrier spray technique resulted in an extraordinary reduction of many of the local cutworm populations, especially the springtime climbing species such as the spotted cutworm and *A. vetusta*.

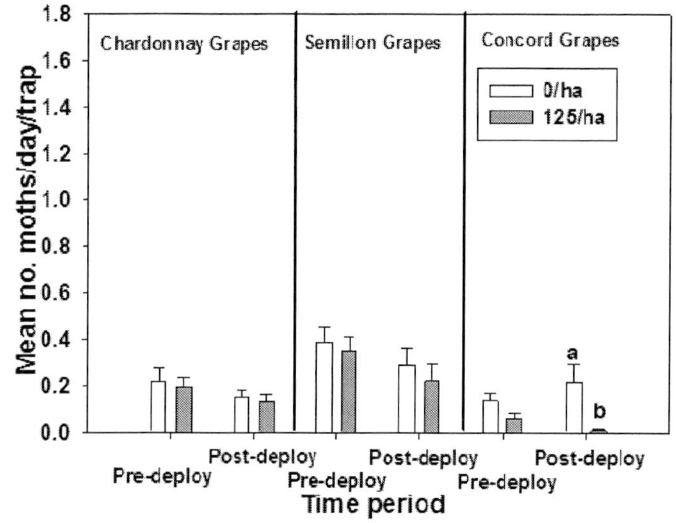

Figure 7. Effect of feeding attractant bait stations on climbing cutworm moth trap captures in wine and juice grape blocks, Prosser and Paterson, WA, 2005. Means grouped within block with the same or no letter above them are not significantly different (P > 0.05).

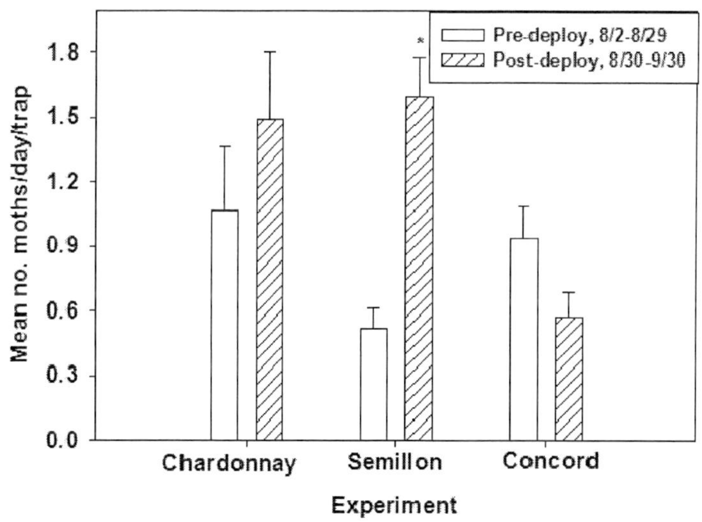

Figure 8. Effect of time period on captures of spotted cutworm male moths in control plot pheromone traps in wine and juice grape blocks, Prosser and Paterson, WA, 2005. Asterisks denote significant difference from pre- to post-deployment (P < 0.05).

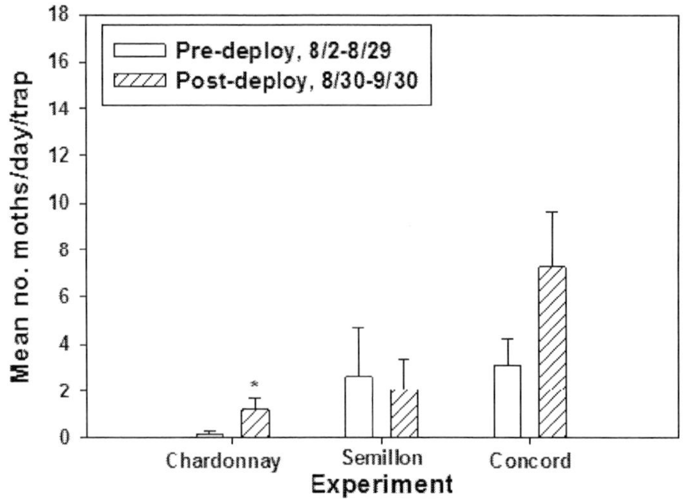

Figure 9. Effect of time period on captures of climbing cutworm moths in control plot light traps in wine and juice grape blocks, Prosser and Paterson, WA, 2005. Asterisks denote significant difference from pre- to post-deployment ($P < 0.05$).

Recently, a study of cutworms in south central Washington vineyards was published that clarified the species composition of climbing cutworms on grapevines [20]. As mentioned before, the cutworm species usually blamed for bud injury in Washington vineyards were the spotted cutworm and the redbacked cutworm. However, the more recent study revealed that two different species, *Abagrotis orbis* (Grote) and *A. vetusta*, were the most abundant grapevine feeders. Wright et al. did find spotted cutworm, *X. c-nigrum*, in low numbers but redbacked cutworm, *E. ochrogaster*, was not detected during the 5-year study. We did not find *A. orbis* in any trap sample from any year; however, based on a larval collection study that was conducted in the same region of Washington State as our study, we believe that it was present at our study sites but undetected. Males and females of *A. orbis* are known to be attracted to ultraviolet light and fermenting fruit baits [36]; even if the AAMB lure derived from fermenting molasses was not particularly attractive to *A. orbis*, we should have found this species in the light traps. In 2003, our collection period was very short, extending only to the middle of September. That year we may have missed the flight period of *A. orbis*, which reportedly occurs in late summer and early autumn [37, 38]. Early spring barrier spray applications in vineyards could have caused localized extinctions of this species, which has been collected on grapevines as early as March 13 [20]; this may explain the absence of *A. orbis* in our 2004 and 2005 trap samples.

Bait Station Effects on Nontarget Cutworms in Grapes

As evidenced by the grape trap data from 2003 to 2005, most of the moths captured in the AAMB and light traps were from nontarget species. It is beyond the scope of this chapter to describe AAMB bait station effects on every nontarget cutworm. However, results for true armyworm (TAW), *P. unipuncta*, in 2005 are worth mentioning because this is a known pest species and because those results provide additional evidence of bait station efficacy in reducing moth populations. We analyzed the 2005 AAMB trap data for TAW, a key pest of corn, small grains, and grasses [39]. TAW moths caught in vineyard AAMB traps probably came from neighboring corn and/or hay fields. Previous studies have suggested that the TAW flight occurring in late summer and fall is a premigratory flight to find food for additional energy reserves needed for long-range dispersal and subsequent reproduction; this would explain the strong attraction of TAW to AAMB observed in a previous trapping study in southeastern Washington State [17] and in the study described in this chapter.

For the Chardonnay wine grape block, AAMB trap captures of TAW were lower in the treated plots compared with control plot trap counts post-deployment and significantly reduced compared with treated plot counts pre-deployment (Fig. 10, $P < 0.05$). A noteworthy 71% reduction of TAW capture was achieved in the treated plots of the Chardonnay block, from pre-deployment to post-deployment. TAW moth captures were not significantly different between control and treated plots pre-deployment in the Chardonnay block nor were they different in control plots pre-deployment compared with post-deployment captures (Fig. 10). In contrast, TAW trap captures were lower in the Semillon block and much lower in the Concord block, and no significant differences were found due to bait station treatment (Fig. 10).

Springtime climber cutworm results from 2003 and TAW results from 2005, both from wine grape blocks, provide evidence that the attract-and-kill bait stations were effective at reducing certain local cutworm moth populations. Interestingly, similar experiments were conducted in peppermint (Labiatae: *Mentha* X *piperita*) near Paterson, WA, in 2003, and while large numbers of cutworm moths were captured in AAMB, light, and pheromone traps, no bait station treatment effect was observed on cutworm numbers post-deployment (data not shown). The apparent failure of AAMB-based attract-and-kill technology in peppermint may be explained by a basic botanical difference between the two perennial crops. While peppermint plants are primarily pollinated by insects and possess an abundance of floral and extrafloral nectaries to attract insects, grape plants (both wine and juice) are primarily self-pollinating, with little assistance from wind and insects, and do not have nectaries that would attract insects. Most likely, AAMB-based bait stations did not work in peppermint fields due to odor competition from nectaries, as has been noted as a reason for failure in previous attract-and-kill studies [2]. Put another way, the same bait station design may have worked better in grapes due to the lack of odor competition from nectaries.

Efficacy and Longevity of Bait Station Killing Agents in Grapes

The key objective when developing a novel insect management technique is the demonstration of efficacy. For the attract-and-kill system in grapes, this meant determining whether a reduction in the adult population of the pest cutworm species resulted in a subsequent reduction in the overwintering larval population. It is very difficult and time-consuming to search for larvae on the grapevines [20], and even more burdensome to find a meaningful relationship between the low number of larvae that would be found on the grapevines and the high number of moths found in the traps. Alternatively, at the time of budbreak in 2004, 2005, and 2006, we conducted systematic searches for bud damage caused by cutworm larvae in the grape blocks that had been treated the summer before. However, no bud damage was found, even in the control plots. We suspect that the grape growers had implemented prophylactic barrier spray applications targeted against cutworms prior to our searches.

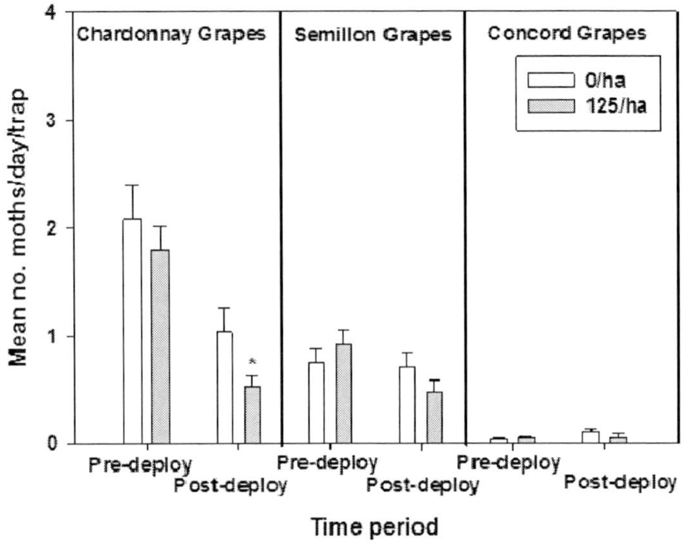

Figure 10. Effect of feeding attractant bait stations on trap captures of true armyworm trap captures in wine and juice grape blocks, Prosser and Paterson, WA, 2005. Asterisk denotes a significant difference from pre- to post-deployment ($P < 0.05$).

A separate, small-scale bioassay study to determine longevity of candidate killing agents for bait stations was conducted at the Irrigated Agriculture Research and Extension Center in Prosser, WA, during the summers of 2004 and 2005. Insecticides were mixed with PTFE grease at the following active ingredient concentrations: 2% bifenthrin in Capture® 2EC, 25% chlorpyrifos in Lorsban™ 4E, 2% (2004) and 2.5% (2005) fenpropathrin in Danitol® 2.4EC, and 2% pyrethrins in Pyganic® EC 1.4. These insecticides were all registered for use in grapes in Washington State. Insecticide treatments were applied to compact discs (CDs) with a paintbrush; duplicate CDs were prepared with each insecticide treatment. Control CDs were prepared with PTFE grease only. In an outdoor location, a series of wires were strung between fence posts upon which treated CDs were hung. A wooden hutch was constructed to provide shade over half of the wire lengths. Half of the CDs were hung on wires in full sunlight conditions and the other half were hung on wires in the shade of the hutch. Cutworm moths used in bioassays were collected from walk-in style light traps located nearby just prior to testing. Care was taken to collect only those moths that had been in the trap overnight and appeared active and healthy. Efficacy of each insecticide treatment was tested on the day of field installment and at weekly intervals for up to seven weeks. This was accomplished by briefly touching the legs of each moth to a CD and quickly placing the moth in a container to monitor for mortality over time.

As moths were not fed in their holding containers, our 24-hr mortality scores showed 100% mortality for nearly all of the moths, including the controls. Therefore, the mortality scores at 3 hr post exposure were chosen as an earlier time point to compare among treatments (Table 1). Generally, the bifenthrin and fenpropathrin shade treatments showed 100% efficacy as far out as 5 weeks after deployment in 2004; the fenpropathrin shade treatment in 2005 showed 100% efficacy further out at 7 weeks after deployment. The 2004 bifenthrin sun treatment was 100% efficacious at 5 weeks after deployment but not efficacious after 6 weeks in the sun. Efficacy began to decline in the fenpropathrin sun treatment after the second week in 2004 and after the first week in 2005, though 50-75% mortality was observed even after 6 weeks of sun exposure. In sharp contrast, both the sun and shade chlorpyrifos treatments lost efficacy after only 1 week in the field. Both the sun and shade pyrethrins treatments showed efficacy on the day of treatment but demonstrated no efficacy by 1 week after deployment. The results indicated that bifenthrin and fenpropathrin are the best candidates for killing agents in the bait stations because their residues demonstrated the greatest longevity in toxicity for cutworm moths.

TARGETING HOP LOOPER IN WASHINGTON HOPS

In 2009, approximately 74% of U.S. hops were grown in Washington State; total production was valued at $265 million [40]. The hop looper, *H. humuli*, is a sporadic economic pest in Washington hop yards. Hop looper larvae feed on the leaves and cones (flowers) of the lower portion of the bine (binding vine of hop). The cones contain the hops used in beer production. It is the direct feeding on the cones by the third and largest generation of larvae appearing in mid to late summer that causes the economic damage [21]. The hop looper has re-emerged as an economic pest as growers have shifted away from using broad-spectrum insecticides to more selective compounds to control aphids and spider mites. However, these "softer" compounds are often not effective against lepidopteran pests. In addition, with three overlapping generations during the summertime, all life stages are present in the field simultaneously, making it difficult to time pesticide treatments to target the more vulnerable first and second instar larvae [21]. Larger hop looper larvae require more disruptive insecticides like bifenthrin, a pyrethroid, for control, but pyrethroid applications can wreak havoc on beneficial arthropod populations and contribute to outbreaks of spider mites. A low-input, low-risk alternative pest management strategy was needed for hop looper.

As suggested previously in this chapter, our attract-and-kill bait stations with acetic acid/3-methyl-1-butanol (AAMB) feeding attractant failed to work in peppermint because of odor competition with abundant nectaries in the insect-pollinated mint. In self-pollinating grapes lacking nectaries, the same AAMB bait stations proved to be successful in reducing certain cutworm moth populations. Hop plants are wind-pollinated [41] and do not possess floral nectaries, which makes them more similar to grapes in this regard than mint. Therefore, it seemed likely that our bait station technology would be effective against noctuid pests, specifically hop looper, in hops.

Table 1. Noctuid moth percent mortality (n) following brief exposure to pesticide-treated bait stations treated with pesticide/PTFE grease mixtures and field-aged in sunlight or shade conditions, Prosser, WA, July to August, 2004

Weeks after	Bifenthrin[a]		Chlorpyrifos		Fenpropathrin		Pyrethrins		PTFE grease (control)	
deployment	Shade	Sun	Shade	Sun	Shade	Sun	Shade	Sun	Shade	Sun
					3 hr post-exposure					
0	100 (3)	100 (3)	100 (3)	100 (3)	100 (5)	80 (5)	100 (4)	75 (4)	0 (4)	0 (4)
1	100 (4)	100 (3)	100 (6)	100 (4)	83 (6)	100 (4)	25 (4)	0 (4)	25 (4)	0 (4)
2	25 (4)	75 (4)	25 (4)	25 (4)	100 (4)	100 (4)	0 (4)	0 (4)	0 (4)	0 (4)
3	100 (4)	100 (4)	75 (4)	50 (4)	100 (4)	50 (4)	n.d[b]	n.d	0 (4)	0 (4)
4	100 (4)	100 (4)	50 (4)	50 (4)	100 (4)	25 (4)	n.d.	n.d	0 (4)	0 (4)
5	100 (4)	100 (4)	0 (4)	0 (4)	100 (4)	50 (4)	n.d.	n.d	0 (4)	0 (4)
6	75 (4)	0 (4)	0 (4)	0 (4)	75 (4)	50 (4)	n.d.	n.d	0 (4)	0 (4)
					5 hr post-exposure					
0	100 (3)	100 (3)	100 (3)	100 (4)	n.d.	n.d	100 (4)	100 (4)	n.d.	n.d.
1	100 (4)	100 (4)	100 (6)	100 (4)	83 (6)	100 (4)	50 (4)	0 (4)	n.d.	n.d.
2	75 (4)	100 (4)	75 (4)	25 (4)	75 (4)	100 (4)	0 (4)	0 (4)	25 (4)	n.d.
3	100 (4)	100 (4)	75 (4)	75 (4)	100 (4)	75 (4)	n.d.	n.d	0 (4)	0 (4)
4	100 (4)	100 (4)	50 (4)	50 (4)	50 (4)	25 (4)	n.d.	n.d	0 (4)	25 (4)
5	75 (4)	100 (4)	0 (4)	25 (4)	100 (4)	50 (4)	n.d.	n.d	0 (4)	0 (4)
6	75 (4)	50 (4)	25 (4)	0 (4)	75 (4)	50 (4)	n.d.	n.d	0 (4)	0 (4)

[a] Insecticides were mixed with PTFE grease at the following active ingredient concentrations: 2% bifenthrin in Capture® 2EC, 25% chlorpyrifos in Lorsban® 4E, 2% fenpropathrin in Danitol® 2.4EC, and 2% pyrethrins in Pyganic® EC 1.4.

[b] n.d. = no data.

In 2009, three large (>30 ha) commercial hop yards with a previous history of hop looper infestations were chosen for the attract-and-kill study. The first one was about 6 km northwest of Moxee, the second about 13 km southwest of Toppenish, and the third one about 6 km southwest of Granger, WA. A randomized complete block design of 2-ha plots with three treatments and four replicates was set up for each hop yard. The three treatments were control (no bait stations), 25 bait stations/ha (50/plot), and 150 bait stations/ha (300/plot). The lower rate of 25/ha was used to prove the system worked; the higher rate of 150/ha was conveniently equivalent to one bait station on every hop pole in the plot. Two AAMB-baited universal bucket traps were placed diagonally across from each other in each plot on the following dates: Moxee, 27 May; Toppenish, 28 May; and Granger, 8 June, 2009. A large, walk-in style light trap was positioned outside of each hop yard to sample the local moth diversity. PVC bait stations treated with β-cyfluthrin/food-grade grease were deployed on 3 June in the Toppenish hop yard, 4 June in the Moxee hop yard, and 8 June in the Granger hop yard. Stations were affixed with screws to the north-facing side of the hop poles at a height of approximately 1.5 m. In the interest of conserving financial and human resources, only the bait stations in the proof-of-concept, low-rate plots were recharged with fresh insecticide/grease and feeding attractant on 14 July. In addition to moth sampling, lepidopteran larvae and other arthropods on the hop plants were sampled by shaking three bines into a large wooden funnel and collecting the insects in a 296 ml container with 75% ethyl alcohol below the funnel. Shake sampling commenced on 25 June and was done biweekly until the middle of July, after which sampling occurred on a weekly basis until 24 August. Also, during the early part of the summer, hop leaves were collected and examined for hop looper eggs and larvae. While universal bucket traps and light traps were removed prior to harvest, bait stations remained in the field through the winter of 2010-2011.

In 2010, bait stations already in the field were recharged with fresh β-cyfluthrin/grease mixture and fresh vials containing AAMB on 3 June in the Moxee hop yard and on 8 June in the Toppenish hop yard; the Granger hop yard was dropped from the study. Bait stations were again recharged with fresh toxicant/grease mixture and AAMB in the Moxee hop yard on 20 July 2010 but not in the Toppenish hop yard because moth populations remained low at that site. Sampling for moth populations was accomplished weekly using universal bucket traps baited with AAMB, beginning on 6 May and ending mid-August. In addition, a large walk-in style light trap was deployed at each location to sample the local moth diversity. Lepidopteran larvae and other arthropods on hop plants were sampled as in 2009 starting on 14 June 2010 and continuing on a biweekly basis until the middle of July, after which sampling occurred on a weekly basis until 17 August. Also, during the early part of the summer, hop leaves were collected and examined for hop looper eggs and larvae.

Trap samples were frozen, then sorted and counted according to lepidopteran species, when the species could be properly identified. Larvae in shake samples preserved in ethanol were sorted and counted according to species, when the species could be properly identified. 2009 trap data for moths from similar-length pre-deployment and post-deployment periods and sample data for lepidopteran larvae from each month post-deployment were pooled and subjected to chi-square analysis and ANOVA followed by means separation with Tukey's correction, α=0.05, using SAS [28]. 2010 trap data for moths were pooled for three similar periods of time: pre-deployment, first month post-deployment and second month post-deployment. Likewise, larval sample data were pooled for June, July, and August, all post-deployment. Pooled data were compared among bait station treatments using ANOVA followed by means separation with Tukey's correction, $\alpha=0.05$ [28].

Species Composition in Hops

In 2009 and 2010, the most abundant moth caught in AAMB traps was the hop looper, followed by spotted cutworm, Bertha armyworm, a geometrid called the common gray moth, *Anavitrinella pampinaria*, and *L. subjuncta*. The larvae of all of these species are known to feed on hops and are considered to be potential pests [23]. In 2009, 98% of the AAMB trap capture in three hop yards (n=19,248) was hop looper (n=18,795), with 80% captured in the Moxee hop yard bucket traps. Included in the remaining 2% of the catch were 156 spotted cutworms, 137 Bertha armyworms, 97 common gray moths, and 61 *L. subjuncta*. In 2010, moth population densities were greatly reduced; only 3,565 hop loopers were caught in AAMB traps, representing 72% of the total moth catch (n=4,980). Considerably more spotted cutworms were trapped (956 (19%)) as well as more Bertha armyworms (199 (4%)), common gray moths (180 (nearly 4%)), and *L. subjuncta* (80 (2%)).

In the shake samples, larvae were separated based on the number of proleg pairs. The noctuid hop looper has four pairs of prolegs (three mid-abdominal pairs and one terminal pair) and the geometrid common gray larva has only two pairs of prolegs, one abdominal pair and one terminal pair. All other noctuid larvae of interest in the samples (spotted cutworm, Bertha armyworm, *L. subjuncta*, and others), have five pairs of prolegs. In the interest of saving time, the other noctuid larvae were not separated by species but counted as one group. In 2009, a total of 9,083 larvae were found in the shake samples from the three study sites; seventy percent were hop looper larvae, 29% were common gray, and 1% fell into the group with five pairs of prolegs. Approximately 99% of the hop looper larvae were found in shake samples taken in the Moxee hop yard, while the majority of the common gray larvae (68%) were found in samples from the Granger hop yard. In 2010, numbers of larvae in shake samples were greatly reduced; a total of 1,112 larvae were identified and counted in samples from the Moxee and Toppenish hop yards. Approximately 57% of the larvae were identified as common gray while only 25% were hop looper larvae. Nearly all of the common gray larvae (94%) were collected from the Toppenish hop yard, while the majority of the hop looper larvae (71%) were found in the Moxee hop yard.

Bait Station Effects on Hop Looper in Hops

For 2009, seasonal abundance of hop looper larvae and adults in the Moxee hop yard appears in Fig. 11. Very few eggs were found in the leaf samples; these data were not analyzed. Numbers of adults and larvae remained low until the middle of July, when adults and larvae began to increase exponentially. ANOVA results from the Moxee trap data indicated that our attract-and-kill bait stations did not exert a reducing effect on moth or larval numbers (data not shown). These results suggest that there is a pest density level above which the attract-and-kill management system shows a progressively smaller and smaller reducing effect on the numbers of moths. In other words, the bait stations cannot attract and remove moths from the population fast enough to keep up with their emergence. These results agree with what was found in previous pheromone-based attract-and-kill field trials involving codling moth, in which they noted poor control with high initial pest density (see review [2]). The Moxee hop grower applied a rescue treatment of bifenthrin in blocks under severe caterpillar pressure during late July/early August, which decimated the damaging larval populations but apparently did not affect the moths because we continued to collect them in the AAMB traps (Fig. 11).

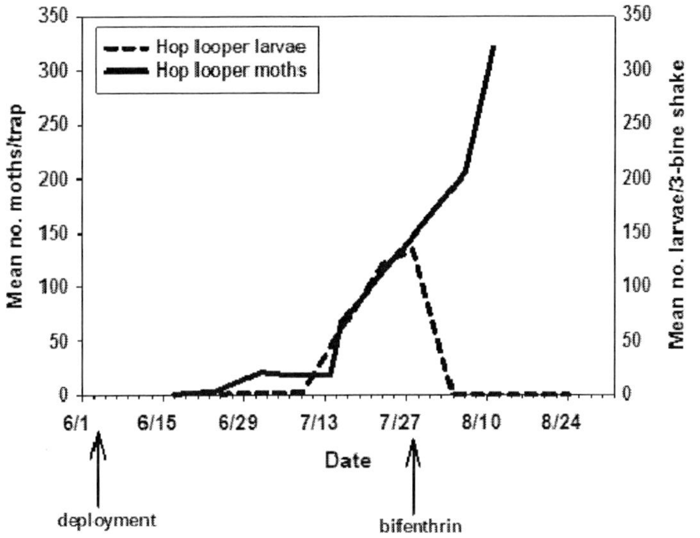

Figure 11. Seasonal abundance of hop looper adults and larvae, Moxee, WA, 2009. Bait station deployment occurred on 3 June and bifenthrin was applied to hop yard during the last few days of July.

The reduction in Moxee hop looper moth trap captures from 2009 to 2010 was probably due primarily to the grower- implemented bifenthrin treatment in late July 2009, which lowered but did not eradicate the local moth populations. Also, some moths may have been killed when they contacted the toxicant in the bait station, but we had no means of directly verifying that. Since moths presumably fly some distance away from the bait station they contact before succumbing to the toxicant, collecting dead moths at the station is not possible and no proof of bait station efficacy can be obtained that way. Finally, differences in weather patterns, especially the springtime temperatures, may have influenced trap catches [42]. The warm temperatures experienced in the spring of 2009 promoted rapid growth in moth populations. In contrast, growth in moth populations in 2010 was mitigated by colder than average temperatures during the spring.

In 2010, numbers of hop looper moths and larvae remained low at both study sites early in the season but showed an exponential increase in abundance starting in early July. Again, very few eggs were found in the leaf samples; these data were not analyzed. During the pre-deployment period prior to recharging the bait stations from early May to early June, there were no significant differences in AAMB trap captures of hop looper among treatments for either hop yard (Figs. 12 and 13). During the first month after recharging, there were no significant differences in hop looper trap counts found among treatments either. Moth numbers were too low to detect significant differences. However, when the moth populations grew rapidly in July, trap capture data indicated that bait stations were significantly reducing hop looper moths in the treated plots compared with numbers in control plots, even at the low rate of 25/ha (Figs. 12 and 13, $P<0.05$). Hop looper moth trap captures at the Moxee site were reduced by 31% in the 25/ha treated plots and by 80% in the 150/ha treated plots, relative to trap captures in the control plots (Fig. 12). Likewise, hop looper moth trap captures at the Toppenish site were reduced by 23% in the 25/ha treated plots and 67% in the 150/ha treated plots, relative to trap captures in the control plots (Fig. 13).Unfortunately, the same effects could not be observed in the Moxee hop looper larval populations because the grower had made two insecticide applications for caterpillar control during the season. *Bacillus thuringiensis*, subsp. *kurstaki* (Dipel® DF) was applied in late June and spinosad (Entrust®) was applied in mid July, so that by August, very few larvae of any species were found in Moxee shake samples. However, there was a noteworthy trend of lower hop looper larval counts in the treated plots compared with the control plots during June and July (Fig. 12). In the Toppenish hop yard, hop looper larval numbers were quite variable and no bait station effects were detected (Fig. 13).

In 2010, AAMB bait stations were effective in reducing hop looper moth numbers in hops. However, we were not able to demonstrate that reduced moth populations would translate to reduced larval populations due to confounding of the data by grower-implemented pesticide applications.

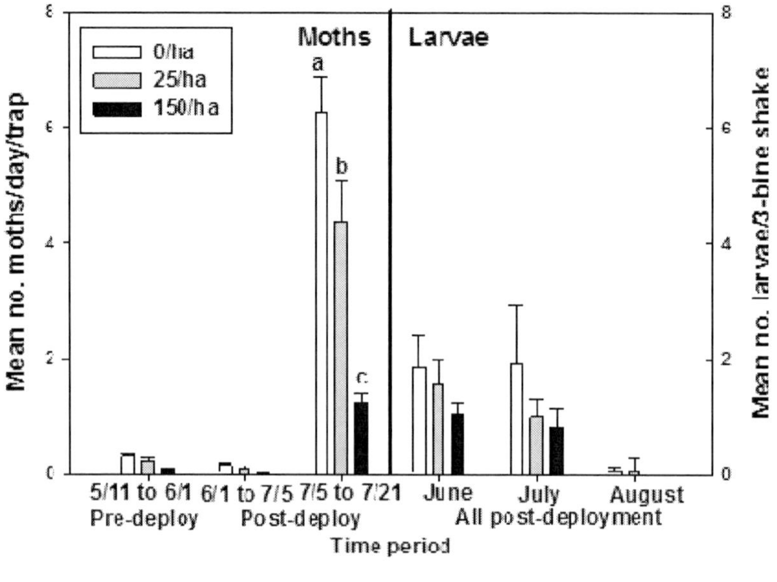

Figure 12. Effect of feeding attractant bait stations on hop looper moth trap captures in Moxee, WA, 2010. Means grouped within time period with the same or no letter above them are not significantly different ($P > 0.05$).

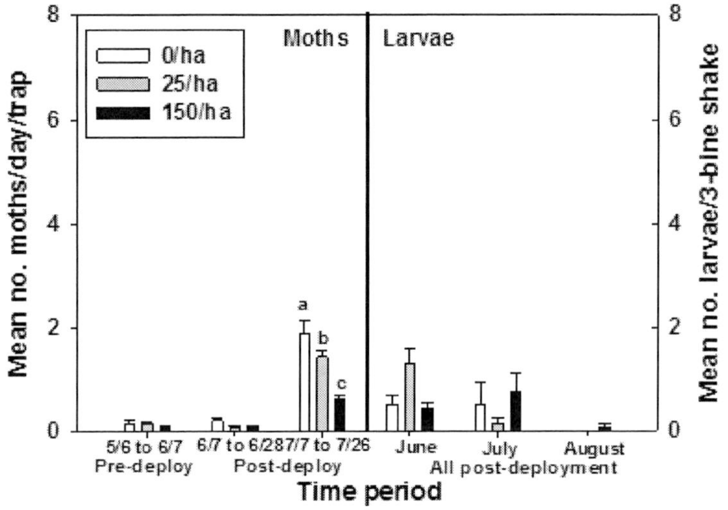

Figure 13. Effect of feeding attractant bait stations on hop looper moth trap captures in Toppenish, WA, 2010. Means grouped within time period with the same or no letter above them are not significantly different (P > 0.05).

Efficacy and Longevity of Bait Station Killing Agent in Hops

To determine the field efficacy and longevity of the killing agent, β-cyfluthrin, used on the hop bait stations, small-scale bioassay studies in the laboratory were conducted in August 2010, at the end of our trapping period. For positive controls, PVC bait stations similar to those used in the field were treated with fresh 2.5% β-cyfluthrin (in the formulation Baythroid® XL) in food-grade grease. Field-aged bait stations were collected from a Moxee plot 3 weeks after recharging with fresh pesticide/grease mixture and from a Toppenish plot 9 weeks after recharging. A negative control treatment consisted of a bait station daubed with food-grade grease only. Moths used in bioassays were collected just prior to testing from walk-in style light traps located near hop yards. Care was taken to collect only those moths that had been in the trap for one night and appeared active and healthy. Moths were identified to species and, when possible, moths from the same species were tested as a group during the same session. Testing protocol consisted of briefly touching the legs of each moth to a treated bait station and quickly placing the moth in a container to monitor for mortality over time.

As moths were not fed in their holding containers, our 24-hr mortality scores showed 25 to 42% mortality of the controls (Table 2). Therefore, 3 hr post exposure was chosen as an earlier time point to compare mortality scores among treatments (Table 2). At 3 hr after exposure, the fresh β-cyfluthrin/grease mixture showed 100% efficacy against hop loopers, common grays, and the common moth group (spotted cutworm, Bertha armyworm, and Western yellow-striped armyworm). In addition, the insecticide residue on bait stations that had been field-aged for 9 weeks showed 100% efficacy against hop looper, while the 3-week-old field-aged residue showed only 80% efficacy (Table 2). In sharp contrast, field-aged residues at 3 weeks showed no efficacy against common gray moth and 50% efficacy against moths from the common moth group; field-aged residues at 9 weeks showed no efficacy against these species either. Our data indicate that the hop looper is relatively more susceptible to β-cyfluthrin than the other commonly found pest moth species. These results suggest that if the target pest is hop looper, then β-cyfluthrin would be a good choice for a bait station killing agent, but if the target is spotted cutworm, Bertha armyworm, western yellow-striped armyworm, and/or common gray moth, perhaps another pyrethroid such as bifenthrin (longevity data in Table 1) should be considered.

Table 2. Efficacy testing of β-cyfluthrin in Baythroid® mixed with food-grade grease against hop looper and other moths in freshly treated versus field-aged bait stations, 2010

Bait station treatment	3 hr percent mortality (n)			
(in food-grade grease)[a]	Hop looper	Common gray	Common moths[b]	Other moths[c]
Negative control	0 (8)	0 (3)	0 (1)	17 (12)
Positive control	100 (3)	100 (5)	100 (3)	77 (13)
Field-aged 3 weeks, Moxee hop yard	80 (5)	50 (4)	0 (3)	46 (11)
Field-aged 9 weeks, Toppenish hop yard	100(3)	0 (3)	0(1)	39 (13)
	24 hr percent mortality (n)			
Negative control	25 (8)	33 (3)	0 (1)	42 (12)
Positive control	100 (3)	100 (5)	100 (3)	100 (13)
Field-aged 3 weeks, Moxee hop yard	100 (5)	75 (4)	66(3)	91 (11)
Field-aged 9 weeks, Toppenish hop yard	100 (3)	33 (3)	100 (1)	85 (13)

[a] Negative control was food-grade grease only. Positive control was freshly mixed 2.5% β-cyfluthrin in Baythroid® with food-grade grease. Moths were exposed to grease mixture for 1 second.

[b] Includes spotted cutworm, Bertha armyworm, western yellow striped armyworm.

[c] Includes alfalfa looper, cabbage looper, *Caradrina* sp., *Choristoneura* spp., other geometrids, other noctuids, *Spilosoma*, clover cutworm.

Bait Station Effects on Nontarget Cutworms in Hops

Similar to what was seen in the previous grape experiments, a fraction of the captured moths in AAMB traps were nontarget cutworms and other moths, in this case, moths that were not hop looper. While the fraction of nontarget moths in 2009 amounted to a little more than 2% of the total catch, in 2010, the fraction increased to 28% nontargets, due to the greatly reduced hop looper population in the Moxee hop yard. The nontarget species frequently (≥10 specimens) found in hop AAMB traps were (in order of abundance): spotted cutworm; Bertha armyworm; common gray moth; *L. subjuncta*; speckled green fruitworm, *Orthosia hibisci* Guenée, and western yellow-striped armyworm, *S. praefica*.

Because spotted cutworm and Bertha armyworm larval populations can reach economically damaging levels occasionally in hop yards [43, 44], 2010 trap data for these species were analyzed for bait station effects. Seasonal abundance of spotted cutworm and Bertha armyworm moths peaked in late May/early June (Figs. 14 and 15); numbers of both of these species were already declining before the bait stations were recharged with fresh toxicant and attractant (recharge dates were 3 June in Moxee hop yard and 8 June in Toppenish hop yard, Figs. 14 and 15). Thus, during the post-deployment period, no significant effects on trap captures of these two species were found that could be attributable to bait station treatment.

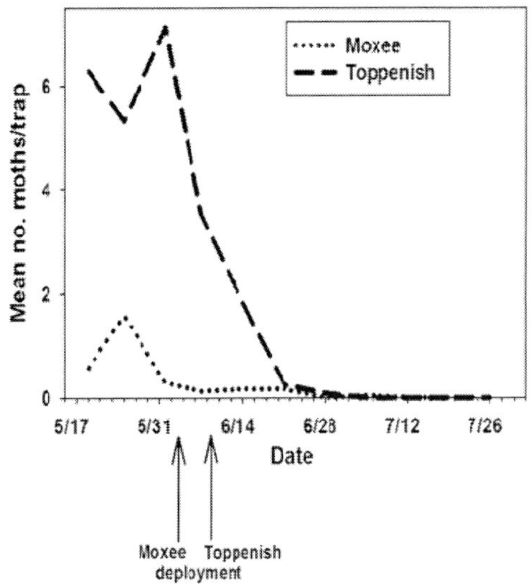

Figure 14. Seasonal abundance of spotted cutworm adults in feeding attractant monitoring traps, Moxee and Toppenish hop yards, Washington State, 2010. Recharging of bait stations occurred on 3 June and 8 June at the Moxee and Toppenish sites, respectively.

For the third most abundant nontarget species, the common gray moth, there is no previous record of their larvae causing economic damage in hop yards. Grasswitz and James [23] indicated that while common gray is quite frequently found feeding on hop foliage, it is heavily parasitized. In our 2010 studies, common gray moth captures were too low in either hop yard to analyze for bait station effects (Toppenish hop yard, Fig. 16). The low numbers of this geometrid in the AAMB traps were not unexpected; in previous studies, geometrid moths seemed to respond poorly to AAMB attractant. In AAMB traps in central Washington, no geometrids were caught [30], and in Alaska, only 15 were caught during a 3-year survey, compared with 5,359 noctuids [22]. Utrio [45] noted a generally greater response of noctuids compared with geometrids to sugar baits and compounds derived from fermented molasses; he proposed that noctuids generally have higher energy demands and thus, different foraging strategies than geometrids. Therefore, if moths are attracted to AAMB because they are foraging for sugar sources in the field [22], then it would follow that trap captures of the more robust noctuids with higher energy demands would be greater in number than captures of the more slender geometrids with lower energy requirements.

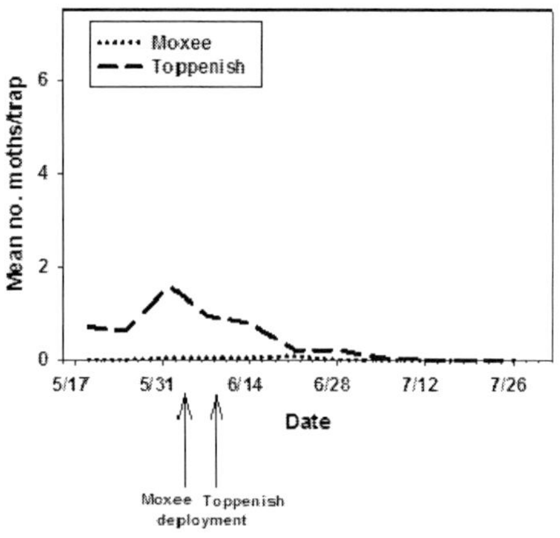

Figure 15. Seasonal abundance of Bertha armyworm adults in feeding attractant monitoring traps, Moxee and Toppenish hop yards, Washington State, 2010. Recharging of bait stations occurred on 3 June and 8 June at the Moxee and Toppenish sites, respectively.

In contrast to the low adult numbers, common gray larvae were quite numerous in Toppenish shake samples (n=591). Numbers of common gray larvae in the Toppenish hop yard were observed to peak in late June and again in August (Fig. 16). Very few common gray larvae were found in Moxee samples (n=38); more than likely the pesticide applications targeting hop looper larvae reduced larval numbers of this species as well. No significant bait station effects were found on common gray larval populations in either hop yard. Our data show low captures of common gray moths in the AAMB traps, indicating a weak response to the feeding attractant, which more than likely explains the absence of a bait station effect as well.

CONCLUSION

In this chapter, we have presented evidence that an attract-and-kill system involving a bait station containing acetic acid+3-methyl-1-butanol as an attractant and a pyrethroid insecticide as a toxicant, can significantly reduce pest noctuid moth populations in grapes and hops. In grapes, we showed that moth populations of springtime climber species (i.e., spotted cutworm, *A. vetusta*, dingy cutworm, variegated cutworm, and black cutworm) were reduced by bait station treatment. In hops, we demonstrated that moth populations of hop looper were effectively reduced by bait station treatment. However, we were not able to demonstrate a subsequent decrease in damaging larval populations in either grapes or hops, because grower-implemented insecticide applications in both cases reduced or eliminated the larval populations before we were able to gather larval data. While the bait station design went through a series of changes through the years, we have no evidence to conclude that the basic design of the bait station affected moth trap captures. Rather, it appeared to be the attractant and toxicant components of the bait stations that had the greatest impact on moth trap captures. The AAMB feeding attractant proved to be useful in attracting our target species; the main drawback we observed was its broad-spectrum attraction to non-target moths in both grapes and hops. Future studies should examine potential environmental risks of AAMB-based attract-and-kill technology. From our bioassay studies, we concluded that choice of toxicant should depend on susceptibility of the target species to a particular toxicant. Preliminary bioassays to determine susceptibility of the target moths should be conducted prior to large-scale field implementation.

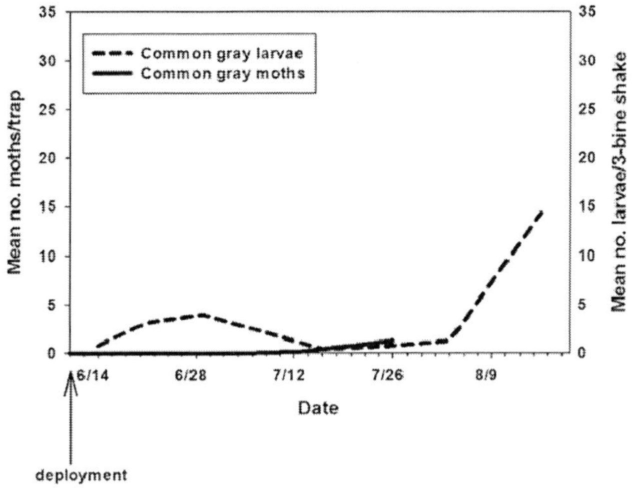

Figure 16. Seasonal abundance of common gray moths and larvae, Toppenish hop yard, 2010. Recharge of bait stations occurred in Toppenish on 8 June.

Because our results showed that moth populations were never reduced to near zero after bait station deployment, we conclude that AAMB-based attract-and-kill should not be considered as a stand-alone management technique for the specific lepidopteran pests we studied. In grapes, we learned, through serendipity, that a two-pronged approach might be advisable involving (1) a prophylactic barrier spray at bud break to kill climbing cutworm larvae followed by (2) a clean-up strategy using bait station deployment in the late summer to eliminate moths that give rise to overwintering larvae. In hops, we discovered, again by chance, that bait stations could be deployed to kill the moths while a soft insecticide could be applied to kill the larvae at the same time. A feasible three-year scheme in hops might consist of a knockdown spray the first year to reduce the local pest noctuid density, followed by a second and if needed, a third year of a combination of attract-and-kill bait station deployment to kill hop looper moths during the second flight period coupled with strategically timed applications of a soft insecticide targeted at first and second instar larvae. Future research should involve testing these potential management schemes under controlled conditions.

ACKNOWLEDGEMENTS

We are very grateful for the grape and hop growers in Washington State who allowed us to conduct research on their land. Many thanks go out to the technical staff of the Environmental and Agricultural Laboratory at the WSU Irrigated Agriculture and Extension

Center for their countless hours of assistance in field implementation, monitoring, and sample processing. We extend our gratitude to the funding sources for the grape work: Washington State Commission on Pesticide Registration (WSCPR), American Farmland Trust/U.S. E.P.A. Region 10, Washington Association of Wine Grape Growers, Concord Grape Research Council, Wine Advisory Committee, and for the hop work: WSCPR, Hop Research Council and Washington Hop Commission.

REFERENCES

[1] National Coalition on Integrated Pest Management. (1994). Toward a goal of 75 percent cropland under IPM by 2000. National Coalition on IPM, Jan. Austin, TX.

[2] El-Sayed, A. M., Suckling, D. M., Byers, J. A., Jang, E. B., & Wearing, C. H. (2009). Potential of "lure and kill" in long-term pest management and eradication of invasive species. *J. Econ. Entomol.*, *102*, 815-835.

[3] Beasley, C. A. & Henneberry, T. J. (1984). Combining gossyplure and insecticides in pink bollworm control. *Calif. Agric.*, *38*, 22-24.

[4] Butler, G. D. & Las, A. S. (1983). Predaceous insects: effect of adding permethrin to the sticker used in gossyplure applications. *J. Econ. Entomol.*, 76, 1448.

[5] Charmillot, P. J., Hofer, D. & Pasquier, D. (2000). Attract and kill: a new method for control of the codling moth *Cydia pomonella. Entomol. Exp. Appl.*, *94*, 211-216.

[6] Lösel, P. M., Penners, G., Potting, R. P. J., Ebbinghaus, D., Elbert, A. & Scherkenbeck, J. (2000) Laboratory and field experiments towards the development of an attract and kill strategy for the control of the codling moth, *Cydia pomonella. Entomol. Exp. Appl.*, *95*, 39-46.

[7] Krupke, C. H., Roitberg. B. D. & Judd, G. J. R. (2002). Field and laboratory responses of male codling moth (Lepidoptera: Tortricidae) to a pheromone-based attract-and-kill strategy. *Environ. Entomol.*, *31*, 189-197.

[8] Evenden, M. L. & McLaughlin, J. R. (2004). Initial development of an attracticide formulation against the Oriental fruit moth, *Grapholita molesta* (Lepidoptera: Tortricidae). *Environ. Entomol.*, *33*, 213-220.

[9] Suckling, D. M. & Brockerhoff, E. G. (1999). Control of light brown apple moth (Lepidoptera: Tortricidae) using an attracticide. *J. Econ. Entomol.*, *92*, 367-372.

[10] Weihman, S. W. & Liburd, O. E. (2006). Mating disruption and attract-and-kill as reduced-risk strategies for control of grape root borer *Vitacea polistiformis* (Lepidoptera: Sesiidae) in Florida vineyards. *Fla. Entomol.*, *89*, 245-250.

[11] Camelo, L. D. A., Landolt, P. J. & Zack, R. S. (2007). A kairomone based attract-and-kill system effective against alfalfa looper (Lepidoptera: Noctuidae). *J. Econ. Entomol.*, *100*, 366-374.

[12] Landolt, P. J., Adams, T., Reed, H. C. & Zack, R. S. (2001). Trapping alfalfa looper moths (Lepidoptera: Noctuidae) with single and double component floral chemical lures. *Environ. Entomol.*, *30*, 667-672.

[13] Landolt, P. J., Adams. T. & Zack, R. S. (2006). Field response of alfalfa looper and cabbage looper moths (Lepidoptera: Noctuidae, Plusiinae) to single and binary blends of floral odorants. *Environ. Entomol.*, *35*, 276-281.

[14] Landolt, P. J. (2000). New chemical attractants for trapping *Lacanobia subjuncta, Mamestra configurata,* and *Xestia c-nigrum* (Lepidoptera: Noctuidae). *J. Econ. Entomol., 93,* 101-106.

[15] Norris, M. J. (1936). The feeding habits of the adult lepidopteran Heteroneura. *Trans. R. Entomol. Soc. Lond., 85,* 61-90.

[16] Landolt, P. J. & Alfaro, J. F. (2001). Trapping *Lacanobia subjuncta, Xestia c-nigrum,* and *Mamestra configurata* (Lepidoptera: Noctuidae) with acetic acid and 3-methyl-1-butanol in controlled release dispensers. *Environ. Entomol., 30,* 656-662.

[17] Landolt, P. J. & Higbee, B. S. (2002). Both sexes of the true armyworm (Lepidoptera: Noctuidae) trapped with the feeding attractant composed of acetic acid and 3-methyl-1-butanol. *Fla. Entomol., 85,* 182-185.

[18] Utrio. P. & Eriksson, K. (1977). Volatile fermentation products as attractants for Macrolepidoptera. *Ann. Zool. Fenn., 14,* 98-104.

[19] Howell, J. F. (1979). Phenology of the adult spotted cutworm in the Yakima Valley. *Environ. Entomol., 8,* 1065-1069.

[20] Wright, L. C., James, D. G., Reyna, V., Del Conte, S. C., Gingras, S., Landolt, P. & Brooks, T. (2010). Species composition of cutworm (Lepidoptera: Noctuidae) larvae in south central Washington vineyards. *Ann. Entomol. Soc. Am., 103,* 592-596.

[21] Grasswitz, T. R. & James, D. G. (2008). Biology, phenology and control of *Hypena humuli* (Lepidoptera: Noctuidae), an emerging pest of hops (*Humulus lupulus*) (Cannabaceae) in the USA. *Int. J. Pest Manag., 54,* 333-338.

[22] Landolt, P. J., Pantoja, A., Hagerty, A., Crabo, L. & Green, D. (2007). Moths trapped in Alaska with feeding attractant lures and the seasonal flight patterns of potential agricultural pests. *Can. Entomol., 139,* 278-291.

[23] T. R. Grasswitz, T. R. & James, D. (2008). *Biology of hop looper and its natural enemies.* Washington State University EB2037E. Pullman, WA: Washington State University Extension Publishing and Printing. Available from: http://cru.cahe.wsu.edu/CEPublications/eb2037e/eb2037e.pdf

[24] Landolt, P. J. (2002). Lacanobia moths: why monitor them when you can just kill them? In *Proc. WA State Hort. Soc.* (98, pp. 119-123). Wenatchee, WA: WSHA.

[25] USDA, NASS. (2006). *Washington annual statistical bulletin.* Available from: http://www.nass.usda.gov/Statistics_by_State/Washington/Publications/Annual_Statistical_Bulletin/2006/ab87-91.pdf.

[26] Ferguson, H., O'Neal, S. & Walsh, D. (2007). *Survey SAYS: great grapes!* Washington State University EB2025E. Pullman, WA: Washington State University Extension Publishing and Printing. Available from: http://cru.cahe.wsu.edu/CEPublications/eb2025e/eb2025e_new.pdf

[27] Wright, L. C. & Cone, W. W. (1980). Economic damage and vine response from simulated cutworm damage to Concord grape buds. *J. Econ. Entomol., 73,* 787-790.

[28] SAS Institute Inc. (2003). SAS for Windows, v. 9.1.3.

[29] Wright, L. C. & Cone, W. W. (1983). Barriers for control of cutworm (Lepidoptera: Noctuidae) damage to Concord grape buds. *J. Econ. Entomol., 76,* 1175-1177.

[30] Landolt, P. J. & Hammond, P. C. (2001). Species' composition of moths captured in traps baited with acetic acid and 3-methyl-1-butanol, in Yakima County, Washington. *J. Lepid. Soc., 55,* 53-58.

[31] Bournier, A. (1976). Grape insects. *Annu. Rev. Entomol., 22,* 355-376.

[32] Lafontaine, J. D. (2004). Noctuoidea: Noctuidae (part). In *The moths of America north of Mexico: Fascicle 27.1* (p. 214). Boca Raton, FL: The Wedge Entomological Research Foundation.

[33] Lowery, D. T. & Mostafa, A. M. (2010). Cutworm species (Lepidoptera: Noctuidae) feeding on grapevines in the Okanagan Valley, British Columbia, with notes on rearing. *Can. Entomol.*, *142*, 173-180.

[34] Landolt, P. J., Guédot, C. & Zack, R. S. (2010). Spotted cutworm, *Xestia c-nigrum* (L.) (Lepidoptera: Noctuidae) responses to sex pheromone and blacklight. *J. Appl. Entomol.* Available from: http://onlinelibrary.wiley.com/doi/10.111/j.1439-0418.2010. 01571. x/pdf

[35] Walsh, D., Ferguson, H. & Wight, R. (2003). Cutworms can climb no more! Barrier technique shows promise, conserves beneficials in vineyards. *Agrichem. Environ. News, 207*. Available from: http://aenews.wsu.edu/july03AENews/July03AENews.htm#Cutworm

[36] Rings, R. W. (1972). Contributions to the bionomics of climbing cutworms: the distribution and developmental biology of *Abagrotis barnesi*. *J. Econ. Entomol.*, *65*, 397-401.

[37] Lafontaine, J. D. (1998). Noctuiodea: Noctuidae (part). In *The moths of America north of Mexico: Fascicle 27.3* (pp. 230-231). Boca Raton, FL: The Wedge Entomological Research Foundation.

[38] Landolt, P. J., Lowery, D. T., Wright, L. C., Smithhisler, C., Guédot, C. & James, D. (2010). Trap response of cutworm moths (*Abagrotis orbis*) to a sex-attractant lure in grape vineyards. *Can. Entomol.*, *142*, 135-142.

[39] Pfadt, R. E. (1978). Insect pests of small grains. In R. E. Pfadt (Ed.), *Fundamentals of applied entomology* (pp. 261-301). New York: MacMillan Publishing Co.

[40] USDA, NASS. (2010). National hop report. Available from: http://usda.mannlib.cornell.edu/usda/current/hops/hops-12-17-2010.pdf

[41] Small, E. (1997). Cannabaceae. In Flora of North America Editorial Committee (Eds.), *Flora of North America north of Mexico* (pp. 381-387). New York: Oxford University Press.

[42] Washington State University. (2011). AgWeatherNet. Available from: http://weather.wsu.edu

[43] George, A. (2001). *Crop profile for hops in Washington*. Available from: www.ipmcenters.org/cropprofiles/docs/wahops.html

[44] Barbour, J. & James, D. (2011). Hop pests. In Hollingsworth, C. S. (Ed.), *Pacific Northwest insect management handbook*. Available from: http://uspest.org/pnw/insects

[45] Utrio, P. (1983). Sugaring for moths: Why are noctuids attracted more than geometrids? *Ecol. Entomol.*, *8*, 437-445.

INDEX

B

C

D

E

F

G

H

I

J

K

L

O

P

Q

R

S

T

U

V

W

Y

Z